Planetary Geodesy
and
Remote Sensing

Planetary Geodesy
and
Remote Sensing

EDITED BY **Shuanggen Jin**

CRC Press
Taylor & Francis Group
Boca Raton London New York

CRC Press is an imprint of the
Taylor & Francis Group, an **informa** business

CRC Press
Taylor & Francis Group
6000 Broken Sound Parkway NW, Suite 300
Boca Raton, FL 33487-2742

First issued in paperback 2019

© 2015 by Taylor & Francis Group, LLC
CRC Press is an imprint of Taylor & Francis Group, an Informa business

No claim to original U.S. Government works

ISBN-13: 978-1-4822-1488-8 (hbk)
ISBN-13: 978-0-367-86886-4 (pbk)

Library of Congress Cataloging-in-Publication Data

Planetary geodesy and remote sensing / edited by Shuanggen Jin.
    pages cm
  Includes bibliographical references and index.
  ISBN 978-1-4822-1488-8 (hardback : acid-free paper) 1. Planetary geographic
  information systems. 2. Outer space--Exploration--Technological innovations. I. Jin,
  Shuanggen, editor.

  QB600.33.P53 2015
  559.9--dc23   2014027310

Visit the Taylor & Francis Web site at
http://www.taylorandfrancis.com

and the CRC Press Web site at
http://www.crcpress.com

Contents

Preface...vii

Editor..ix

Contributors..xi

1. **Lunar Geodesy and Sensing: Methods and Results from Recent Lunar Exploration Missions** 1
   Shuanggen Jin, Sundaram Arivazhagan, and Tengyu Zhang

2. **Improvement of Chang'E-1 Orbit Accuracy by Differential and Space VLBI** 19
   Wei Yan, Shuanggen Jin, and Erhu Wei

3. **Laser Altimetry and Its Applications in Planetary Science** 51
   Hauke Hussmann

4. **Photogrammetric Processing of Chang'E-1 and Chang'E-2 STEREO Imagery for Lunar Topographic Mapping** 77
   Kaichang Di, Yiliang Liu, Bin Liu, and Man Peng

5. **Integration and Coregistration of Multisource Lunar Topographic Data Sets for Synergistic Use** 97
   Bo Wu, Jian Guo, and Han Hu

6. **Estimates of the Major Elemental Abundances with Chang'E-1 Interference Imaging Spectrometer Data** 119
   Yunzhao Wu

7. **Lunar Clinopyroxene Abundance Retrieved from M³ Data Based on Topographic Correction** 157
   Pengju Guo, Shengbo Chen, Jingran Wang, Yi Lian, Ming Ma, and Yanqiu Li

8. **Martian Minerals and Rock Components from MRO CRISM Hyperspectral Images** 175
   Yansong Xue and Shuanggen Jin

9. **Anomalous Brightness Temperature in Lunar Poles Based on the SVD Method from Chang'E-2 MRM Data** 209
   Yi Lian, Sheng-bo Chen, Zhi-guo Meng, Ying Zhang, Ying Zhao, and Peng-ju Guo

10. **Mercury's Magnetic Field in the MESSENGER Era**..........................223
    Johannes Wicht and Daniel Heyner

11. **Lunar Gravity Field Determination from Chang'E-1 and Other
    Missions' Data**..263
    Jianguo Yan, Fei Li, and Koji Matsumoto

12. **Martian Crust Thickness and Structure from Gravity and
    Topography**...293
    Tengyu Zhang, Shuanggen Jin, and Robert Tenzer

13. **Theory of the Physical Libration of the Moon with a
    Liquid Core**..311
    *Yuri Barkin, Hideo Hanada, José M. Ferrándiz, Koji Matsumoto,
    Shuanggen Jin, and Misha Barkin*

Index...375

Preface

Planetary science is dedicated to exploring the origin, formation, and evolution of Mercury, Venus, Earth, Moon, Mars, Saturn, and Jupiter, etc., and seeking life beyond Earth. Planetary exploration provides the most important direct observations and constraints on planetary structure and dynamics, as well as evolution, particularly planetary geodesy and remote sensing, for example, very long baseline interferometry, laser ranging, laser altimetry, microwave radiometers, Mineralogy Mapper, and other sensors. In the 1960s, the United States made its first attempt to obtain closer images of the lunar surface with the Ranger series, and particularly the successful landing of the lunar Apollo 11 mission in 1969 was a scientific milestone. After that, many more explorations on the moon, Mars, Venus, Jupiter and elsewhere have been conducted from all over the world, such as the recent lunar SMART-1, SELENE, ChangE-1/2/3, Chandrayaan-1, LRO/LCROSS and GRAIL, Mars Global Surveyor, Mars Express, Mars Odyssey, Mars Reconnaissance Orbiter, Venus Express, Phoenix, and other missions. These explorations provided new understanding and insights on planetary atmosphere, space environments, surface processes, evolution and interior structure, as well as dynamics.

However, the recent results from various missions are challenging our previous understanding of the moon and other planets, such as the identification of ice, OH/H_2O, and new mineral components. For example, the early results showed that the moon and some planets have practically no atmosphere and lost their thermal energy in the initial stages of formation, so they have undergone little change since its early formation, unlike Earth, which has undergone drastic changes. Therefore, Moon and other planets have lots of long-standing questions, such as planetary environments, origin, formation and evolution, magnetization of crustal rocks, internal structure, and possible life. Furthermore, the high-resolution topography, gravity and magnetic field, surface processes, and interior activities of planets are not clear. One of the main factors is the lack of high-precision and high-resolution geodetic and remote-sensing techniques. Planetary geodesy and remote sensing from recent planetary missions, with higher spatial and spectral resolution, provided new opportunities to explore and understand Moon and planets in more detail. In this book, the methods and techniques of planetary geodesy and remote sensing are presented, as well as scientific results on probe orbit, topography, gravity field, crustal thickness, mineral components, major elements, clinopyroxene, and physical libration of planets.

This book provides the main techniques, methods, and observations of planetary geodesy and remote sensing, and their applications in planetary science for planetary explorers and researchers who have geodetic and remote-sensing background and experiences. Furthermore, it is also useful for planetary probe designers, engineers, and other specialists, for example, planetary geologists and geophysicists.

This work is supported by the National Basic Research Program of China (973 Program) (Grant no. 2012CB720000) and Main Direction Project of Chinese Academy of Sciences (Grant no. KJCX2-EW-T03). Also, we would like to gratefully thank Taylor & Francis/CRC Press for their efforts and cordial cooperation to publish this book.

Shuanggen Jin
Shanghai Astronomical Observatory
Chinese Academy of Sciences
Shanghai, China

Editor

Shuanggen Jin is a professor at the Shanghai Astronomical Observatory, Chinese Academy of Sciences. He completed his B.Sc. degree in geodesy/geomatics at Wuhan University in 1999 and his Ph.D. degree in GNSS/geodesy at the Chinese Academy of Sciences in 2003. His main research areas include satellite navigation, remote sensing, satellite gravimetry, and space/planetary sensing. He has written over 200 papers in various journals, five books/monographs, and he has five patents/software copyrights. He is the president of the International Association of Planetary Sciences (IAPS) (2013–2015), chair of the IAG Sub-Commission 2.6 (2011–2015), editor-in-chief of International Journal of Geosciences, associate editor of *Advances in Space Research* (2013), and editorial board member of *Journal of Geodynamics* and other six international journals. He has received many awards during his career, including the Special Prize of Korea Astronomy and Space Science Institute (2006), 100-Talent Program of Chinese Academy of Sciences (2010), Fellow of International Association of Geodesy (IAG) (2011), Shanghai Pujiang Talent Program (2011), Fu Chengyi Youth Science and Technology Award (2012), Second Prize of Hubei Natural Science Award (2012), Second Prize of National Geomatics Science & Technology Progress Award (2013), and Liu Guangding Geophysical Youth Science & Technology Award (2013).

Contributors

Sundaram Arivazhagan
Department of Geology
Periyar University
Salem, India

Misha Barkin
Moscow Aviation Institute
Moscow, Russia

Yuri Barkin
Sternberg Astronomical Institute
Moscow State University
Moscow, Russia

Shengbo Chen
College of Geoexploration Science
   and Technology
Jilin University
Changchun, China

Kaichang Di
Institute of Remote Sensing and
   Digital Earth
Chinese Academy of Sciences
Beijing, People's Republic of China

Jose Ferrandiz
University of Alicante
Alicante, Spain

Jian Guo
Department of Land Surveying and
   Geo-Informatics
The Hong Kong Polytechnic
   University
Kowloon, Hong Kong

Pengju Guo
College of Geoexploration Science
   and Technology
Jilin University
Changchun, China

Hideo Hanada
National Astronomical Observatory
   of Japan
Mizusawa, Japan

Daniel Heyner
Institute for Geophysics and Extra
   Terrestrial Physics
TU Braunschweig
Braunschweig, Germany

Han Hu
Department of Land Surveying and
   Geo-Informatics
The Hong Kong Polytechnic
   University
Kowloon, Hong Kong

and

Mapping and Remote Sensing
Wuhan University
Wuhan, People's Republic of China

Hauke Hussmann
DLR Institute of Planetary Research
Berlin, Germany

Shuanggen Jin
Shanghai Astronomical Observatory
Chinese Academy of Sciences
Shanghai, China

Fei Li
Mapping and Remote Sensing
Wuhan University
Wuhan, China

Yanqiu Li
College of Geoexploration Science
    and Technology
Jilin University
Changchun, China

Yi Lian
College of Geoexploration Science
    and Technology
Jilin University
Changchun, China

Bin Liu
Institute of Remote Sensing and
    Digital Earth
Chinese Academy of Sciences
Beijing, People's Republic of China

Yiliang Liu
Institute of Remote Sensing and
    Digital Earth
Chinese Academy of Sciences
Beijing, People's Republic of China

Ming Ma
College of Geoexploration Science
    and Technology
Jilin University
Changchun, China

Koji Matsumoto
RISE Project
National Astronomical Observatory
    of Japan
Oshu, Japan

and

National Astronomical Observatory
    of Japan
Mizusawa, Japan

Zhi-guo Meng
College of Geoexploration Science
    and Technology
Jilin University
Changchun, China

Man Peng
Institute of Remote Sensing and
    Digital Earth
Chinese Academy of Sciences
Beijing, People's Republic of China

Robert Tenzer
School of Geodesy and Geomatics
Wuhan University
Wuhan, China

Jingran Wang
College of Geoexploration Science
    and Technology
Jilin University
Changchun, China

Erhu Wei
School of Geodesy and Geomatics
Wuhan University
Wuhan, China

Johannes Wicht
Max Planck Institute for Solar
    System Research
Kaltenburg-Lindau, Germany

Bo Wu
Department of Land Surveying and
    Geo-Informatics
The Hong Kong Polytechnic
    University
Kowloon, Hong Kong

Yunzhao Wu
School of Geographic and
    Oceanographic Sciences
Nanjing University
Nanjing, China

Yansong Xue
Shanghai Astronomical Observatory
Chinese Academy of Sciences
Shanghai, China

Jianguo Yan
RISE Project
National Astronomical Observatory
    of Japan
Oshu, Japan

and

Mapping and Remote Sensing
Wuhan University
Wuhan, China

Wei Yan
National Astronomical
    Observatories
Chinese Academy of Sciences
Beijing, China

Tengyu Zhang
Shanghai Astronomical Observatory
Chinese Academy of Sciences
Shanghai, China

Ying Zhang
College of Geoexploration Science
    and Technology
Jilin University
Changchun, China

Ying Zhao
College of Geoexploration Science
    and Technology
Jilin University
Changchun, China

1

Lunar Geodesy and Sensing: Methods and Results from Recent Lunar Exploration Missions

Shuanggen Jin, Sundaram Arivazhagan, and Tengyu Zhang

CONTENTS

1.1 Introduction 2
1.2 Selenodetic Methods and Techniques 2
    1.2.1 Very Long Baseline Interferometry 2
    1.2.2 Lunar Laser Ranging 3
    1.2.3 Lunar Laser Altimetry 5
    1.2.4 Satellite Gravimetry 5
    1.2.5 Lunar Remote Sensing 6
1.3 The Recent Lunar Missions and Explorations 6
1.4 Selenodetic Results from Recent Lunar Missions 7
    1.4.1 Lunar Topography 7
    1.4.2 Lunar Gravity Field 8
    1.4.3 Major Lunar Minerals 9
    1.4.4 Surface Components Mapping 9
    1.4.5 Long-Lived Volcanism on the Lunar Far Side 10
    1.4.6 Lunar Interior Dynamics 11
1.5 Future Selenodetic Questions and Missions 12
    1.5.1 Tectonics of the Moon..12
    1.5.2 Lunar Quakes 13
    1.5.3 Lunar Dynamics..13
    1.5.4 Future Lunar Missions 13
1.6 Conclusion 14
Acknowledgments 15
References 15

1.1   Introduction

Early results showed that the Moon has practically no atmosphere and that it lost its thermal energy in the initial stages of formation. As a result, it has undergone minimal changes from its earlier formation unlike Earth, which has undergone drastic changes. In addition, the Moon is more heterogeneous and underwent prolonged igneous activity spanning hundreds of millions of years. This process led to varied and large-scale differentiation and fractionation during its early history. Therefore, long-standing questions such as lunar environments, origin, formation and evolution, magnetization of crustal rocks, internal structure, and possible life are of interest to the researcher.

Since 1960, the United States made its first attempt to obtain closer images of the lunar surface with the Ranger series. After that, several explorations of the Moon have been conducted from all over the world. These explorations provided a new understanding and further insights about the Moon. However, the recent results from various missions are challenging our previous understanding of the Moon, including identification of ice on polar regions, OH/H_2O, and new mineral identification. Furthermore, the far-side gravity field, high-resolution topography, surface processes, and interior activities are not clear. One of the main factors is the lack of high precision and resolution mapping. The selenodesy from recent lunar missions provided new opportunities to explore and understand the Moon in more detail with higher spatial and spectral resolution. In this chapter, the selenodetic methods, techniques, and results are presented and some scientific questions about the Moon are discussed along with future lunar missions.

1.2   Selenodetic Methods and Techniques

1.2.1   Very Long Baseline Interferometry

Very long baseline interferometry (VLBI) is one of the main technologies in deep space probe orbit determination because of its high precision and high angular resolution (Liu et al. 2009). For example, the unified S-band (USB) monitoring system and VLBI have been successfully used for CE-1's orbit determination. In order to improve the precision of VLBI, differential VLBI observations have been developed to eliminate the common errors of the probe's radio signals and radio source's signals during the propagation path, such as the station location errors and transmission media delays (Wei et al. 2013). The principle of ΔVLBI is illustrated in Figure 1.1a (Sekido et al. 2005), in which $\bar{K}$ is the direction vector of the CE-1 signal, $\bar{I}$ is the direction vector of an alternating observed radio source signal, and $\bar{B}$ is the baseline vector

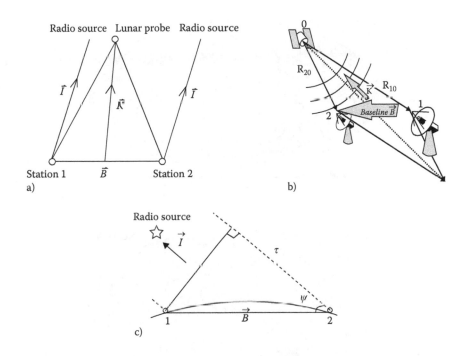

FIGURE 1.1
Principle of ΔVLDI observations a), where R_{10} and R_{20} represent the distance of the CE-1 to ground stations 1 and 2, respectively b), which are not equal and so $\bar{K}$ is neither parallel to R_{10} nor R_{20}, ψ is the angle between $\bar{I}$ and $\bar{B}$ and c), τ is the time delay of the radio source signals.

of two ground stations. The propagation paths of the CE-1 and radio source signals are different (Figure 1.2b and c).

1.2.2 Lunar Laser Ranging

The basic theory of lunar laser ranging (LLR) is to measure the distance from Earth to the Moon as

$$D = \frac{ct}{2} \tag{1.1}$$

where c is the speed of light and t is the time for the laser to reflect back. The round-trip time is affected by a host of factors, such as the relative motion of Earth and the Moon, rotation of the Earth, lunar libration, polar motion, and propagation delay through Earth's atmosphere. The very first LLR experiment was developed in the late 1950s for the gravitational program at Princeton University. The distance was measured between Earth and the Moon using the laser round-trip time during the Apollo program. Smullin and Fiocco (1962) at the Massachusetts Institute of Technology observed the

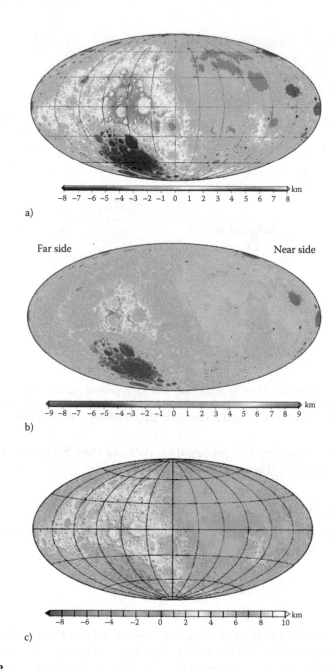

FIGURE 1.2
Topographic maps of the Moon: a) Chang'E-1. (From Su, X., et al., 42nd Lunar and Planetary Science Conference, Abstract # 1077, 2011.) b) SELENE laser altimetry. (From Araki, H., et al., *Science*, 323, 897–900, 2009.) c) LRO laser altimetry. (From Smith, D.E., et al., *Geophys. Res. Lett.*, 37, L18204, 2010.)

laser light pulse reflecting from the lunar surface using a laser with millisecond pulse length. The ranging accuracy was greatly improved to tens of centimeters after the installation of a retroreflector array by the Apollo Mission in 1969. In the later missions, Apollo 14 and 15, two more retroreflectors were installed on the lunar surface. Both the Lick observatory and the McDonald observatory received the return laser light pulse, and some results from LLR have been obtained concerning Earth's rotation, precession, and coordinate system as well as the gravitational phenomenology and understanding of the lunar interior.

1.2.3 Lunar Laser Altimetry

The theory of laser altimetry measures the two-way time of laser pulse flight to determine the distance between the surface and the orbiter. The laser altimeter is an instrument that measures the distance from an orbiting spacecraft to the surface of the planet or an asteroid to produce a precise global lunar topographic model. The lunar orbiter laser altimeter (LOLA), with a payload on the lunar reconnaissance orbiter, can measure the lunar surface topography and provide more detailed information on landing sites and a precise global geodetic grid on the Moon. Meanwhile, the elevations can be used to estimate the slope over a range of baselines, and the root mean square roughness of the surface at the scale of the laser footprint can be measured by the time of backscattered pulses. In addition, the high precision and resolution global topography of the Moon can be used to investigate the lunar origin and evolution related to the precise figure of the Moon, the nature of gravity anomalies, and the thermal and loading histories of major impact basins.

1.2.4 Satellite Gravimetry

The lunar gravity field plays an important role in understanding the interior structure and evolution of the Moon. Since the first Russian lunar mission of Luna 10 in 1966 (Akim 1966), many efforts for more precise lunar gravity field models have been made. However, various factors restrict the precision and resolution of the lunar gravity field. Recently, the Gravity Recovery and Interior Laboratory (GRAIL) mission was designed to determine the lunar gravity field (Zuber et al. 2012). GRAIL is a spacecraft-to-spacecraft tracking mission of the Moon, in principle similar to the gravity recovery and climate experiment (GRACE) mission. Each GRAIL spacecraft has a single science instrument, the lunar gravity ranging system (LGRS), which measures the change of distance between the twin orbiting spacecrafts flying above the lunar surface. The spacecrafts are perturbed by the gravitational attraction of topography and subsurface mass variations. The satellite-to-satellite tracking observations from GRAIL can determine with high precision, the lunar gravity field with up to degree-and-order 660, which is a great improvement when compared with the previous lunar gravity models. The GRAIL field revealed new features and internal structure of the Moon, including tectonic structures, volcanic landforms, basin rings, crater central peaks, and numerous simple craters.

1.2.5 Lunar Remote Sensing

Advanced remote sensing techniques, such as the lunar radar and long-wave infrared camera, play an important role in lunar exploration. A set of bistatic radar observations was carried out on the Clementine spacecraft in 1994 and an enhancement in echoes with the same sense circular polarization from regions near the South Pole in a near backscatter was detected, which demonstrated the possible existence of large quantities of water ice near the pole (Nozette et al. 1994). In addition, the spectral data from the Clementine mission together with data from the new 12.5-cm Mini-RF radar of the Lunar Reconnaissance Orbiter were used to map TiO_2 at higher resolution.

The long-wave infrared camera (LWIR) on the Clementine spacecraft was a single passband at a wavelength of 8.75 μm, which systematically mapped the lunar surface at visible, near-infrared, and thermal-infrared wavelengths. The Clementine LWIR camera was designed to study the thermal properties of the lunar regolith and some results on various lunar thermophysical properties were obtained. During the day, the surface temperature of the Moon is controlled by albedo at low incidence angles. However, at higher angles, the lunar temperature is more controlled by the surface roughness (Lawson et al. 2000).

1.3 The Recent Lunar Missions and Explorations

Lunar geologic mapping started in the late 1960s when five lunar orbiter (LO) missions were launched in 1966 and 1967, providing an excellent photographic image at ~99% coverage (Hansen 1970). Exploration of lunar geology was intensified after the successful landing of the Apollo 11 mission in 1969. Since then, explorations have been carried out to understand the chemistry, mineralogy, and rock types of the Moon. The Moon has not been geologically mapped in a systematic fashion for more than 25 years, and some major advances in lunar science have occurred in the last 15 years.

Recently, several countries have launched lunar missions to explore the lunar surface and to understand the Moon (Jin et al. 2013). The important and recent lunar exploration missions are Japan's Kaguya (SELENE) that was launched on September 14, 2007, to obtain scientific data of the lunar origin and evolution and to develop the technology for the future lunar exploration; China's Chang'E-1 lunar orbiter that was launched on October 24, 2007, to study the

TABLE 1.1

Scientific Instruments for Chang'E-1, SELENE, Chandrayaan-1, and LRO/LCROSS Mission

Instruments	Chang'E-1 (China, 2007)	SELENE (Japan, 2007)	Chandrayaan-1 (India, 2008)	LRO (USA, 2009)
Stereo imager	X	X	X	X
VNIR camera	X	X	X	
UV imager	—			X
IR spectrometer	—	X	X	
Magnetometer	—	X		
X-ray spectrometer	X	X	X	
γ-ray experiment	X	X		
Neutron detector				X
Laser altimeter	X	X	X	X
Plasma/ion experiments	X	X	X	X
Sub-Kev atom reflecting analyzer			X	
Microwave sounder	X	X	X	X
Thermal emission radiometer				X
Radiation dose monitor			X	
Penetrator/impactor			X	X

Abbreviation: VNIR camera, visible and near-infrared camera.

lunar environment and the three-dimensional surface topography; India's orbital satellite Chandrayaan-1 launched on October 22, 2008; and USA's Lunar Reconnaissance Orbiter/Lunar Crater Observation and Sensing Satellite (LRO/ LCROSS) launched on June 18, 2009, to search for water ice in a permanently shadowed crater near one of the Moon's poles. In addition, on September 10, 2011, the NASA's Discovery Program GRAIL was launched to get a high-quality gravitational field mapping of the Moon to determine its interior structure. The scientific instruments used in various missions are shown in Table 1.1.

1.4 Selenodetic Results from Recent Lunar Missions

1.4.1 Lunar Topography

The global lunar topographic model and lunar digital elevation model with 3-km spatial resolution were obtained using laser altimeter data from Chang'E-1 (Li et al. 2010). The first 2-months data of Chang'E-1 with careful calibration were used to produce a global topographic map of the Moon with a vertical accuracy of approximately 30 m and a spatial resolution of ~7.5 km (Su et al. 2011). Topographic maps of the Moon from Chang'E-1 and SELENE laser altimetry are shown in Figure 1.2. This Chang'E-1-derived topographic map has a little higher spatial resolution than the one from SELENE.

1. Topographic map from Chang'E-1
2. Topographic map from SELENE
3. Topographic map from LRO LOLA

LOLA is a payload element on the LRO. The objective of the LOLA investigation is to characterize potential future robotic or human landing sites and to provide a precise global geodetic grid of the Moon. In addition, LOLA offered surface slope, surface roughness, and reflectance as ancillary measurements. The maps produced using high-resolution LOLA data are of definite help to future lunar projects in selecting precise landing locations.

1.4.2   Lunar Gravity Field

The lunar gravity field plays a pivotal role in understanding the structure and the evolution of our celestial neighbor. As the Moon is in a state of synchronous rotation, it is difficult to obtain detailed information of the "global" lunar gravity field. Lunar surface gravity at the equator is 5.32 ft/sec^2 (1.622 m/sec^2), nearly 1/6th of Earth's gravity of 32.174 ft/sec^2 (9.806 m/sec^2). The lunar gravity field is uneven due to mass concentration variations in the near and far sides. The far-side lunar surface has never been directly observed by lunar missions. The near-side lunar gravity is measured through two-way Doppler tracker orbiting the Moon, which is of low sensitivity. SELENE measured the far-side gravity field using a four-way Doppler measurement with a round trip communication between the Earth and the main orbiter on the far side through a relay satellite. Matsumoto et al. (2010) determined the SELENE lunar gravity field model with degree-and-order 100 (SGM100h), including 14.2 months of SELENE Doppler range data with all the usable four-way Doppler data and the historical tracking data of the LOs, Apollo subsatellites, Clementine, SMART-1, and Lunar Prospector. The SGM100h revealed in greater detail the far-side, free-air gravity anomalies, which have larger correlation with the SELENE and laser altimeter-derived topography (Araki et al. 2009). Ishihara et al. (2009) computed the lunar crustal thickness globally and confirmed that the crust below the Moscoviense basin on the far side is extremely thin. Ishihara et al. (2011) proposed a double impact hypothesis for the origin of the Moscoviense basin using the SELENE topographic map of the area, which has triple and offset ring structure.

Recently, the high-precision lunar gravity field with up to degree-and-order 660 was determined from the GRAIL mission, which is a great improvement in comparison with the previous lunar gravity models (Zuber et al. 2012). The GRAIL field reveals new features and internal structure of the Moon, including tectonic structures, volcanic landforms, basin rings, crater central peaks, and numerous simple craters.

1.4.3 Major Lunar Minerals

The γ-ray (GRS) and x-ray spectrometer (XRS) onboard Chang'E-1 can retrieve the abundance of rock-forming elements and map the major rocks and minerals to evaluate the important resources of the Moon (Ogawa et al. 2008; Grande et al. 2009). A multichannel microwave radiometer is used to measure the brightness temperature (TB) of the Moon to derive the thickness of lunar regolith and estimate the amount of helium-3 (Haruyama et al. 2008; Ohtake et al. 2009). The Chang'E-1 Interference Imaging Spectrometer (IIM) completed 84% coverage of the lunar surface between 70°S and 70°N. Chang'E-1's GRS, IIM, and XRS have determined the abundance of some key elements and distribution of major minerals of the Moon (Ouyang et al. 2010). Liu et al. (2010) characterized the Ti spectral features including full width at half maximum (FWHM), absorption position, depth, area, and symmetry using Chang'E-1 IIM images, Reflectance Experiment Laboratory (RELAB) spectra, and Lunar Soil Characterization Consortium (LSCC) data. Ling et al. (2011) developed a new algorithm to map TiO_2 abundance using Chang'E-1 IIM data, and then validated it with the results of Apollo 16 landing site samples and Clementine UVVIS-derived TiO_2 data. Highland and mare regional studies suggested that the TiO_2 map has a good correlation with the Clementine UVVIS results in corresponding areas. The global inventory of helium-3 in lunar regolith was estimated using data from the Chang'E-1 multichannel microwave radiometer (Fa et al. 2010).

1.4.4 Surface Components Mapping

The Lyman Alpha Mapping Project (LAMP) ultraviolet spectrograph onboard LRO observed the plume generated by the Lunar Crater Observation Sensing Satellite (LCROSS) impact as far-ultraviolet emissions from the fluorescence of sunlight by molecular hydrogen and carbon monoxide and resonantly scattered sunlight from atomic mercury, with contributions from calcium and magnesium (Gladstone et al. 2010). The light curve from LAMP is completely consistent with the immediate formation of H_2, and substantial contribution of H_2 from the photolysis of water is exempted by the strict upper limits on other photolysis products, for example, H and O. The LCROSS mission was designed to provide evidence of water ice, which may be present in permanently shadowed craters of the Moon. On October 9, 2009, a spent Centaur rocket struck the Permanently Shadowed Region (PSR) within the lunar south pole crater Cabeus, ejecting debris, dust, and vapor, which was observed by a second "shepherding" spacecraft. Water vapor, ice, and ultraviolet emissions attributable to hydroxyl radicals support the presence of water in the debris recognized by Near Infrared (NIR) absorbance. Light hydrocarbons, sulfur-bearing species, and carbon dioxide were observed along with water (Colaprete et al. 2010).

The Diviner Lunar Radiometer Experiment (DLRE) on the LRO provided global coverage maps of thermal-infrared derived compositions on the Moon for the first time in lunar exploration history. The DLRE compositional investigation relied primarily on the three shortest wavelength thermal infrared channels near 8 μm: 7.55–8.05 μm, 8.10–8.40 μm, and 8.38–8.68 μm. These observations were consistent with previous regional lunar surface observations and comparable with laboratory measurements of returned lunar rocks and soils in a simulated lunar environment (Murcray et al. 1970).

Mini-RF is a lightweight synthetic aperture radar (SAR) on NASA's LRO operated at either 12.6- or 4.2-cm wavelength with 7.5 m/pixel spatial resolution and ~18 and ~6 km swath, respectively. Mini-RF has acquired good quality data with radar backscatter properties of the lunar surface. Thomson et al. (2011) reported Mini-RF observations of the 20-km-diameter Shackleton crater situated at the inner rim massif of the much higher South Pole-Aitken Basin. LCROSS impacts nearby Cabeus region (Colaprete et al. 2010) and Lunar Prospector as well as LRO's lunar exploration neutron detector suggested polar concentrations of excess hydrogen (Lawrence et al. 2006; Mitrofanov et al. 2010). Mini-RF remains a viable candidate site of polar volatile accumulation when viewed in conjunction with other data like that from Diviner (Paige et al. 2010).

In addition, parts of the basaltic regions from Mare Orientale, Lacus Veris, and Lacus Autumni of the Orientale basin were studied with the Moon Mineralogical Mapper (M^3) data of the Chandrayaan-1 orbiter. The spectral profiles and LSU techniques were used to characterize the various basalts such as low-, medium-, and high-Ti basalts in the Mare Orientale basin and compared with RELAB basaltic spectra. The RELAB averaged chemistry and spectral data of low (15071), medium (12030), and high (71501) Ti basalts were taken as the end member for analysis. The distribution and nature of TiO_2 basalts in the Orientale basaltic regions were analyzed in a quantitative manner. The low-, medium-, and high-Ti basalt's concentration map was established for the Orientale region (Figure 1.3). The results are consistent with RELAB and conventional analysis. It was been found that Mare Orientale has high-Ti basalts compared to the nearest Lacus Autumni and Lacus Veries. With the fully calibrated M^3 data, we would be able to characterize basalts globally in the lunar surface. This task provides a new challenge in lunar studies.

1.4.5 Long-Lived Volcanism on the Lunar Far Side

The volcanic history of the Moon is important to understand the lunar origin and evolution. However, it is very difficult to get the precise age of mare deposits due to few lunar samples returned by Apollo and the lack of high spatial

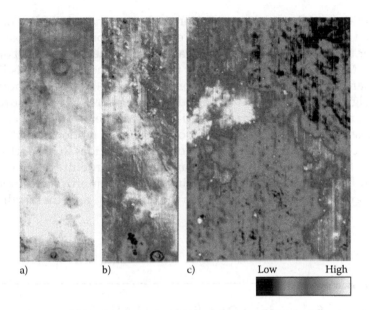

FIGURE 1.3
Low and high Ti basalt characterization using M^3 and RELAB spectral data. a) Mare Orientale; b) Lacus Veries; c) Lacus Autumni.

resolution images, particularly in the far side. SELENE provided a unique opportunity to map the whole lunar surface using terrain camera (TC) with 10 m/pixel to determine the model ages of mare deposits on the Moon based on the crater size frequency distributions (CSFD). From high-resolution TC images, it was found that most mare volcanism on the lunar far side stopped before 3 billion years, while a few mare deposits stopped before ~2.5 billion years (Haruyama et al. 2009). Recent missions Chandrayaan-1 (TMC—5 m/pixel) and LRO (LROC-NAC—0.5 m/pixel) provided higher spatial resolution images than those from the TC. These fully calibrated TMC and LROC images will delineate further interpretation of the lunar surface features.

1.4.6 Lunar Interior Dynamics

The LLR can determine precise lunar orbit and ephemeris (Dicky et al. 1994). The accuracy of lunar orbit determined with laser ranging data had dramatic improvement compared with the classical optical data, while the orientation was improved at least two orders of magnitude and the radial distance variations were improved to 2~3 cm range-accuracy. Therefore, a continuous supply of high-quality measurements and more analysis of data will greatly contribute to maintaining and enhancing accuracy. Since the Sun has a strong influence on the lunar orbit, it somehow permits the range data to be used to determine the mass ratio of the Sun/(Earth+Moon) and the relative orientation of the Earth–Moon system.

In addition, LLR can test gravitational physics and relativity. For example, the LLR can be used to test the equivalence principle by examining whether the Moon and Earth accelerate alike in the Sun's field. For Earth and the Moon accelerated by Sun, if the equivalence principle is broken, the lunar orbit will be displaced along the Earth–Sun line, producing a range signal with a 29.53-day period. Furthermore, LLR measurements can observe the Earth's rotation variations and determine the constants of precession and nutation, station coordinates and motions, Earth's gravitational coefficient, and tides. Therefore, LLR provides an opportunity to obtain information on the dynamics and interior structure of the Moon. In the near future, the expected increasing data and improved accuracy will permit greater understanding of Earth, the Moon, and the Earth–Moon system.

1.5 Future Selenodetic Questions and Missions

1.5.1 Tectonics of the Moon

The Moon is the only extraterrestrial object studied extensively in all aspects in the twenty-first century including structural geology. The finite deformation of the lunar lithosphere was visualized by tectonic features such as wrinkle ridges and straight rilles and interpreted as an anticlinal fold (Watters and Schultz 2010). These features imply that the secular and spatial changes of the surface and the boundary condition of stress have been controlled by endogenic and exogenic processes. Such information not only limits our understanding of the mechanical properties of the lunar lithosphere, but also on the regional and global thermal histories and even on the origin of the Earth–Moon system (Atsushi et al. 1998). Impact basins are found in tectonic landforms in and around the near side lunar maria, such as wrinkle ridges, rilles, and narrow troughs, which exclusively occur in mare basalts and margins of basins adjacent to highlands. The Lee–Lincoln scarp located on the Taurus–Littrow valley near the Apollo 17 landing site exemplifies an important class of lunar faults, the lobate scarp thrust faults. The vast majority of the Moon's large-scale tectonic features are found in the basalt-filled impact basins and the adjacent highlands (Watters and Johnson 2010).

There are a few outstanding questions to be addressed including the tectonic history of the Moon. High-resolution imaging is required for the determination of global distribution and ages of the lobate scarp thrust faults. The next critical question is, what is the source mechanism for the shallow and deep-seated moonquakes? The present mission's data disclosed detailed information about the critical understanding of lunar tectonics.

1.5.2 Lunar Quakes

Scientists study moonquakes to understand the structure, interior, and dynamic activity of the Moon. Seismologists study seismic waves in the subsurface of the Moon to understand the interior and its behavior. With the successful installation of a geophysical station at Hadley Brille, on July 31, 1971, on the Apollo 15 mission and the continued operation of stations 12 and 14 approximately 1100 km SW, the Apollo program for the first time achieved a network of seismic stations on the lunar surface (Latham et al. 1971). A recent reevaluation of the 1970s Apollo data suggests that "the moon is seismically active," and the possibility of four different models of moonquakes are suggested. The vibrations from bigger earthquakes cease within 2 min, but vibrations from moonquakes exist for more than 10 min like those of a tuning fork. This nature of the moonquake is attributed to the presence of water-depleted stone, which has a different mineral structure expansion and a dry, cool, and rigid surface and interiors. In order to establish long-standing lunar bases, it is mandatory to understand the occurrences and nature of moonquakes. This understanding could be achieved by deploying network seismometers designed to collect precise seismic data from different regions of the Moon.

The current understanding of the Moon is based on existing seismic data and as more questions arise, they need to be answered by future missions. Compared to Apollo and LROC images, the wrinkles have grown in the past few decades, but the Moon's diameter has shrunk by just 200 m in the last few billion years, suggesting that recent cooling is relatively small (Watters et al. 2010). In the near future, if we wish to set up a base station and think of habitation on the lunar surface, we ought to understand the lunar dynamics, moonquakes, and tectonics through geophysical methods such as gravity and seismic study of the Moon.

1.5.3 Lunar Dynamics

Analysis of LLR data provides information on the lunar orbit, rotation, and solid-body tides. The lunar dynamical and thermal evolution may be connected since heat is generated by both tidal and core and mantle boundary (CMB) dissipation, and the Moon's former molten interior affects its tidal response. Information pertaining to the past magnetic field and thermal history is of great interest. The gravity field determined from orbiting spacecraft has been used in combination with LLR results to determine the lunar moment of inertia and rotation. The more accurate gravity field from GRAIL and future missions can detect the lunar inner core and properties, which are still not completely understood.

1.5.4 Future Lunar Missions

The past and present lunar missions have provided rock samples and high-resolution imaging data to understand the lunar surface features and evolution, which have resulted in a lot of curiosity among the lunar scientific community. Hence, almost all leading space agencies have planned their own missions to the Moon and the results are available in the literature and news-letters. These lunar missions include the United States' Lunar Atmosphere and Dust Environment Explorer (LADEE-2013) and International Lunar Network (2018), United Kingdom's the Moon Lightweight Interior and Telecoms Experiment (MoonLITE-2014), India's Chandrayaan-2 (2014), Russia's Lunar Glob 1 and 2 (2014/2015), which will enable us to answer the unresolved questions on lunar exploration and science.

1.6 Conclusions

The Chang'E-1, SELENE-1, Chandrayaan-1, and LRO missions provided a good source of information about the lunar gravity field, surface geomor-phology, and major components of the Moon. These data are sufficient for basic understanding, but when the question extends to the possibility of human habitability and setting up a permanent base station, more data are required for further understanding of lunar exploration activities. New missions like Chang'E-3 & 4, SELENE-2, and Chandrayaan-2 are expected to provide photogeological, mineralogical, and chemical data with which the elemental level chemical mapping can be done. Through that we can understand the stratigraphy and nature of the Moon's crust and thereby assess in certain aspects of magma ocean hypothesis. The improved spatial and spectral images will allow us to determine the compositions of impac-tors that bombarded the Moon during its early evolution, which is also rel-evant to the understanding of Earth's formation. These images help us to determine the compositions of the projectiles (asteroid and comet material) from the fresh impact craters and impact ejecta, which are not affected by impact melts, topography, structural features, morphological characters, and surface roughness. The major problem in the lunar surface is that the original crust is blanketed by regolith and dust due to impact and space weathering. The central peaks of the older craters and recent impact (fresh) craters allow us to study the compositional information of the original crust of the lunar surface.

Based on the current knowledge of the geologic setting map of the lunar surface, we have to identify the potential and promising test sites for further exploration with high spatial/spectral resolution and sample collection. The images from the above-mentioned missions are still at the calibration level. Once the calibrated images are available to the public domain, we could expect much more striking results and new findings from the current and forthcoming lunar mission's data sets, particularly from the recent missions, namely, Chang'E-3, LADEE, MoonLITE, Chandrayaan-2, and SELENE-2.

Acknowledgments

The authors are grateful for contributions from the SELENE (Kaguya), Chang'E-1, Chandrayaan-1, and LRO/LCROSS programs. This work was supported by the National Basic Research Program of China (973 Program) (Grant No. 2012CB720000) and the Main Direction Project of Chinese Academy of Sciences (Grant No. KJCX2-EW-T03).

References

Akim, E.L., (1966), Determination of the gravitational field of the Moon from the motion of the artificial satellite "Lunar-10," *Dokl. Akad. Nauk SSSR*, 170, 799–802.

Araki, H., Tazawa, S., Noda, H., et al., (2009), Lunar global shape and polar topography derived from Kaguya-LALT laser altimetry, *Science*, 323, 897–900.

Atsushi, Y., Sho, S., Sushi, Y.Y., Takayuki, O., Junichi, H., Tatsuaki, O., (1998), Lunar tectonics and its implications for the origin and evolution of the moon, *Memoirs Geol. Soc. Jpn.*, 50, 213–226.

Chin, G., Brylow, S., Foote, M., et al., (2007), Lunar reconnaissance orbiter overview: The instrument suite and mission, *Space Sci. Rev.*, 129(4), 391–419.

Colaprete, A., Schultz, P., Heldmann, et al., (2010), Detection of water in the LCROSS ejecta plume, *Science*, 330(6003), 463–468.

Dickey, J.O., Bender, P.L., Faller, J.E., et al., (1994), Lunar laser ranging: A continuing legacy of the Apollo program, *Science*, 265(5171), 482–490.

Fa, W.Z., Jin, Y.Q., (2010), Global inventory of helium-3 in lunar regoliths estimated by a multi-channel microwave radiometer on the Chang'E 1 lunar satellite, *Chin. Sci. Bull.*, 55(35), 4005–4009.

Gladstone, G.R., Hurleu, D.M., Retherford, K.D., et al., (2010), LRO-LAMP observations of the LCROSS impact plume, *Science*, 330(6003), 472–476.

Grande, M., Maddison, B.J., Howe, C.J., et al., (2009), The C1XS x-ray spectrometer on Chandrayaan-1, *Planet. Space Sci.*, 57, 717–724.

Hansen, T.P., (1970), Guide to lunar orbiter photographs: Lunar orbiter photographs and maps missions 1 through 5, *NASA SP-242*, N71-36179, 254.

Haruyama, J., Ohtake, M., Matsunaga, T., et al., (2009), Long-lived volcanism on the lunar farside revealed by SELENE terrain camera, *Science*, 323(5916), 905–908.

Haruyama, J., Ohtake, M., Matsunaga, T., et al., (2008), Lack of exposed ice inside lunar south pole Shackleton crater, *Science*, 322(5903), 938–939.

Ishihara, Y., Goossens, S., Matsumoto, K., Noda, H., Araki, H., Namiki, N., Hanada, H., Iwata, T., Tazawa, S., Sasaki, S., (2009), Crustal thickness of the moon: Implications for farside basin structures, *Geophys. Res. Lett.*, 36(19), L19202.

Ishihara, Y., Morota, T., Nakamura, R., Goossens, S., Sasaki, S., (2011), Anomalous Moscoviense basin: Single oblique impact or double impact origin? *Geophys. Res. Lett.*, 38(3), L03201.

Jin, S.G., Arivazhagan, S., Araki, H., (2013), New results and questions of lunar exploration from SELENE, Chang'E-1, Chandrayaan-1 and LRO/LCROSS, *Adv. Space Res.*, 52(2), 285–305.

Lawson, S.L., Jakosky, B.M., Park, Hye-Sook, Mellon, M. T. 2000. Brightness temperatures of the lunar surface: Calibration and global analysis of the Clementine long-wave infrared camera data, *J. Geophys. Res.*, 105(E2), 4273–4290.

Latham, G., Ewing, M., Dorman, J., Lammlein, D., Press, F., Toksoz, N., Sutton, G., Duennebier, F., Nakamura, Y., (1971), Moonquakes and lunar tectonism, *The Moon*, 4(3–4), 373–382.

Lawrence, D.J., Feldman, W.C., Elphic, R.C., Hagerty, J.J., Maurice, S., McKinney, G.W., Prettyman, T.H., (2006), Improved modeling of Lunar Prospector neutron spectrometer data: Implications for hydrogen deposits at the lunar poles, *J. Geophys. Res.*, 111, E08001.

Li, C., Ren, X., Liu, J., et al., (2010), Laser altimetry data of Chang'E-1 and the global lunar DEM model, *Sci. China Earth Sci.*, 53(1), 1582–1593.

Ling, Z., Zhang, J., Liu, J., et al., (2011), Preliminary results of FeO mapping using Imaging Interferometer data from Chang'E-1, *Chin. Sci. Bull.*, 56(4–5), 376–379.

Liu, Q.H., Kikuchi, F., Goossens, S., (2009), S-band same-beam VLBI observations in SELENE (Kaguya) and correction of atmospheric and ionospheric delay, *J. Geodet. Soc. Jpn.*, 55(2), 243–254.

Liu, F., Shi, J., Le, Q., Rong, Y., (2010) Lunar titanium characterization based on Chang'E (CE-1) interference imaging spectrometer (IIM) imagery and RELAB spectra, 41st Lunar and Planetary Science Conference, Abstract # 1642.

Matsumoto, K., Goossens, S., Ishihara, Y., et al., (2010), An improved lunar gravity field model from SELENE and historical tracking data: Revealing the farside gravity features, *J. Geophys. Res.*, 115(E6), E06007.

Mitrofanov, I.G., Sanin, A.B., Boynton, W.V., et al., (2010), Hydrogen mapping of the lunar south pole using the LRO neutron detector experiment LEND, *Science*, 330(6003), 483–486.

Murcray, F.H., Murcray, D.G., Williams, W.J., (1970), Infrared emissivity of lunar surface features 1. Balloon-borne observations. *J. Geophys. Res.*, 75(14), 2662–2669.

Nozette, S., Rustan, P., Pleasance, L.P., et al., (1994), The Clementine mission to the moon: Scientific overview, *Science*, 266(5192), 1835–1839.

Ogawa, K., Okada, T., Shira, K., Kato, M., (2008), Numerical estimation of lunar x-ray emission for x-ray spectrometer onboard SELENE. *J. Earth Planets Space* 60(4), 283–292.

Ohtake, M., Matsunaga, T., Haruyama, J., et al., (2009), The global distribution of pure anorthosite on the moon, *Nature*, 461(7261), 236–240.

Ouyang, Z., Li, C., Zhou, Y., et al., (2010), Chang'E-1 lunar mission: An overview and primary science results. *Chin. J. Space Sci.*, 30(5), 392–403.

Paige, D.A., Siegler, M.A., Zhang, J., et al., (2010), Diviner lunar radiometer observations of cold traps in the moon's south polar region. *Science*, 330(6003), 479–482.

Sekido, M., Fukshima, T., (2005), Relativistic VLBI delay model for finite distance radio source. In: *International Association of Geodesy Symposia*: vol. 128, Springer, Berlin, pp. 141–145.

Smullin, L.D., Fiocco, G., (1962), Project Luna Sea. *Inst. Elec. Electron. Eng. Proc.*, 50 (7), 1703–1704.

Smith, D.E., Zuber, M.T., Jackson, G.B., et al., (2010a), The lunar orbiter laser altimeter investigation on the lunar reconnaissance orbiter mission, *Space Sci. Rev.*, 150(1–4), 209–241.

Smith, D.E., Zuber, M.T., Neumann, G.A., et al., (2010b), Initial observations from the Lunar Orbiter Laser Altimeter (LOLA), *Geophys. Res. Lett.*, 37(18), L18204.

Su, X., Huang, Q., Yan, J., Ping, J., (2011), The improved topographic model from Chang'E-1 mission, 42nd Lunar and Planetary Science Conference, Abstract # 1077.

Thomson, B.J., Bussey, D.B.J., Cahill, J.T.S., Neish, C., Patterson, G.W., Spudis, P.D., (2011). The Interior of Shackleton crater as revealed by Mini-RF orbital radar, 42nd Lunar and Planetary Science Conference, Abstract # 1626.

Watters, T.R., Schultz, R.A., (2010), *Planetary Tectonics: Introduction, Planetary Tectonics* (Eds); Watters, T.R., Schultz, R.A., Cambridge University Press, pp. 1–14.

Watters, T.R., Johnson, C.L., (2010), *Lunar Tectonics, Planetary Tectonics* (Eds); Watters, T.R., Schultz, R.A., Cambridge University Press, pp. 121–182.

Watters, T.R., Robinson, M.S., Beyer, R.A., Banks, M.E., Bell, J.F., Pritchard, M.E., Hiesinger, H., Bogert, C.H.V., Thomas, P.C., Turtle, E.P., Williams, N.R., (2010), Evidence of recent thrust faulting on the moon revealed by the Lunar Reconnaissance Orbiter Camera, *Science*, 329(5994), 936–940.

Wei, E., Yan, W., Jin, S.G., Liu, J., Cai, J., (2013), Improvement of earth orientation parameters estimate with Chang'E-1 ΔVLBI observations, *J. Geodyn.*, 72, 46–52.

Zuber, M., Smith, D., Watkins, M., Asmar, S., Konopliv, A., Lemoine, F., Melosh, H., Neumann, G., Phillips, R., Solomon, S., Wieczorek, M., Williams, J., Goossens, S., Kruizinga, G., Mazarico, E., Park, R., Yuan, D., (2012), Gravity field of the moon from the Gravity Recovery and Interior Laboratory (GRAIL) mission, *Science*, 339, 668–671.

2

Improvement of Chang'E-1 Orbit Accuracy by Differential and Space VLBI

Wei Yan, Shuanggen Jin, and Erhu Wei

CONTENTS

2.1   Introduction 20
2.2   Differential VLBI 21
    2.2.1   Relativistic ΔVLBI Model for the CE-1 Transfer Orbit 22
    2.2.2   Adjustment Model 24
    2.2.3   Observations and Processing Strategies 24
        2.2.3.1   Observations 24
        2.2.3.2   Orbit Determination Strategy 25
        2.2.3.3   Strategies and Methods 25
    2.2.4   Results and Discussion 26
        2.2.4.1   Parameter Estimation Using Time Delay
                Observations of CE-1 on October 31, 2007 27
        2.2.4.2   Overall Estimation of the Entire Transfer Orbit 30
    2.2.5   Summary 32
2.3   Space VLBI 34
    2.3.1   Relativistic SVLBI Model for the CE-1 Transfer Orbit 35
    2.3.2   Adjustment Model 37
    2.3.3   Estimation of Unknown Parameters 39
    2.3.4   Experiment 39
        2.3.4.1   Parameter Estimation under the Actual
                Observation Conditions of CE-1 40
        2.3.4.2   Optimal Conditions of Observation for SVLBI 43
        2.3.4.3   Parameter Calculation under Optimal Conditions
                for Observation 46
    2.3.5   Summary 47
2.4   Conclusions 49
Acknowledgments 49
References 49

2.1 Introduction

Deep space exploration has become a topic of great interest in recent years. Following the exciting soft-landing and extraordinary roving exploration around the Moon, the Chang'E-3 probe (CE-3), equipped with the Lander and the Rover vehicles, has established a new milestone in China's history of deep space exploration programs. CE-3's roving exploration of the lunar surface was accomplished using power supply; thermal control; telemetry, tracking, and control (TT&C); and communications support during lunar days. Lunar radars, infrared spectrometers, panoramic cameras, and particle excitation x-ray spectrometers were used to conduct scientific exploration tasks and accumulate rich experience for the future development of China's lunar exploration programs. CE-3 is the major component of the second phase of China's lunar exploration program, following the deployment of Chang'E-1 (CE-1) and Chang'E-2 (CE-2) in the first phase. CE-1's flight orbit included a phasing orbit, transfer orbit, and mission orbit, which are typical of a lunar probe orbit. Thus, a study on the orbit determination of CE-1 had important practical significance and application value.

CE-1 was China's first lunar probe; it was successfully launched on October 24, 2007. CE-1 has provided abundant information on various parameters such as the lunar landscape, soil, gravity field, and atmosphere (Ouyang et al. 2010; Jin et al. 2013; Hering et al. 1991). To overcome the inadequacy of traditional geodesy technology, very long baseline interferometry (VLBI) has become the technology of choice for deep space probe orbit determination in recent years because of its high precision and high angular resolution. Liu et al. (2009) have demonstrated the orbit determination accuracy of a lunar probe using VLBI technology to a level of tens of meters, showing that VLBI has great potential in the field of deep space exploration. In a practical engineering application, the unified S-band (USB) monitoring system and VLBI have been used for CE-1's orbit determination, where VLBI also played an important role when the probe entered the transfer orbit.

CE-1 orbit determination has been reported using VLBI time delay and delay rate data (e.g., Huang 2006; Li et al. 2009; Chen et al. 2011; Wang 2005). However, current results show that the orbit determination accuracy of CE-1, which is just 1–2 km for the transfer orbit and several hundred meters for the mission orbit, would not meet the application requirements of a high-precision field such as geodesy. Therefore, improving the probe orbit determination accuracy has become the main challenge for China's lunar exploration program. In addition, Earth orientation parameters (EOPs) were still not well estimated or clearly understood during the CE-1 orbit determination due to the complex geophysical mechanisms (Jin et al. 2010, 2011, 2012), which are a part of space geodetic observations and essential for the interconnection of various reference systems. Because predicted EOP values are used in practical engineering applications, the influence of their precision on orbit determination accuracy needs to be studied further. Furthermore, the relativistic effect correction is also relevant to orbit determination because the current accuracy of the CE-1 VLBI time delay observations has achieved the level of the relativistic effects of the time delay observations at the nanosecond level (Lin 1985; Zheng 1999) and will affect estimated parameter accuracy. Therefore, some new method, such as differential VLBI (ΔVLBI) and combining space VLBI observations, must be developed to solve these problems.

2.2 Differential VLBI

Currently, the CE-1 orbit was determined without estimation of EOPs. Because CE-1 time delay observations contain the components of the probe orbital parameters and EOPs, it is possible to estimate these parameters simultaneously. Here, the USB observations are not used, which may affect the accuracy of CE-1 orbit determination using VLBI time delay observations. So the ΔVLBI technology is used to determine the CE-1 orbit parameters and EOPs with high accuracy. ΔVLBI observations are obtained by differencing time delay observations of the probe's radio signals and a nearby alternating observed radio source's signals. The differential technology allows the elimination of common errors in the probe's radio signals and radio source's signals during propagation, such as station location errors and transmission media delays, to obtain high-precision observations that can guarantee high accuracy of the estimated parameters. This chapter describes the mathematical model of orbit determination using ΔVLBI technology. As explained previously, EOPs and relativistic effect corrections are also introduced into the CE-1 orbit determination.

CE-1 was launched on October 24, 2007, from Xichang Satellite Launch Center. The entire flight orbit of CE-1 is divided into the phasing, transfer, and mission orbits (Figure 2.1) (Huang 2006). USB and VLBI technologies have been used for CE-1 orbit determination during its flight, where VLBI plays an important role because of its high measurement accuracy, especially when the probe is in the transfer orbit. When the probe is in the mission orbit, its motion is under the reference frame of the selenocenter, the determination of the probe orbit involves transformation between the geocentric coordinate

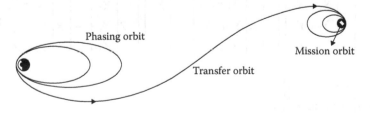

FIGURE 2.1
The orbit of CE-1.

system and selenocentric coordinate system, and the impact of EOP accuracy is not significant. So estimation of the orbital parameters and EOPs in the mission orbit is not discussed here.

In this section, a relativistic ΔVLBI time delay mathematical model including orbital parameters for the CE-1 transfer orbit and EOPs is derived and unknown parameters are estimated simultaneously using measured time delay data in this model. The accuracies of the CE-1 orbital parameters and EOPs are then assessed and evaluated (Wei et al. 2013).

2.2.1 Relativistic ΔVLBI Model for the CE-1 Transfer Orbit

The principle of ΔVLBI is illustrated in Figure 2.2a, where $\bar{k}$ is the direction vector of the CE-1 signal, $\bar{I}$ is the direction vector of an alternating observed radio source signal, and $\bar{B}$ is the baseline vector of two ground stations.

Two factors that must be considered for derivation of mathematical model are:

1. The propagation paths of the CE-1 signals and radio source signals are different (Figure 2.2b and c). The former encloses a small angle at the vertex of the probe, while the latter are parallel to each other. As a result, $\bar{k}$ and $\bar{I}$ are different as shown in Figure 2.2 (Sekido et al. 2005).
2. The relativistic effects must be considered because the accuracy of the CE-1 VLBI time delay observations is similar to the relativistic effects of the time delay observations and will affect the accuracy of the estimated parameter.

It can be seen from Figure 2.2b that R_{10} and R_{20}, which represent the distance of CE-1 to ground stations 1 and 2, respectively, are not equal, so $\bar{k}$ is neither parallel to R_{10} nor R_{20}. In Figure 2.2c, ψ is the angle between $\bar{I}$ and $\bar{B}$ and τ is the time delay of the radio source signals; it indicates that $\bar{k}$ and $\bar{I}$ are not parallel to each other.

Based on the above description a ΔVLBI time delay mathematical model is derived as follows:

$$L = c\Delta\tau + c\tau' = c(\tau_{CE} - \tau_S) + c\tau' = \bar{B} \cdot (\bar{k} - \bar{I}) + c\tau'$$

$$= \left\{ -\left(R \begin{bmatrix} X_1 - X_2 \\ Y_1 - Y_2 \\ Z_1 - Z_2 \end{bmatrix} \right)^T \cdot \left(R \begin{bmatrix} X_1 + X_2 \\ Y_1 + Y_2 \\ Z_1 + Z_2 \end{bmatrix} - 2 \begin{bmatrix} X_S \\ Y_S \\ Z_S \end{bmatrix} \right) \left| 2 \begin{bmatrix} X_S \\ Y_S \\ Z_S \end{bmatrix}^T - \left(R \begin{bmatrix} X_1 + X_2 \\ Y_1 + Y_2 \\ Z_1 + Z_2 \end{bmatrix} \right)^T \right|^{-1} \right.$$

$$\left. -a \begin{bmatrix} \cos\delta\cos a \\ \cos\delta\sin a \\ \sin\delta \end{bmatrix} \right\} + c\tau' \tag{2.1}$$

where τ_{CE} is the time delay observations of CE-1 signals, τ_S is the time delay observations of radio source signals, and τ' is the residual time delay including the differential residual parts of the time delay observations caused by the atmosphere, station errors, clock parameters, and so on. The last parameter is too small to affect the estimated accuracy of unknown parameters determined using VLBI technology, so it is ignored in this calculation. For the expansion of $\vec{k}$, $\vec{I}$ and $\vec{B}$, (X_i, Y_i, Z_i), $i = 1,2$, are the coordinates of ground VLBI stations fixed in the ITRS. (X_s, Y_s, Z_s) are the coordinates of CE-1 in the J2000.0 ICRS. (α,δ) are the right ascension and declination coordinates of the radio source in the J2000.0 International Celestial Reference System (ICRS). $a = c^2 / (c^2 - U(X_\oplus))$ is the correction coefficient for the general relativistic effect, where c is the light speed of light and $U(X_\oplus)$ is the gravitational potential of the Sun, Moon, and other planets (excluding Earth) in the barycentric celestial reference system (BCRS), with the first-order term expressed as $U(X_\oplus) = \Sigma_{A \neq M} GM_A / r_A$ (M_A indicates the lunar mass and r_A is the distance between the Earth and the Moon). $\vec{K}$ is the rotation matrix containing the EOPs, specifically x_p, y_p, $UT1-UTC$ [Earth rotation parameters (ERPs)], and $\Delta\psi$ and $\Delta\varepsilon$ (nutation parameters).

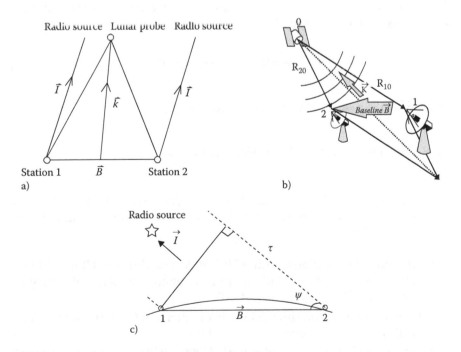

FIGURE 2.2

Principle of ΔVLBI. a) the propagation paths of CE-1 signals and radio source signals; b) the propagation paths of CE-1 signals; c) the propagation paths of radio source signals.

2.2.2   Adjustment Model

According to the mathematical model outlined in Section 2.2.1, the error equation of the ΔVLBI time delay observations can be written as:

$$V = Ax - l \tag{2.2}$$

where A is the design matrix formed by the partial derivatives of the unknown parameters, x is the correction for the unknown parameters (including the CE-1 orbital parameters and EOPs), and l is the difference between the observations and the model values. Accounting for the a priori accuracy of the EOP $(D(X) = \sigma_{X0}^2 P_X^{-1})$, the estimate of x, denoted as $\hat{x}$, can be determined as follows after least squares adjustment (Cui et al. 2005):

$$(A^T P A + \sigma_0^2 \cdot \sigma_{X0}^{-2} \cdot P_X) \hat{x} = A^T P l$$
$$\hat{x} = (A^T P A + \sigma_0^2 \cdot \sigma_{X0}^{-2} \cdot P_X)^{-1} (A^T P l) \tag{2.3}$$

where P is the weight matrix of observations and P_X is the a priori weight matrix of the unknown parameters, σ_0^2 is the a priori unit weight variance of observations, and σ_{X0}^2 is the a priori unit weight variance of the unknown parameters. So the accuracy of $\hat{x}$ can be estimated as:

$$D\{\hat{x}\} = \hat{\sigma}_0^2 (A^T P A + \sigma_0^2 \cdot \sigma_{X0}^{-2} \cdot P_X)^{-1} \tag{2.4}$$

where $\hat{\sigma}_0^2$ is the posteriori variance of unit weight.

2.2.3   Observations and Processing Strategies

2.2.3.1   Observations

The CE-1 transfer orbit lasted from October 31, 2007, to November 5, 2007. The data used for the simultaneous estimation of the CE-1 orbital parameters and EOPs are as follows:

1. Ground VLBI stations with *ITRS 2000* coordinates: China VLBI Net (CVN), including VLBI antennas in Shanghai (SH), Beijing (BJ), Kunming (KM), and Urumqi (UM).
2. Epochs of measured time delay observations of CE-1 in the transfer orbit with a sample interval of about 5 s.
3. A priori values of CE-1 orbit with *J2000.0* ICRS coordinates: 1-min sample interval and interpolated every 5 s by Chebyshev polynomials (Yu et al. 2004).

4. The radio source with *J2000.0* ICRS coordinates: the radio source is selected by the average direction of CE-1 during the selected orbital arc.

5. A priori values of EOPs: predicted values and accuracy from the International Earth Rotation and Reference Systems Service (IERS).

2.2.3.2   Orbit Determination Strategy

In real-time applications of lunar exploration missions, predicted EOP values are used for the CE-1 orbit determination. Therefore, further studies need to be undertaken to determine whether the prediction accuracy meets the demands of the CE-1 orbit determination. For this purpose, the differences in orbit determination results between estimating orbital parameters and estimating orbital parameters and EOPs simultaneously are analyzed using the measured time delay observations of CE-1 on October 31, 2007, to determine the importance of EOPs for the CE-1 orbit determination and to formulate the orbit determination strategy for this section. The results are illustrated in Figure 2.3.

It can be seen that the difference in the accuracy of each estimated CE-1 coordinate component between the two strategies is more than 40 m, with the difference for the Y_s component being more than 100 m. Thus, it can be concluded that EOPs are critical for the determination of the CE-1 orbit using ΔVLBI time delay observations and they should be treated as unknown parameters during the estimation. It also reflects the importance of EOPs for interconnection between the Earth coordinate system and the celestial reference system involved in the VLBI time delay observations. So the CE-1 orbit parameters and EOPs should be estimated simultaneously (Yan et al. 2012; Wei et al. 2013).

2.2.3.3   Strategies and Methods

Using the orbit determination strategy outlined in Section 2.2.3.2, the unknown parameters cannot be estimated with least squares adjustment

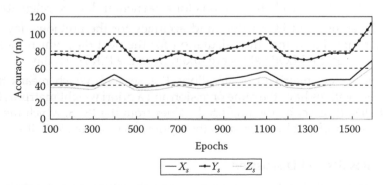

FIGURE 2.3
The difference in orbit determination results between the two strategies.

using a single epoch algorithm because there are 11 unknown parameters (six orbital parameters and five EOPs) but just six observations (six ground baselines) for each epoch. Therefore, an overall adjustment of multiple epochs is used. Using this method, CE-1 can be treated as being in one orbit for a selected orbital arc so that the orbital parameters and EOPs can be estimated. The specific method is as follows:

1. The time delay observations must be selected before the orbit determination to ensure that there are six time delay observations per epoch.

2. The time delay observations that contain gross errors must be removed because they will affect the estimated parameter accuracy. The principle of data rejection is that the time delay residuals of each baseline at an epoch must be between three times the negative and positive values of accuracy of time delay observations ($\pm 3\sigma$). Otherwise, the observations will be treated as problematic data and are removed. Another principle is that the amount of data removed cannot exceed 10% of the total observations.

3. The entire CE-1 transfer orbit is calculated per day, with each day's orbit being divided into several subarcs. The subarc's length is increased from 100 to 900 epochs with an interval of 100 or 50 epochs, which is determined by data quality.

4. The orbital elements of each subarc's initial epoch are calculated by the interpolation of CE-1's coordinates and velocities. Then the coordinates at subsequent epochs can be obtained by using Kepler's equation for the calculation of the ΔVLBI time delay model values.

5. The optimal length of the subarc is determined by analyzing the internal and external agreement of each unknown parameter.

6. When the optimal length of one subarc is determined, the adjustment values and accuracy of the EOPs are considered to be as a priori information for the next subarc to determine its optimal orbital arc length until all observations for a particular day are calculated.

7. The adjustment EOP values and accuracy for the previous day are not introduced into the calculation for the next day, which means that predicted EOP values are taken from the IERS each day.

8. The weight matrix of observations P is set as I, which means that observations have the same precision and a priori weight matrix as the EOPs. P_X is set by the predicted values from the IERS, which means that $P_X = \sigma^2_{XO}/\sigma^2_X$, where σ_X is the a priori accuracy of the EOPs.

2.2.4 Results and Discussion

This section will first focus on parameter estimation using the time delay observations of CE-1 on October 31, 2007 to introduce the data processing methods. Then the overall estimation results of the entire transfer orbit will be discussed.

2.2.4.1   *Parameter Estimation Using Time Delay Observations of CE-1 on October 31, 2007*

2.2.4.1.1   *Time Delay Residuals Analysis*

In the parameter estimation process, the time delay residuals will affect the value of using posteriori variance of unit weight $\hat{\sigma}_0^2$ and the estimated parameter accuracy when the number of time delay observations is constant. So the time delay residuals of the baselines need to be analyzed.

The time delay residuals of six baselines using the time delay observations of CE-1 on October 31, 2007, and a different number of epochs are illustrated in Figure 2.4.

It can be seen from Figure 2.4 that the time delay residuals of each baseline are distributed between −1.5 and 1.5 m (±3σ), which represents an advantage of ΔVLBI technology. Moreover, with the increase in the subarc length, the time delay residuals become larger, which will result in a larger $\hat{\sigma}_0^2$ and, consequently, decrease the estimated accuracy of the unknown parameter according to Equation 2.4. Therefore, the orbital arc length is important for the accuracy of the unknown parameters and must be selected carefully.

2.2.4.1.2   *Optimal Orbital Arc Length*

The estimated accuracy of the unknown parameters for the first subarc of CE-1 on October 31, 2007, using different orbital arc lengths is shown in Figures 2.5 to 2.7.

Figure 2.5 illustrates the estimated accuracy of CE-1 orbital parameters for the initial epoch. The accuracy improves from tens of kilometers to several hundred meters with the increase in the subarc length. So it can be concluded that the estimated accuracy of CE-1 coordinates can be improved by increasing the number of observations, and does not change significantly after 600 epochs. In addition, Figures 2.6 and 2.7 show that the estimated EOP accuracy decreases with increase in the subarc length, which still maintains the same order of magnitude. The accuracy stops changing significantly after 600 epochs. So the optimal orbital arc length of the first subarc of CE-1 on October 31, 2007, will be larger than 600 epochs according to the estimated parameter accuracy.

Finally, the externally coincident accuracy of EOPs is calculated based on the difference between the IERS final EOP values and the adjusted values of this section to determine the optimal orbital arc length using the optimal estimated accuracy of the parameters. The results are illustrated in Figure 2.8.

From Figure 2.8, it is evident that the externally coincident accuracy of EOPs increases with increase in the subarc length and then decreases. It reaches

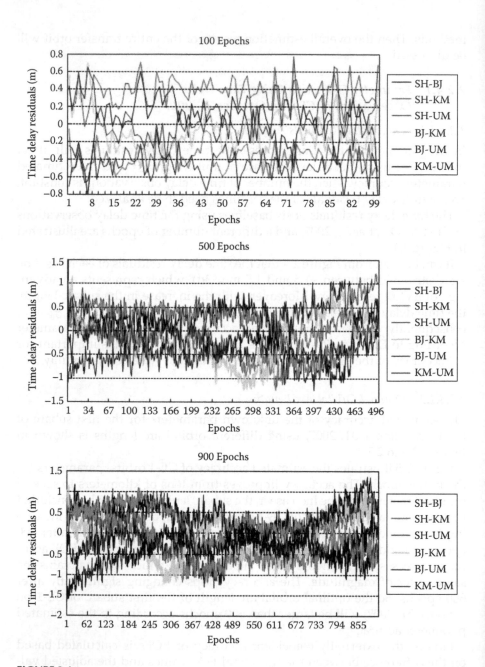

FIGURE 2.4
Time delay residuals of each baseline. *Abbreviations:* SH-BJ, baseline from Shanghai to Beijing; SH-KM, baseline from Shanghai to Kunming; SH-UM, baseline from Shanghai to Urumqi; BJ-KM, baseline from Beijing to Kunming; BJ-UM, baseline from Beijing to Urumqi; KM-UM, baseline from Kunming to Urumqi.

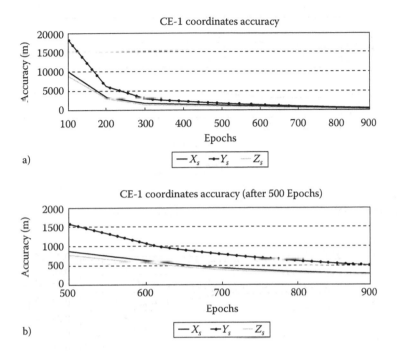

FIGURE 2.5
Accuracy of CE-1 coordinates for the initial epoch.

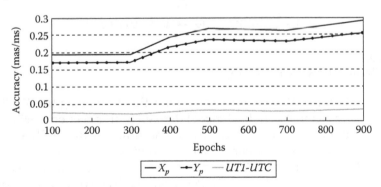

FIGURE 2.6
Accuracy of ERPs.

an optimal value when the subarc length is about 600 epochs; by contrast, the estimated accuracy of certain parameters, such as *UT1–UTC*, begins to diverge. Unfortunately, the externally coincident accuracy of CE-1 coordinates is not calculated because there are no accurate orbit determination results for CE-1. Taking into account the estimated accuracy and the external agreement of the unknown parameters, the optimal length of the first subarc

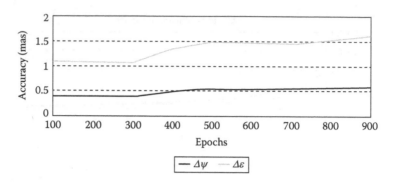

FIGURE 2.7
Nutation accuracy.

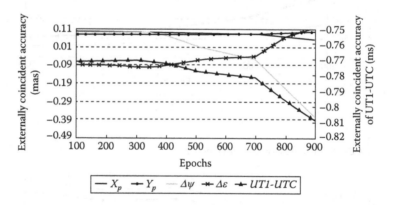

FIGURE 2.8
Externally coincident accuracy of EOPs.

of CE-1 on October 31, 2007, for parameter estimation using the ΔVLBI time delay observation mathematical model derived in this section is 600.

By considering the adjusted values and accuracy of the EOPs as a priori information for the next subarc, the entire orbit's observations for CE-1 on October 31, 2007, can be calculated. Based on the adjustment results, the entire orbit of CE-1 on October 31, 2007, can be divided into three subarcs that have optimal lengths of 600, 800, and 184, respectively. The externally coincident accuracy of the EPOs for each subarc is listed in Table 2.1.

2.2.4.2 Overall Estimation of the Entire Transfer Orbit

For the observations on the following days in the CE-1 transfer orbit, problematic data with gross errors, which affect the estimated parameter accuracy, should be rejected using the principle outlined in Section 2.2.3. Figure 2.9 gives a comparison of the time delay residuals of each baseline before and after data rejection. It can be seen that a reasonable distribution of residuals

TABLE 2.1

Externally Coincident Accuracy of EOPs

	A Priori Accuracy	Adjusted Accuracy (1st Subarc)	Adjusted Accuracy (2nd Subarc)	Adjusted Accuracy (3rd Subarc)
x_p (mas)	0.1	0.083	0.076	0.076
y_p (mas)	0.08	0.079	0.077	0.077
UT1–UTC (ms)	−0.77	−0.78	−0.77	−0.77
$\Delta\psi$ (mas)	0.091	−0.025	−0.031	−0.030
$\Delta\varepsilon$ (mas)	−0.083	−0.032	−0.151	−0.150

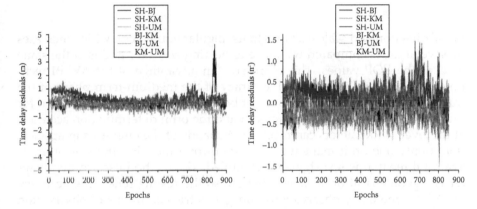

FIGURE 2.9
The effect of data rejection.

is obtained after data rejection to ensure the estimated parameter accuracy. The observations used in Figure 2.9 are from the fourth subarc in November 2, 2007, with 900 epochs.

According to the data processing methods described previously, the optimal estimated accuracy of the CE-1 orbital arc parameters and EOPs in the transfer orbit using an optimal orbital is shown in Table 2.2.

It can be seen from Table 2.2 that the estimated accuracy of the daily EOPs in the transfer orbit all achieve or are better than the IERS accuracy level, which is at the level of 0.01 mas for diagonal components and 0.001 ms for UT1–UTC. The results of the observations on October 31 are slightly worse than the results on the other days because of its short orbital arc, which is roughly 2 hours. Therefore, the estimated accuracy is considered to be reasonable.

For the CE-1 orbital parameters, using the optimal orbital arc an optimal estimated accuracy of the order of several hundred meters can be achieved for each coordinate component in the beginning and end phases of the transfer orbit, where the results for November 5 are within 100 m. It means that CE-1 can achieve an orbit determination accuracy level of about 100 m for the

TABLE 2.2

Estimated Parameter Accuracy in the CE-1 Transfer Orbit

	σ_{xp} (mas)	σ_{yp} (mas)	$\sigma_{UT1-UTC}$ (mas)	$\sigma\Delta_{\psi}$ (mas)	$\sigma\Delta_{\varepsilon}$ (mas)	σ_{Xs} (m)	σ_{Ys} (m)	σ_{Zs} (m)
31 Oct.	0.133	0.116	0.012	0.269	0.736	371.79	631.71	330.22
1 Nov.	0.070	0.070	0.024	0.005	0.002	459.32	717.64	430.66
2 Nov.	0.035	0.035	0.012	0.027	0.009	778.47	966.63	571.95
3 Nov.	0.011	0.011	0.004	0.011	0.003	1506.07	1208.95	712.99
4 Nov.	0.013	0.013	0.004	0.024	0.004	128.13	132.96	155.33
5 Nov.	0.035	0.035	0.002	0.027	0.008	80.4	75.67	105.14

radial direction and scale 82.7 mas to its angular equivalent, which increases by several times compared with its engineering accuracy and is at the same level as the EOP values. This represents an advantage of the ΔVLBI technology. However, the estimated accuracy of the medium-term phase of the transfer orbit is worse, with the worst being roughly 1.5 km. The reason for this is that the probe's flight is smooth and the orbit determination accuracy at the level of kilometers can meet the demands of the engineering application if only one orbit maneuver is made during the CE-1 transfer orbit, as against three that were originally scheduled. The method used in this chapter is a geometrical orbit determination, so the results reflect the accuracy of ΔVLBI time delay observations and geometric conditions of observation but cannot make use of orbital constraints or forecasting, which is reflected in the results. Although the results in this chapter show that the CE-1 orbit determination accuracy in the transfer orbit can be better than its engineering precision using the ΔVLBI time delay observations, with the simultaneous estimation of the probe's orbital parameters and the EOPs, they can still meet the demands of the lunar exploration mission.

The external agreement of the daily estimated EOP values in the transfer orbit are listed in Table 2.3. It can be seen that the accuracy of EOPs is improved when compared with their predicted values except for values of Δε and UT1–UTC for some days, from November 2 to 5, 2007, using the method for estimating the CE-1 orbital parameters and EOPs simultaneously introduced in this chapter. The results show that UT1–UTC is sensitive to errors in ΔVLBI time delay observations. Thus, a meaningful attempt to improve EOP estimates using ΔVLBI observations is provided.

2.2.5 Summary

In this section, relativistic ΔVLBI time delay observations have helped in deducing a mathematical model for the CE-1 transfer orbit with the advantage of eliminating some of the significant commonly propagated errors during the signals' propagation path. The orbit determination strategy of

TABLE 2.3

Externally Coincident Accuracy of EOPs in the CE-1 Transfer Orbit

	x_p (mas)		y_p (mas)		UT1–UTC (ms)		$\Delta\psi$ (mas)		$\Delta\varepsilon$ (mas)	
	Predicted	Estimated	Predicted	Estimated	Predicted	Estimated	Predicted	Estimated	Predicted	Estimated
31 Oct.	0.10	0.076	0.08	0.077	−0.7715	−0.77	0.091	−0.031	−0.083	−0.151
1 Nov.	−0.077	0.012	−0.44	−0.411	−0.2211	−0.025	0.772	0.772	0.115	0.115
2 Nov.	0.195	0.194	−0.693	−0.692	−0.1916	0.966	0.934	0.741	0.155	0.152
3 Nov.	1.027	0.993	−0.708	−0.645	−0.1004	−0.043	0.804	0.524	0.121	0.121
4 Nov.	1.773	1.766	−0.44	−0.416	−0.0419	0.709	0.536	0.202	0.022	0.022
5 Nov.	2.334	2.335	−0.418	−0.417	0.0646	1.239	0.331	0.324	−0.067	−0.066

estimating the CE-1 orbital parameters and EOPs simultaneously is formulated and the unknown parameters are estimated using the measured time delay data of the CE-1 transfer orbit.

Based on the results, it is evident that the CE-1 orbital parameters and EOPs can be estimated simultaneously using the derived model and that higher estimated parameter accuracies are obtained. To be specific, the overall accuracy level of the estimated CE-1 orbital parameters be of the order of of a few hundred meters, with a best value of 75 m (Y direction), which is much better than their engineering accuracy. The estimated EOP accuracies are improved when compared with their predicted values except for the values of $\Delta\varepsilon$ and $UT1-UTC$ on certain days. Therefore, it is suggested that $\Delta\varepsilon$ and $UT1-UTC$ be considered as known values to reduce the effects of the predicted EOP values on the estimated accuracy of the CE-1 orbital parameters and EOPs.

To conclude, it can be said that the high-precision time delay observations obtained using ΔVLBI technology are useful for estimating the CE-1 orbital parameters and EOPs. In addition, the radio source is a part of the ΔVLBI observing target such that the observations are more sensitive to EOPs when compared with the engineering survey measurements in terms of the ability to improve predicted EOP values. It provides a new method to estimate EOPs in addition to the traditional observation methods. Thus, ΔVLBI technology is an approach that can improve the accuracy of both CE-1 orbital parameters and EOPs.

2.3   Space VLBI

In the previous section, EOPs and relativistic effect correction are introduced in the orbit determination of CE-1 using a derived differential VLBI (ΔVLBI) time delay mathematical model because they are crucial for lunar probe orbit determination and must be estimated simultaneously. The results of CE-1's measured data show that the accuracy of CE-1 orbit parameters and EOPs can be improved significantly compared with their predicted use under this method.

However, the accuracy of CE-1 ΔVLBI time delay observations, which is of the order of nanoseconds, is still lower than the accuracy of traditional VLBI observations, which can have accuracies of the order of picoseconds, so the estimated accuracy of unknown parameters will be affected. In addition, the geometrical structure of CE-1 observations gradually deteriorates with increasing orbital altitude, and some other problems are exposed when the probe enters the mission orbit, such as the influence of the lunar gravitational field, precise coordinate conversion between the Earth system and the Moon system, and so on. These factors make it difficult to obtain a more accurate orbit using the ΔVLBI geometric observations under the current observation conditions, which will restrict the further development of the lunar exploration mission and so should be solved.

An intuitive idea is to place an antenna on a satellite in the sky, which is referred to as SVLBI. SVLBI is an extension of ground-based VLBI and its time delay observations can be obtained by observing stable extragalactic radio sources using VLBI telescopes placed on the probe and on the ground. Because the baseline of SVLBI is longer than the diameter of Earth, more accurate observations can be obtained (Kulkarni 1992). Furthermore, SVLBI has the advantage of improving the geometrical structure of time delay observations, interconnecting three coordinate systems involved in geodesy and geodynamics directly (Earth reference system, celestial reference system, and dynamic reference system), calculating orbital parameters and EOPs simultaneously, and so on. So SVLBI can be used for deep space probe orbit determination.

In fact, although it is difficult to place a large antenna on a satellite, some useful attempts have been made until now with the development of related theory and techniques, such as VLBI Space Observatory Programme (VSOP), RaidoAstron, Advanced Radio Interferometry between Space and Earth (ARISE), and so on. These projects provided useful experiments for the application of SVLBI. In addition, many researchers have studied the application of SVLBI in the field of geodesy. Ádám (1990) studied the estimation of geodetic parameters from SVLBI observations. Following this, Kulkarni (1992) researched the feasibility of SVLBI for geodesy and geodynamics. SVLBI time delay observations were simulated using VSOP design orbit and model parameters, such as satellite orbital parameters, and ERPs were estimated. The results show that the estimated accuracy of ERPs can be precisely obtained using SVLBI compared with other geodetic technologies with precise a priori information on the parameters and precise modeling of systematic influences. Zheng et al. (1993) studied the establishment of reference systems and their connection using SVLBI. Wei (2006) researched the design of a Chinese SVLBI system and some computation simulation and showed that SVLBI has great potential for application in geodesy. Its advantage is that it provides an effective method to overcome the problems posed by the orbit determination strategy of CE-1 using ground-based ΔVLBI geometric observations.

Therefore, it is not hard to imagine that there will be a deep space probe with SVLBI antenna in the near future and that the application of SVLBI for orbit determination will further improve the estimated accuracy of unknown parameters. This section will focus on the improvement of orbit accuracy for the CE-1 transfer orbit using SVLBI observations. Similar to the previous section on the ΔVLBI, CE-1 mission orbit is not discussed here.

2.3.1 Relativistic SVLBI Model for the CE-1 Transfer Orbit

The principle of SVLBI is illustrated in Figure 2.10. The propagation paths of signals from the radio source to the two VLBI telescopes can be seen as

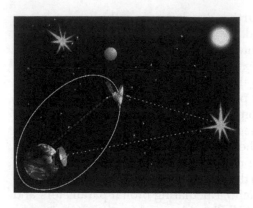
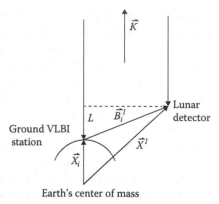

FIGURE 2.10
Principle of SVLBI.

parallel lines because the radio source is far from two telescopes. So the SVLBI time delay observations can be written as:

$$\tau = \tau_g + \Delta\tau = -\frac{1}{c}(\vec{B}\cdot\vec{K}) + \Delta\tau \tag{2.5}$$

where τ_g indicates the geometrical time delay observations and $\Delta\tau$ denotes the nongeometrical time delay observations including the effects of random error and systematic errors, such as solar radiation pressure, atmospheric refraction, and so on. In the expression, $\vec{B}$ is the baseline vector from the ground VLBI station to the lunar probe, $\vec{K}$ is the direction vector of the radio source signals, and c is the light speed. Similar to the significant influence on the time delay observations, EOPs and relativistic effect correction are also introduced into this model (Wei et al. 2013). So a relativistic SVLBI time delay observable mathematical model for the CE-1 transfer orbit can be derived as Equation 2.6 by unifying each component's coordinate system into *J2000.0* ICRS.

$$L = -a\left\{\left(\left[R\begin{bmatrix}X_i\\Y_i\\Z_i\end{bmatrix}\right]^T - \begin{bmatrix}X^I\\Y^I\\Z^I\end{bmatrix}^T\right)\cdot\left\{\begin{bmatrix}\cos\delta\cos\alpha\\\cos\delta\sin\alpha\\\sin\delta\end{bmatrix}\right\} + c\Delta\tau\right\} \tag{2.6}$$

where (X_i, Y_i, Z_i) are the coordinates of the ground VLBI station fixed in the Earth-fixed system, such as *ITRS 2000*; (X^I, Y^I, Z^I) are the coordinates of CE-1 defined under *J2000.0* ICRS. (α, δ) are the coordinates of the radio source fixed in *J2000.0* ICRS. *a* is a relativistic effect correction and is considered to be a constant here with its first-order expansion expressed as $a = c^2 \cdot (c^2 - U(X_\oplus))$, where $U(X_\oplus)$ is the gravitational potential of the Sun, Moon, and other planets (excluding Earth) in BCRS, $X_\oplus$ is the coordinate of geocenter in *J2000.0* ICRS; and *R* is the rotation matrix containing EOPs, specifically x_p, y_p, dUT, ERP, $\Delta\psi$, and $\Delta\varepsilon$ (nutation parameters). In Equation 2.6, (X_i, Y_i, Z_i) can be precisely obtained by GPS or other geodetic technologies and (α, δ) can be obtained from astronomical ephemeris, so (X^I, Y^I, Z^I) and EOPs are considered to be unknown parameters in this study.

Compared with the mathematical models in Ádám (1990) and Kulkarni (1992), the mathematical model derived in this section expands the interconnection parameters of the reference system from ERPs to EOPs while considering the influence of relativistic effects on SVLBI time delay observations. The mathematical model seems to be more reasonable by ensuring the theoretical rigor of the probe parameters estimation.

2.3.2 Adjustment Model

According to the mathematical model derived in Section 2.3.1, the error equation for SVLBI time delay observations can be written as:

$$V = Ax - l \tag{2.7}$$

where *A* is the design matrix formed by the partial derivatives of the unknown parameters, *x* is the correction for the unknown parameters (including the CE-1 orbital parameters and EOPs), and *l* is the difference between the observations and the model values. On linearization, Equation 2.7 can be rewritten as follows:

$$v = A_1 dX_S + A_2 dY_S + A_3 dZ_S$$
$$+ A_4 dx_p + A_5 dy_p + A_6 dUT + A_7 d(\Delta\psi) + A_8 d(\Delta\varepsilon) \tag{2.8}$$

where A_i ($i = 1, \ldots, 8$) are components of the design matrix *A* whose specific expressions are as follows:

1. Partial derivatives of probe orbital elements

$$\begin{pmatrix} A_1 \\ A_2 \\ A_3 \end{pmatrix} = a \begin{bmatrix} \cos\delta\cos\alpha \\ \cos\delta\sin\alpha \\ \sin\delta \end{bmatrix} \tag{2.9}$$

Based on the formula for the derivatives of the orbital elements of the coordinate components (Ádám 1990), the of the orbital elements derivatives to the model can also be obtained for further study.

2. Partial derivatives of ERPs: R can be rewritten as:

$$R = R_{ERP} \cdot R_N \cdot R_P \tag{2.10}$$

where R_{ERP} is the *ERP* matrix, R_N is the nutation matrix, and R_P is the precession matrix. Then partial derivatives of *ERP* can be written as:

$$\begin{pmatrix} A_4 \\ A_5 \\ A_6 \end{pmatrix} = -a \begin{pmatrix} \dfrac{dR_{ERP}}{dx_p} \\ \dfrac{dR_{ERP}}{dy_p} \\ \dfrac{dR_{ERP}}{ddUT} \end{pmatrix} \begin{bmatrix} X_i \\ Y_i \\ Z_i \end{bmatrix}^T \cdot R_N R_P \cdot \begin{bmatrix} \cos\delta\cos a \\ \cos\delta\sin a \\ \sin\delta \end{bmatrix} \tag{2.11}$$

The expressions $\dfrac{dR_{ERP}}{dx_p}$, $\dfrac{dR_{ERP}}{dy_p}$, and $\dfrac{dR_{ERP}}{ddUT}$ can be seen in Ádám (1990).

3. Partial derivatives of nutation parameters

$$\begin{bmatrix} A_7 \\ A_8 \end{bmatrix} = -a \begin{bmatrix} \dfrac{dR_N}{d\Delta\psi} \\ \dfrac{dR_N}{d\Delta\varepsilon} \end{bmatrix} \cdot \begin{bmatrix} X_i \\ Y_i \\ Z_i \end{bmatrix}^T \cdot R_{ERP} R_P \cdot \begin{bmatrix} \cos\delta\cos a \\ \cos\delta\sin a \\ \sin\delta \end{bmatrix} \tag{2.12}$$

The expressions $\dfrac{dR_N}{d\Delta\psi}$ and $\dfrac{dR_N}{d\Delta\varepsilon}$ can be seen in Ádám (1990).

Taking into account the a priori accuracy of EOP ($D(X) = \sigma_{X0}^2 P_X^{-1}$), the estimate of x, denoted as $\hat{x}$, and its accuracy can be determined as follows after weighted least squares adjustment (Huang 1990):

$$\hat{x} = (A^T P_\Delta A + \sigma_0^2 \cdot \sigma_{X0}^{-2} \cdot P_X)^{-1}(A^T P_\Delta l) = N^{-1}(A^T P_\Delta l) \tag{2.13}$$

$$\hat{\sigma}_0^2 = \frac{V^T P_\Delta V}{r} = \frac{l^T P_\Delta l - (A^T P_\Delta l)^T \hat{x}}{n-t} \tag{2.14}$$

$$D_x = \hat{\sigma}_0^2 (N^{-1} A^T P_\Delta D_\Delta P_\Delta A N^{-1}) = \hat{\sigma}_0^2 N^{-1} \tag{2.15}$$

where P_Δ is the weight matrix of the time delay observations, P_X is the a priori weight matrix of the unknown parameters, σ_0^2 is the a priori unit weight variance of observations, σ_{X0}^2 is the a priori unit weight variance of unknown parameters, $\hat{\sigma}_0^2$ is the posteriori variance of unit weight, N is the coefficient matrix of the normal equation, n is the number of observations, t is the number of unknown parameters, and r is the freedom. P_X is considered here for better estimated accuracy of the parameters because of the sensitivity of EOPs to the lunar probe.

2.3.3 Estimation of Unknown Parameters

Estimation of unknown parameters can be analyzed by studying the linear correlation of each column of matrix A. If a linear correlation exists between columns of A, then the matrix will be column rank deficient and the normal matrix will consequently be singular, which means that not all of the parameters can be estimated. According to Equations 2.9–2.12, there is no evidence of linear correlation, so the estimated parameters of the derived model are as follows:

$$(X_S, Y_S, Z_S, x_p, y_p, dUT, \Delta\psi, \Delta\varepsilon) \tag{2.16}$$

It should be noted that there is linear correlation between the ground station coordinates, ERPs, radio source ascension, and probe orbit right ascension of the ascending node (RAAN) in the mathematical model described by Ádám (1990), which influences the estimated accuracy of the unknown parameters because the nutation parameters are not involved. The estimability analysis in our study illustrates the function of nutation parameters for improving the linear correlation of the unknown parameters. The model derived in this section is suitable in studying the influence of EOPs on probe orbit parameters and the orbit strategy of estimating EOPs and probe orbit parameters simultaneously using SVLBI time delay observations.

2.3.4 Experiment

Because a real VLBI antenna is not present on CE-1, simulated SVLBI time delay observations are used here for the estimation of orbital parameters and EOPs. To compare with CE-1's engineering orbit accuracy, SVLBI time delay observations are first simulated under CE-1's actual observation conditions.

2.3.4.1 Parameter Estimation under the Actual Observation Conditions of CE-1

2.3.4.1.1 Actual Observation Conditions of CE-1

CE-1's actual observation conditions are consistent with the description in Section 2.2.3.1.

2.3.4.1.2 Simulation of SVLBI Time Delay observations

According to Equation 2.5, SVLBI time delay observations comprise both geometrical (τ_g) and nongeometrical time delay observations $\Delta\tau$, of which τ_g can be easily calculated under observation conditions using the mathematical model described in Equation 2.6.

Nongeometrical time delay $\Delta\tau$ in SVLBI observations is mainly caused by random error and systematic errors, of which the former is a type of stochastic quantity with the characteristics of a normal distribution. It is also called white noise and is very important for SVLBI-simulated time delay observations because of the authenticity of simulation and the mathematical characteristics of least squares adjustment. In addition, the effects of solar radiation pressure, atmospheric refraction, clock errors, and relativistic effect will be considered as major systematic errors in the simulated observations because they are significant in the SVLBI system.

2.3.4.1.3 Simulation Steps

1. *Visibility test.* The purpose of this test is to check whether CE-1 and the ground station can receive the signals from the target radio source at some epoch simultaneously.
2. If visibility test is passed, the geometrical time delay τ_g can be simulated. Otherwise, the next epoch is subjected to the visibility test.
3. Simulation of random error at some epoch.
4. Simulation of systematic errors by calculating the residuals of solar radiation pressure, atmospheric refraction, clock errors, relativistic effect, and so on. Then nongeometrical time delay $\Delta\tau$ can be generated.
5. Simulation of SVLBI time delay observations by τ_g and $\Delta\tau$.

2.3.4.1.4 Estimation Results

The CE-1 transfer orbit lasted from October 31, 2007, to November 5, 2007. Since EOPs are very sensitive to the fast flying probe, the whole orbit of CE-1 is divided into several subarcs and the optimal length of each subarc is determined in addition to considering the a priori accuracy of EOPs to ensure that the estimated accuracy of the CE-1 orbit and EOPs is optimal (Wei et al. 2013). Figure 2.11 illustrates an example of the parameter's estimated accuracy under different subarc lengths of the CE-1 transfer orbit on October 31, 2007.

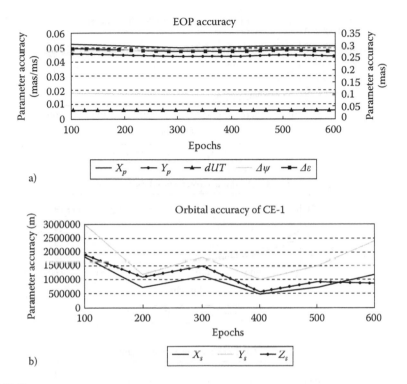

FIGURE 2.11
Parameter accuracy using the simulated SVLBI observables on October 31, 2007.

It can be seen that the estimated accuracy of EOPs for different lengths of the test orbital arc is quite good and constant, and all of variables achieved the accuracy level of the IERS. This is attributed to the high accuracy of SVLBI observations and the a priori accuracy of EOPs. However, the estimated accuracy of the CE-1 orbit is not reasonable; it is more than 3000 km and therefore cannot meet the demands of practical application. This is attributed to the poor geometric structure of the SVLBI system under CE-1's actual observable conditions, which include observations for just one target radio source. This in turn will yield an ill-conditioned normal equation and impact parameter estimation accuracy.

In order to improve CE-1 orbit accuracy, the number of ground stations and radio sources is increased during the simulation and the orbital parameters are estimated again. The estimated accuracy is illustrated in Figure 2.12.

The results indicate that the estimated accuracy of the CE-1 orbit showed distinct improvement with improvement in observation conditions. The orbit determination accuracy can be as high as 100 m at 600 epochs for each observation condition. So it is necessary to improve the observation conditions of CE-1 to obtain a more accurate orbit.

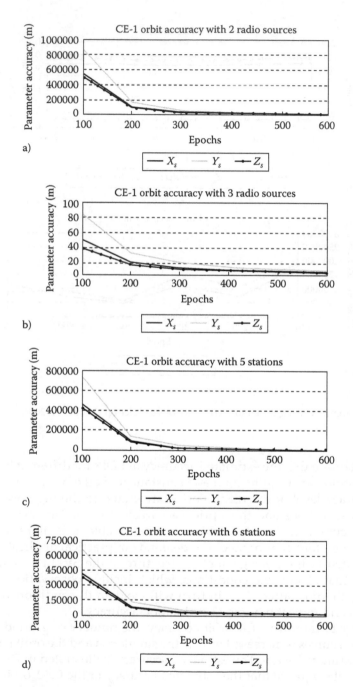

FIGURE 2.12
Parameter accuracy using simulated SVLBI under different observation conditions.

2.3.4.2 Optimal Conditions of Observation for SVLBI

According to the previous analysis, good observation conditions are essential for the estimation of parameters using SVLBI time delay observations. Therefore, optimal observation conditions for CE-1, including number of stations, observing epochs, and numbers of target radio sources, must be studied to obtain optimal estimated accuracy of orbital parameters and EOPs. Because the estimated accuracy of CE-1 orbital parameters increases with improvement in the observation conditions and the change in the estimated accuracy of EOPs is irregularly, this section will focus on the change in the estimated accuracy of EOPs under different observation conditions to confirm the optimal conditions for observations.

The different observation conditions for CE-1 are listed in Table 2.4. By computing the coefficient matrix of the normal equation A under different observable conditions, the change in the estimated accuracy of EOPs can be obtained from Equation 2.15 to analyze the optimal observable conditions for CE-1. In this chapter, P_Δ is the unit matrix and σ_0^2 is determined using a value of 0.05 ns for the accuracy of the SVLBI observable (Kulkarni 1992).

It should be mentioned that the value of 0.05 ns for the accuracy of the SVLBI observable is just suitable for the simulated estimation in this chapter. For practical application, more complex situations should be considered to ensure the authenticity and accurateness of the SVLBI time delay observations, such as signal receiving ability of a fast-moving probe from a radio source, signal-to-noise ratio of the SVLBI time delay observations, and so on (Yan et al. 2010).

2.3.4.2.1 Impact of Number of Stations on Estimated Parameter Accuracy

In order to analyze the impact of number of stations on estimated EOP accuracy, the observable epochs are set as 1000, the number of radio sources are set as nine, and station numbers are set to range from 4 to 10 with an interval of one station. The selection of stations is done from CVN and an additional six stations from the International VLBI Service (IVS). The results are illustrated in Figure 2.13.

Because the accuracy of dUT is stable at about 0.005 ms and is better than the IERS accuracy level, no statistical analysis was done in this chapter (the following experiments show the same pattern). The calculation results show that the estimated accuracy of EOPs improves gradually with the number of

TABLE 2.4

Conditions of Observations

Parameter	Range of Variability
Number of stations	4–10
Observable epochs	300–5300
Number of radio sources	1–9

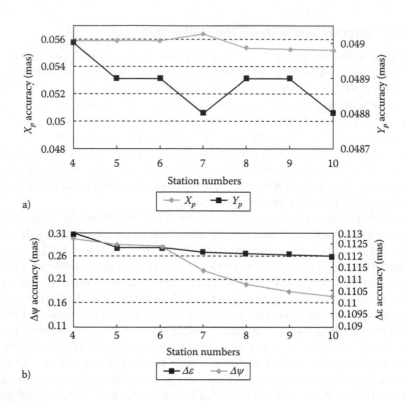

FIGURE 2.13
EOP accuracy for number of stations. a) ERPs' accuracy; b) Nutation parameters' accuracy.

stations, although the variation trends differ from each other. For example, the accuracy of polar motion parameters show variation when the station number is larger than six, while the accuracy of $\Delta\psi$ shows a fast increasing trend under the same conditions. In order to get a stable EOP estimation accuracy, the station numbers should be more than six.

2.3.4.2.2 Impact of Different Observable Epochs on Parameter Calculation Accuracy

In order to analyze the impact of different observable epochs on estimated EOP accuracy, the station numbers are fixed as six, radio sources are fixed as nine, and observable epochs are set to range from 300 to 5300. The results are shown in Figure 2.14.

It can be seen from Figure 2.14 that the estimated accuracy of EOPs improves with an increase in the observable epochs. The estimated accuracy of each parameter shows a huge variation when the observable epochs are shorter than 1100, whereas after that a declining trend is observed, although it does not yield any stable resolutions. Therefore, in order to get a stable and reasonable estimate of EOP values, observable epochs must be greater than 1100.

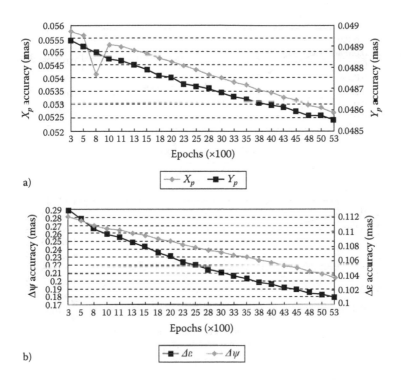

a)

b)

FIGURE 2.14
EOP accuracy for different epochs. a) ERPs' accuracy; b) Nutation parameters' accuracy.

2.3.4.2.3   *Impact of Different Radio Sources on Parameter Calculation Accuracy*

In order to analyze the impact of different radio sources on estimated EOP accuracy, the station numbers are set as six, observable epochs are set as 1100, and radio sources are set to range from one to nine with an interval of two. Radio sources are selected from ICRS and are uniformly distributed in the sky. The results are shown in Figure 2.15.

The results show that the estimation accuracy of EOPs improves gradually with the number of radio sources. When the radio sources are set between five and seven, the estimated EOP accuracy exhibits a comparatively stable trend. To obtain a uniform distribution, it is suggested that the radio sources be set as seven.

In summary, to apply SVLBI in the orbit determination strategy of estimating orbital parameters and EOPs simultaneously and to obtain high estimated accuracy, the following observation conditions are suggested in this chapter: six ground stations, observable epochs as 1100, and seven radio sources for each subarc.

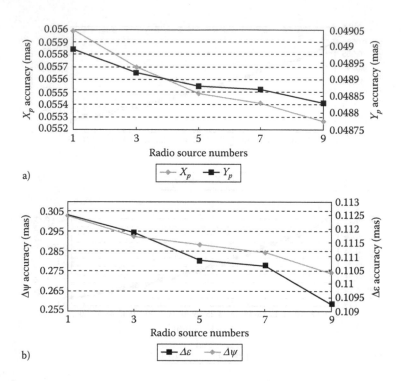

a)

b)

FIGURE 2.15
EOP accuracy for different radio source numbers. a) ERPs' accuracy; b) Nutation parameters' accuracy.

2.3.4.3 Parameter Calculation under Optimal Conditions for Observation

In this section, the SVLBI time delay observations are simulated again under the optimal conditions of observations discussed in Section 2.3.4.2 and the parameters will be reestimated. During the calculation, the observable epochs of each subarc are fixed at 1100, while the ground stations are set as six including CVN, and seven uniform distribution radio sources are selected. In addition, the predicted EOP values and accuracy are selected separately from the IERS for different days, and the previous subarc's estimated EOP results are considered to be the a priori information for the next subarc in the same day until all the simulated SVLBI time delay observations for this day are calculated. The results are listed in Table 2.5.

The results show that the optimal estimated accuracy of each day's orbital parameters can all achieve a level of 2 m, which is much better than their engineering orbit determination accuracy. So SVLBI can obviously improve CE-1 orbit determination accuracy to meet the demands of scientific disciplines, such as geodesy, and engineering applications. Compared with the estimated accuracy of the SVLBI satellite orbit obtained from simulated observations using VSOP-designed orbital elements described by Kulkarni

TABLE 2.5

Parameter Accuracy Using Simulated SVLBI Observables of the CE-1 Transfer Orbit

	σ_{xp} (mas)	σ_{yp} (mas)	σ_{dUT} (ms)	$\sigma_{\Delta\varepsilon}$ (mas)	$\sigma\Delta_{\varepsilon}$ (mas)	σ_{xs} (m)	σ_{ys} (m)	σ_{zs} (m)
31 Oct.	0.046	0.041	0.005	0.093	0.249	0.35	0.51	0.24
01 Nov.	0.047	0.041	0.005	0.092	0.223	0.79	1.16	0.58
02 Nov.	0.047	0.041	0.005	0.093	0.220	1.31	1.55	0.74
03 Nov.	0.039	0.034	0.004	0.076	0.171	1.43	1.39	0.87
04 Nov.	0.043	0.038	0.005	0.085	0.197	1.25	1.18	0.80
05 Nov.	0.050	0.044	0.005	0.101	0.263	0.06	0.07	0.08

(1992), which is at the centimeter level, the estimated accuracy of the CE-1 orbit is credible. SVLBI provides a brand new method for probe orbit determination of China's lunar exploration program.

In addition, the estimated accuracy of EOP's is very steady and is better than the IERS accuracy level. All these results reflect the technological advantages of SVLBI.

The externally coincident accuracy of each day's estimated EOP values in the transfer orbit are listed in Table 2.6. This accuracy is obtained by comparing the predicted values and the estimated values of EOPs with their final values announced by the IERS. The closer the value is to zero, the more accurate the estimated parameters are. It can be seen that the estimated accuracy of some EOP components is improved compared with their predicted values except $\Delta\varepsilon$ and dUT. Thus, it can be concluded that the predicted EOP accuracy can be improved by the orbit determination strategy of estimating orbital parameters and EOPs simultaneously using SVLBI simulated time delay observations. It also reflects the availability of CE-1's estimated orbit accuracy.

2.3.5 Summary

In this section, a relativistic SVLBI observable mathematical model for the CE-1 transfer orbit, which includes probe orbital parameters and EOPs, is derived and the estimability of unknown parameters is discussed. The conclusion that good observation conditions of CE-1 are crucial for the estimation of parameters using SVLBI time delay observations is reached by the analysis of estimated parameter accuracy of simulated SVLBI observations under CE-1's actual observation conditions. Then the optimal observation conditions are analyzed and the unknown parameters are estimated again.

The results show that the estimated accuracy of CE-1 orbital parameters can achieve a level of several meters using SVLBI simulated observations under optimal conditions for observation, which is much better than their engineering orbit determination accuracy. The estimation accuracy of EOPs is achieved or is better than the IERS accuracy level, and the predicted

TABLE 2.6

Externally Coincident Accuracy of EOP by Simulated SVLBI Observables of CE-1 Transfer Orbit

	X_p (mas)		Y_p (mas)		dUT (ms)		$\Delta\psi$ (mas)		$\Delta\varepsilon$ (mas)	
	Predicted	Estimated	Predicted	Estimated	Predicted	Estimated	Predicted	Estimated	Predicted	Estimated
31 Oct.	0.1	0.084	0.08	0.081	-0.7717	-0.192	0.091	0.060	-0.083	-0.151
1 Nov.	-0.077	-0.141	-0.44	-0.329	-0.2211	0.351	0.772	0.451	0.115	-0.399
2 Nov.	0.195	0.1	-0.693	-0.516	-0.1916	0.380	0.934	0.505	0.153	-0.584
3 Nov.	1.027	0.904	-0.708	-0.518	-0.1004	0.468	0.804	0.217	0.121	-0.616
4 Nov.	1.773	1.655	-0.44	-0.282	-0.0419	0.531	0.536	-0.002	0.022	-0.538
5 Nov.	2.334	2.301	-0.418	-0.372	0.0646	0.652	0.331	0.149	-0.067	-0.186

accuracy of certain components can be improved compared with their IERS-predicted values. These results represent the advantage of SVLBI in estimating probe orbital parameters and EOPs simultaneously.

2.5 Conclusions

ΔVLBI and SVLBI were developed from traditional VLBI technology. Their technical features enable relatively high accuracy observations and can compensate for the lack of traditional techniques in the field of deep space exploration. In addition, because EOPs and other useful parameters are involved in ΔVLBI and SVLBI observations, geodetic applications can be implemented during deep space probe orbit determination. This chapter presents a new method for obtaining more accurate orbital parameters, EOPs, and other geodetic parameters using ΔVLBI and SVLBI, and the results obtained are found to be both reasonable and credible.

Acknowledgments

This research was funded by the national "863 Project" of China (No. 2008AA12Z308), National Natural Science Foundation of China (No. 40974003), National Basic Research Program of China (973 Program) (No. 2012CB720000), Main Direction Project of Chinese Academy of Sciences (No. KJCX2-EW-T03), and National Natural Science Foundation of China (NSFC) Project (No. 11173050).

References

Ádám, J. Estimability of geodetic parameters from space VLBI observables. Report No. 406, Dept. of Geodetic Science and Surveying, the Ohio State University, Columbus, Ohio, July. 1990.

Chen, M., Tang, G., Cao, J., et al. Precision orbit determination of CE-1 lunar satellite. *Geomat. Inf. Sci. Wuhan Univ.* 2011, 36(2), 212–217.

Cui, X., Yu, Z., Tao, Benzao., et al. *General Surveying Adjustment*. Wuhan: Wuhan University Press, 2005.

Hering, T.A., Buffett, B., Mathews, P.M., et al. Forced nutations of the earth: Influence of inner core dynamics 3. *J. Geophys. Res.* 1991, 96(B5), 8259–8273.

Huang, W. B. *Modern Adjustment Theory and Its Application*. Beijing: Chinese PLA Press, 1990.

Huang, Y. Orbit determination of the first Chinese lunar exploration spacecraft CE-1. *Shanghai Astron*. Obs. 2006.

Jin, S.G., Arivazhagan, S., Araki, H. New results and questions of lunar exploration from SELENE, Chang'E-1, Chandrayaan-1 and LRO/LCROSS. *Adv. Space Res.* 2013, 52(2), 285–305.

Jin, S.G., Chambers, S., Tapley, B. Hydrological and oceanic effects on polar motion from GRACE and models. *J. Geophys. Res.* 2010, 115, B02403.

Jin, S.G., Hassan, A., Feng, G. Assessment of terrestrial water contributions to polar motion from GRACE and hydrological models. *J. Geodyn.* 2012, 62, 40–48.

Jin, S.G., Zhang, L., Tapley, B. The understanding of length-of-day variations from satellite gravity and laser ranging measurements. *Geophys. J. Int.* 2011, 184(2), 651–660.

Lin, K.X. *Very Long Baseline Interferometry*. Beijing: China Astronautic Publishing House, 1985.

Li, J., Guo, L., Qian, Z., et al., The application of the instantaneous states reduction to the orbital monitoring of pivotal arcs of the Chang'E -1 satellite. *Sci. China* 2009, 39(10), 1393–1399.

Liu, Q.H., Shi, X., Kikuchi, F., et al. High-accuracy same-beam VLBI observations using Shanghai and Urumqi telescopes. *Sci. China Ser. G* 2009, 39(10), 1410–1418.

Kulkarni, M.N. A feasibility study of space VLBI for geodesy & geodynamics. Report No. 420, Dept. of Geodetic Science & Surveying, The Ohio State University, Columbus, Ohio, 1992.

Ouyang, Z.Y., Li, C.L., Zou, Y., et al. Initial scientific results of CE-1. *Chin. J. Nat.* 2010, 32(5), 249–254.

Sekido, M., Fukushima, T. *Relativistic VLBI Delay Model for Finite Distance Radio Source.* International Association of Geodesy Symposia: Vol. 128. Berlin, Springer, 141–145, 2005.

Wang, W. Accuracy analysis of orbit determination for Chinese lunar exploration mission. *Shanghai Astron. Obs.* 2005.

Wei, E. *Research on the Designment of Chinese Space VLBI System and Computation Simulation*. Wuhan University, 2006.

Wei, E., Yan, W., Jin, S.G., Liu, J., Cai, J. Improvement of Earth orientation parameters estimate with Chang'E-1 ΔVLBI observations. *J. Geodyn.* 2013, 72, 46–52.

Yan , W., Wei, E., Liu, J. On the sensitivity of EOPs to the priori precision of Chang'E-1's simulated parameters. *Shanghai China CPGPS* 2010, 321–326.

Yan, W., Wei, E., Liu, J. Determination of CE-1 Orbit and EOPs with ΔVLBI observation in earth-moon transfer orbit. *Geom. Inf. Sci. Wuhan Univ.* 2012, 37(8), 960–962.

Yu, P., Sun, X., Zhao, S. Chebyshev polynomial fitting model for GPS orbit calculation. *Meteoro. Sci. Technol.* 2004, 32(3), 198–201.

Zheng, Y. *VLBI Geodesy*. Beijing: Chinese PLA Press, 1999.

Zheng, Y., Qian, Z. Using space VLBI to establish reference systems and their connection. *J. PLA Inst. Survey. Map.* 1992, 3, 19–22.

3

Laser Altimetry and Its Applications in Planetary Science

Hauke Hussmann

CONTENTS

3.1 Introduction 52
3.2 Measurement Principle and Scientific Objectives 52
3.3 Altimeters on Moon Missions and Planetary Missions.......................57
    3.3.1 The Moon 57
        3.3.1.1 The Apollo 15, 16, and 17 Laser Altimeters.................57
        3.3.1.2 LIDAR on Clementine...58
        3.3.1.3 LALT on Kaguya/SELENE 58
        3.3.1.4 LAM on Chang'E-1 59
        3.3.1.5 LLRI on Chandrayaan-1..60
        3.3.1.6 The Lunar Orbiter Laser Altimeter on the Lunar
                Reconnaissance Orbiter...60
        3.3.1.7 LAM on Chang'E-2 61
    3.3.2 Mars 62
        3.3.2.1 The Mars Orbiter Laser Altimeter on MGS..................62
    3.3.3 Mercury 63
        3.3.3.1 The Mercury Laser Altimeter on MESSENGER 63
    3.3.4 Near-Earth Asteroids...65
        3.3.4.1 NLR on NEAR-Shoemaker..65
        3.3.4.2 Light Detection and Ranging Instrument on
                Hayabusa-1 66
3.4 Future Prospects...67
    3.4.1 BELA on BepiColombo 68
    3.4.2 GALA on the Jupiter Icy Moons Explorer....................................69
    3.4.3 OSIRIS-REx 71
3.5 Concluding Remarks 72
Acknowledgments...72
References...73

3.1 Introduction

Laser altimetry is a powerful tool to address major objectives of planetary physics, geodesy, and geology (e.g., Jin et al. 2013). It can be used to determine the global shape and radius of planetary bodies, global, regional, and local topography of the surface, tidal deformation, rotational states and physical librations, surface roughness and local slopes, seasonal changes of elevations (e.g., polar caps on Mars), atmospheric properties (density by attenuation of the laser beam), and albedo at the laser wavelength. In addition, laser altimetry is used to define accurate reference systems of planetary bodies, which provide the basis for other measurements that have to be referenced to absolute coordinate systems. Furthermore, it can assist in spacecraft orbit and attitude determination. All these methods, except measurements of tidal deformation and physical librations, have been applied to solar system bodies other than Earth.

Since the first applications in planetary explorations on the Apollo missions 15, 16, and 17 to the Moon, Laser altimeters have improved significantly. So far they have been used on planetary missions to the Moon, Mars, Mercury, and the asteroids 433 Eros, and 25143 Itokawa. In this chapter, we review the current status of laser altimetry used in exploration of solar system bodies, other than Earth. We start by describing the measurement principle of laser altimetry including specific technical and operational aspects. In Section 3.3 (Altimeters on Moon Missions and Planetary Missions), we summarize the basic results and technical realizations of laser altimeters that have been flown on planetary and Moon missions. We conclude by briefly describing prospects for future applications.

3.2　Measurement Principle and Scientific Objectives

The basic principle of a range measurements is simple: a laser pulse is transmitted from the instrument onboard the spacecraft to the surface of the planet. The reflected laser pulse (assumed Lambertian) is received in the instrument's telescope. From the measured travel-time Δt of the wave-package the distance d to the surface can be computed by $d = c\,\Delta t/2$, where $c =$ 299,792.458 km/s is the speed of light in vacuum. The factor of $1/2$ is required because the photons pass the distance d twice. Thus, Δt of 1 ns corresponds to approximately 15 cm in range. Despite the simple principle, the technical challenges in realization of such a measurement are substantial. Not only the precision within the instrument (e.g., determination of accurate times for outgoing and received pulse) but also the position and orientation of the spacecraft at the time of the measurement have to be accurately known.

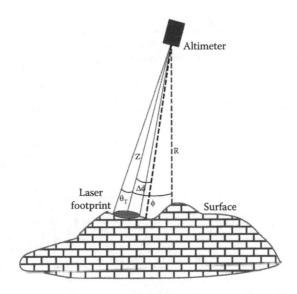

FIGURE 3.1
Basic measurement principle of laser altimetry. (Thomas, N., et al. 2007. The BepiColombo Laser Altimeter (BELA): concept and baseline design. *Planet. Space Sci.* 55, 1398–1413.)

In Figure 3.1, the involved distances and angles are shown. Here we assume a general case with a small off-nadir pointing angle ϕ and a surface slope. The outgoing laser pulse has a beam-divergence of Θ_T. The off-nadir pointing angle may only be known to an accuracy of $\Delta\phi$. The distance to the surface from the altimeter is specified as z. The height h of the surface with respect to a reference surface R_{ref} (usually the geoid) is the actual quantity to be determined. It requires accurate knowledge on the spacecraft distance with respect to the center of mass $R_{s/c}$ and its orientation at the time when the measurement was taken. The height of the laser footprint can then be determined as:

$$h = \sqrt{R_{S/C}^2 + z^2 - 2R_{S/C}z\cos\phi} - R_{ref} \qquad (3.1)$$

Besides the time-of-flight of the photons, the wave-package, which is received by the instrument, contains additional information. Slopes and surface roughness on the planetary surface cause a spreading of the return pulse. Therefore surface characteristics—roughness and slope on the scale of the laser footprint—can be derived from the pulse spread. By filter matching yields, yet more information on the surface on the scale of the laser footprint (e.g., steep slopes at a crater rim) can be extracted from the distortion of the return pulse with respect to the outgoing pulse. Sizes of the footprint range from 5 m (LOLA on LRO) to about 100 m and more. In addition to range, slopes, and surface roughness, the albedo of the surface at the laser-wavelength can be derived

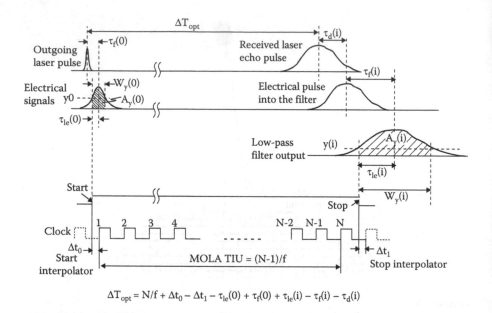

$$\Delta T_{opt} = N/f + \Delta t_0 - \Delta t_1 - \tau_{le}(0) + \tau_f(0) + \tau_{le}(i) - \tau_f(i) - \tau_d(i)$$

FIGURE 3.2
Timing diagram of the MOLA optical and electrical signals. Several contributions have to be taken into account to derive the time-of-flight of the wave-packet and the corresponding error budget. (Abshire, J.B., Sun, X., Afzal, R.S. 2000. Mars orbiter laser altimeter: Receiver model and performance analysis. *J. Appl. Optics* 39, 2449–2460.)

from determining the energy and intensity, or more precisely, the number of photons of the transmitted and received pulses (e.g., Gardner 1982, 1992).

To illustrate the complexity of a range measurements, Figure 3.2 shows the different contributions of timing offsets and the corresponding errors that have to be taken into account using MOLA as an example (Abshire et al. 2000). The duration of each step in the electronic and optical detection chain has to be accurately known to obtain a precise time-of-flight measurement. A time signal (clock counter) is used for reference. In between the time tags an interpolator is used for more precise determination in between two time signals. When the transmitted pulse reaches a predefined threshold, the start time is defined. Note that the electronic signal is slightly delayed with respect to the "real" optical signal. After a certain time (typical order of several milliseconds), the back-scattered optical signal is received from the surface and is converted into an electrical signal. After passing a low-pass filter, the stop-time is determined again by a predefined threshold of the received signal. From the two thresholds of both transmitted and received signal, the pulse spreading and the location of the corresponding peaks can be determined. The time between the two optical peaks is then given as:

$$\Delta t_{opt} = N/f + \Delta t_0 - \Delta t_1 - \tau_{le}(0) + \tau_f(0) + \tau_{le}(i) - \tau_f(i) - \tau_d(i) \qquad (3.2)$$

This example shows that the exact characterization of the signal is not an easy task. Furthermore, each step in the chain has its own error that has to be extremely small to obtain range measurements with an accuracy of a few tens of cm (timing accuracy on the order of ns). Note that the pulse is spread due to surface roughness on the scale of the laser footprint and due to a possible slope in the surface terrain. The natural pulse-spreading due to wave propagation in vacuum is much smaller. Therefore, the pulse spreading can be used to characterize the surface roughness and local slopes.

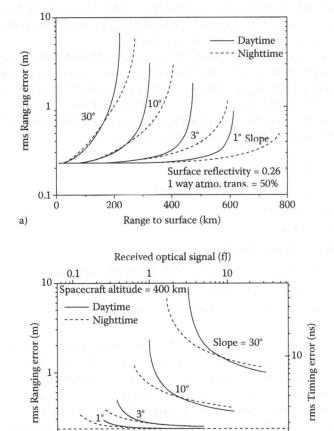

FIGURE 3.3
Ranging error of the MOLA instrument a) as a function of distance to the Martian surface, slopes, illumination conditions, and b) as a function of received number of photons. (From Abshire, J.B., Sun, X., Afzal, R.S. 2000. Mars orbiter laser altimeter: Receiver model and performance analysis. *J. Appl. Optics* 39, 2449–2460.)

In addition to the signal processing, the surface itself and the environmental conditions strongly affect the accuracy of range measurements.
Figure 3.3a shows the error of range measurements for MOLA as a function
of distance for different surface slopes and for day- and night-side measurements on Mars. The error is strongly dependent on the overall distance to the
surface and on the surface slopes. In the case of Mars, the attenuation of the
signal in the atmosphere must also be taken into account. The measurements
are more accurate without the increased noise-level by sunlight. Figure 3.3b
shows the ranging error as a function of number of photons (or energy of the
received optical signal) for different slopes. Only for a very flat surface and
a large number of photons (good signal-to-noise ratio) the limit of 23 cm due
to quantization of the signal can be reached.

All these errors have to be taken into account in the performance budget of
laser altimeters. In addition, the albedo at the laser wavelength is crucial for
the number of returned photons. As shown in Figure 3.3, this not only affects
the probability of detection, but also the ranging error.

Laser altimetry is a powerful tool to address many objectives of planetary
physics and geodesy. It can be used to measure:

1. Global radius and shape
2. Global, regional, and local topography
3. Tidal deformation
4. Rotation and libration
5. Surface roughness and local slopes
6. Seasonal changes of elevations (e.g., polar caps on Mars)
7. Atmospheric properties (e.g., density by attenuation)
8. Albedo (reflectivity) at the laser wavelength

In addition, altimetry can be used to assist in orbit determination and
other navigation purposes (e.g., Shum et al. 1990). All these objectives (except
measuring the tidal deformation by altimetry) have been addressed with different instruments on various missions and planetary targets. We describe
major achievements for specific missions in the next section.

The main task of altimetry is to derive the global and local topography of
planetary bodies. The topography is essential in interpreting the gravity signal
usually obtained from Doppler tracking of the spacecraft. In principle, a gravity anomaly can arise from mass concentrations (or deficiencies) in the interior
or from topography with respect to a reference potential. The latter is called
the geoid for Earth and has to take into account contributions from the planet's
rotation besides the pure gravitational potential (e.g., Rummel 2005). Geoid
anomalies have to be defined with respect to this reference geoid, which is a surface of constant potential. These anomalies have to be corrected for elevations
referenced to the geoid. A simple approach is given by the Bouguer formula,

which corrects the free air gravity anomaly for effects by local topography (see also Chapter 11). It should be noted that the Bouguer anomaly does not take into account the long-wavelength topography. It can only be applied locally. However, local topography can be supported by the lithosphere beneath it and surface loads by large-scale topography deflect the lithosphere downward. The mass associated with the long-wavelength topography can be compensated by a low density "root" because the crust is moved downward. Therefore, for long-wavelength topography, Bouguer anomalies, and topography are not correlated. Topography can be isostatically compensated and has in that ideal case zero free-air gravity anomalies. Therefore, the ratio of Bouguer anomalies and topography as a function of wavelength (admittance) can reveal the state of compensation. From that, crustal thicknesses can be derived if gravity and topography are accurately known. This is an important technique that has been applied to the terrestrial planets to infer the crustal thickness as a function of longitude and latitude. The state of the crust and especially its thickness yields important information about planetary evolution and global and regional processes that have shaped the planetary surfaces. Examples are given in the next section in application to the terrestrial planets.

3.3 Altimeters on Moon Missions and Planetary Missions

In this section the laser altimeters flown so far on missions to the Moon and planetary bodies are briefly described. References to more extensive descriptions, both technically and scientifically, are given. Here, we also summarize the main results that have been achieved based on altimetry data at the Moon, Mars, Mercury, and the asteroids Eros and Itokawa.

3.3.1 The Moon

In the following missions to the Moon carrying altimeters are described. Ordered by the date of arrival in the final science orbit, we describe the technical aspects and main results of the laser altimeter experiments.

3.3.1.1 The Apollo 15, 16, and 17 Laser Altimeters

The first laser altimeters used for exploration of a planetary body other than the Earth were flown on the Apollo 15, 16, and 17 missions to the Moon in 1971 and 1972. Originally designed to provide an altitude measurement for each camera frame taken, the Apollo laser altimeters were part of the metric camera system. However, they could also be used independently from the metric camera providing—combined with precise orbit determination from tracking data—altitudes of the lunar surface with respect to the Moon's center of mass (Roberson and Kaula 1972; Wollenhaupt and Sjogren

1972; Wollenhaupt et al. 1973). Whereas Apollo 15 altimeter data have been reduced for only two ground-tracks on the near and far side, the Apollo 16 and 17 altimeters collected several complete orbits of the lunar command and service module (CSM) and made 2372 and 4026 firings, respectively. All three altimeters used ruby lasers pumped by flash-lamps and a Q-switch to transfer the light pulse to the output resonant reflector. The instruments each had a mass of 22.5 kg. The Apollo 17 altimeter had been modified because of anomalous performances and degrading in the Apollo 15 and 16 campaigns. It operated excellently throughout the mission and near the end of mission, it acquired data for six consecutive revolutions (Wollenhaupt et al. 1973).

The data from the three missions were used to determine the global shape of the Moon yielding evidence for a displacement of the Moon's center of mass with respect to the center of figure of about 2 km, by analysis of the first-degree coefficients of the harmonic expansion of the elevation (Sjogren and Wollenhaupt 1973). Interpretation of the data provided first evidence for a varying thickness of the lunar crust with its maximum at the far side. In addition, depths of basins as well as mountain heights were measured along the ground-tracks and interpreted with respect to the Moon's interior and evolution (Kaula et al. 1973, 1974).

3.3.1.2 LIDAR on Clementine

First global topographic maps of the Moon based on laser altimeter data combined with gravity field data from S-band tracking were derived by the Clementine mission, sponsored by the Ballistic Missile Defense Organization with participation from NASA (Zuber et al. 1994). Analysis of global topography and gravity data revealed that the lunar highlands on the far side are isostatically compensated, while impact basins are in a wide range of compensation states independent from their sizes and ages. Global maps of crustal thickness were derived showing crustal thinning under the basins and the hemispheric differences. The South Pole-Aitken basin was recognized as the largest basin of the solar system from the topographic models based on Clementine data. A 16-km range of elevation has been determined, with the greatest excursions occurring on the lunar far side. The evolution of the Moon turned out to be more complex than previously thought.

The laser ranging instrument (LIDAR) operated for approximately one half hour for each revolution of about 5 h. Ranging was possible at altitudes of 640 km and less. The mapping mission lasted for two months yielding about 72,300 valid ranges out of over half a million shots. The typical shot frequency was 1.67 Hz (Zuber et al. 1994).

3.3.1.3 LALT on Kaguya/SELENE

The laser altimeter LALT onboard the Kaguya/SELENE spacecraft, developed and operated by JAXA, obtained a global lunar topographic map with a spatial resolution finer than 0.5 degrees, corresponding to about 15 km at the equator.

In contrast to the results obtained by the Clementine LIDAR, the data from LALT are not biased because Kaguya was in an almost circular polar 100-km-altitude orbit. Whereas coverage by the Clementine LIDAR suffered from larger orbital distances in the northern high-latitude regions. Comparing the topographic maps obtained by the Clementine LIDAR and LALT, the latter shows a great improvement in accuracy and resolution for features that are less than a few hundred km across (Araki et al. 2009). Polar maps were obtained with complete coverage. These are invaluable data products to determine permanently shadowed regions near the poles (Noda et al. 2009). Those are relevant for energy supply with solar panels at future landing sites, and for the possible presence of ice in polar crater floors.

The lunar topographic range was determined as 19.81 km, 2 km greater than previously thought. The spherical harmonics model of the lunar topography based on LALT data deviates from the Clementine LIDAR model, especially in the high-order range (degree and order >30) indicating that the Moon is rougher at scales smaller than 180 km. Whereas isostatic compensation is the main lithospheric support mechanism at long wavelength, rigid support is suggested to be dominating on smaller scales on the Moon. The latter may be the consequence of a much drier lunar lithosphere (as compared to Earth).

Combined with gravity field data, which has been very accurately determined on the lunar far side by tracking the Kaguya spacecraft from its relay subsatellite, the topography reveals various states of compensation for large impact basins on the lunar far-side (Namiki et al. 2009).

LALT was operational from December 2007 to mid-2009. However, due to a loss of laser power by 5 mJ and a failure of one of the spacecraft's reaction wheels, resulting in less accurate orbit determination, LALT could not be operated as planned from April 2008 to the end of the nominal mission (Araki et al. 2013). In the extended mission, in which the altitude was lowered to 50 km, the number of successful shots increased again. LALT acquired over 22 million successful range measurements over Kaguya's 1.5 year operational phase. Updates of the topographic maps have been obtained using newly available gravity field models. The results are in good agreement with the ones from Lunar Reconnaissance Orbiter and Chang'E-1 (Araki et al. 2013).

3.3.1.4 LAM on Chang'E-1

Range measurements from the laser altimeter LAM onboard the Chinese mission Chang'E-1 have been acquired from November 2007. A global lunar digital elevation model with 3 km spatial resolution similar in accuracy to the global model constructed from SELENE/Kaguya data has been derived from LAM data (Jin et al. 2013). Measurements were taken from a polar orbit at an altitude of about 200 km. LAM was operated at 1 Hz shot frequency and

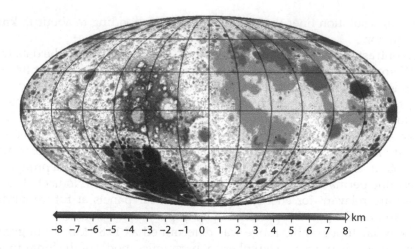

FIGURE 3.4
Global lunar digital elevation model derived from Chang'E-1 LAM data. (Jin, S.G., Arivazhagan, S., Araki, H. 2013. New results and questions of lunar exploration from SELENE, Chang'E-1, Chandrayaan-1, and LRO/LCROSS. *Adv. Space Res.* 52, 285–305.)

yielded a total of 9.12 million measurements between November 2007 and December 2008. A global digital elevation model mapped with Chang'E-1 LAM data is shown in Figure 3.4.

3.3.1.5   LLRI on Chandrayaan-1

Chandrayaan-1 was India's first lunar mission and was successfully launched in October 2008. Originally planned for a two-year operation time, contact with the spacecraft unexpectedly terminated in August 2009, after successful completion of more than 3400 revolutions about the Moon. The lunar laser range instrument (LLRI) was aimed at determining the topography of the lunar surface from a 100-km polar orbit (Kamalakar et al. 2009). Operated at 10 Hz frequency, LLRI's range accuracy between spacecraft and lunar surface was <5 m. Detection and precise profiling of topographic features was proven to be possible with LLRI (Kamalakar et al. 2009). Lunar topography models derived from LLRI data have also been combined with results from the CHACE experiment (Chandra's Altitudinal Composition Explorer), which was part of the Chandrayaan-1's Moon Impact Probe (Sridharan et al. 2013).

3.3.1.6   The Lunar Orbiter Laser Altimeter on the Lunar
Reconnaissance Orbiter

As of this writing (2013), lunar orbiter laser altimeter (LOLA) is still in operation onboard the lunar reconnaissance orbiter (LRO) spacecraft. In combination with data from NASA's GRAIL mission, which has precisely mapped the lunar gravity field, LOLA provides the most accurate geodetic measurements for the Moon. Based on LOLA range measurements and LRO orbit determination, the lunar geodetic grid has been improved to approximately 10 m radial and approximately 100 m spatial accuracy with respect to the Moon's center of mass. LOLA has also provided the highest resolution global maps yet produced of slopes, roughness, and the 1064 nm reflectance of the lunar surface (Smith et al. 2010b).

In contrast to all other altimeters described here, LOLA is a multibeam laser altimeter. A single laser beam is split by a diffractive optical element into five output beams (see also Rowlands et al. 2009). Together with the 28-Hz frequency at which LOLA is operated, this results in a total sampling rate of the lunar surface of 140 measurements per second. Each of the five beams has a 100 mrad divergence and illuminates a 5 m diameter spot from LRO's mapping orbit. The backscattered signals of the five pulses are received on separate APDs. LOLA measurements could also be used to improve the orbit determination of LRO (Mazarico et al. 2010). For this purpose, altimetric cross-over points (intersecting ground-tracks) were used in the inversion to determine the spacecraft trajectory. The total error in spacecraft position could be reduced by several meters or several tens of meters, depending on the a priori gravity field chosen for the inversion (Mazarico et al. 2012). A similar technique has led to an improvement of the orbit determination of the Mars Global Surveyor (MGS) spacecraft using MOLA data (Rowlands et al. 1999). In the case of the Moon, the use of cross-overs is challenging because of the Moon's slow rotation. Ground-tracks intersect with very small angles, making the analysis more difficult. Again the five-beam measurements help to exactly determine the cross-over location.

The main investigation of LOLA is to assist in the selection of future landing sites by providing topographic, surface roughness, surface slope, reflectance at 1064 nm, and a precise global lunar coordinate system (Smith et al. 2010c). However, especially in combination with high-precision gravity field data obtained by NASA's GRAIL mission, LOLA has yielded invaluable information on the lunar topography and processes that have shaped the Moon's surface.

The global data set has an along-track spatial resolution of about 20 m and a cross-track resolution of about 0.07° corresponding to ~1.8 km at the equator and 160 m at 85° latitude, respectively (Smith et al. 2010b). Fundamental parameters of the lunar shape are derived from a spherical harmonic expansion up to degree 720 based on global LOLA data. The highest and lowest elevations with respect to the reference radius are 10.7834 and −9.117 km, respectively.

3.3.1.7   LAM on Chang'E-2

Chang'E-2 was in orbit around the Moon from October 2010 to June 2011. The spacecraft had an altitude of 100 km above the lunar surface. However, the orbit was lowered into a 15 × 100 km orbit on May 2011 (Li et al. 2012). The lowest 15-km altitude point was over Sinus Iridum, the possible landing-site of Chang'E-3. The characterization of this region by high-resolution imaging was one of the main tasks of Chang'E-2. After completing its nominal and extended mission phases, the spacecraft left its lunar orbit for the L2 Sun-Earth Lagrangian point. In December 2012, the space probe completed a close flyby at asteroid Toutatis acquiring high-resolution images during approach. As of November 2013, Chang'E-2 is still in operation as China's first deep space mission. The laser altimeter LAM onboard Chang'E-1 was equivalent to the one on its precursor mission. However, LAM on Chang'E-2 did not operate as expected and could therefore not augment the scientific data obtained by LAM on Chang'E-1.

3.3.2 Mars

3.3.2.1 The Mars Orbiter Laser Altimeter on MGS

Mars orbiter laser altimeter (MOLA) acquired altimetry data in orbit around Mars from 1997 to June 2001. It was operated at 10 Hz frequency and collected more than 600 million laser range measurements. It obtained very precise global topographic maps of the Martian surface and is to date the only laser altimeter to have orbited Mars. The vertical accuracy was of the order of meters. A global topography map derived from MOLA data is

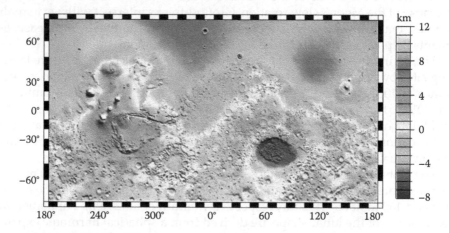

FIGURE 3.5
Map of the global topography of Mars derived from MOLA data (here shown in Mercator projection up to 70° latitude). Note the elevation difference between the northern and southern hemispheres. The Tharsis volcano-tectonic province is centered near the equator in the longitude range 220°E to 300°E and contains the vast east–west trending Valles Marineris canyon system and several major volcanic shields including Olympus Mons, Alba Patera, and others. The Hellas basin is located in the southern highlands. Note that color scale saturates at elevations above 8 km. (Figure slightly modified from Zuber, M.T., Smith, D.E., Lemoine, F.G., Neumann, G.A. 1994. The shape and internal structure of the moon from the Clementine mission. *Science* 266, 1839–1843.)

shown in Figure 3.5. It displays a variety of prominent features including the low northern hemisphere, the Tharsis province, and the Hellas impact basin.

From global shape measurements, a difference of about 20 km between the polar and equatorial axes has been determined indicating a significant flattening due to rotation (Zuber et al. 2000a). The center of mass is shifted by about 3 km along the polar axis toward the south pole with respect to the center of figure. This is consistent with enhanced elevations of about 6 km in the southern highlands, with respect to the northern hemisphere. The topography map clearly shows a global trend of elevations and the dichotomy between the two hemispheres (see also Smith et al. 1999). The latter is also inherent in surface-roughness maps derived from the pulse spreading of MOLA shots (Smith et al. 2001a). The southern highlands display very rough terrain whereas the northern lows are much smoother. The latter is extremely important for the selection of landing-sites on Mars. Combining gravity and topography data, the crustal thickness has been derived from MOLA data (Zuber et al. 2000a).

With MOLA, it was also possible to detect seasonal variations of snow deposits at the polar caps of 1.5–2 m maximum during the Martian winter (Smith et al. 2001b). These measurements resolve changes in heights as little as 10 cm. Topography data from MOLA also contributed to understanding the pathways of liquid water that was flowing at the surface of early Mars. In the case of Mars, laser altimetry data from MOLA was also used to detect cloud structures in the planet's atmosphere. Two different types could be detected: (a) clouds that were reflective at the laser wavelength of 1064 nm and (b) clouds that were opaque omitting the return pulse (Neumann et al. 2003b). Formation and migration of the clouds could be tracked and interpreted with respect to the seasonal cycles on Mars. The very different aspects discussed here for Mars—interior, surface, and atmosphere—show the broad applications of laser altimetry data in different fields of planetary science.

In 1998, the MGS spacecraft had close encounters with the Martian moon Phobos. During the closest flybys with a distance of 265 km two ground-tracks were obtained with MOLA (Banerdt and Neumann 1999). This was the first active ranging experiment to a small solar system body.

3.3.3 Mercury

3.3.3.1 The Mercury Laser Altimeter on MESSENGER

The Mercury laser altimeter (MLA) has been operational in orbit around Mercury since March 2011. Before Mercury orbit insertion, MLA also provided two ground-tracks during two Mercury flybys on January 14 and October 6, 2008 (Smith et al. 2010a). Closest approaches of these flybys were around 200 km. With a beam-divergence of 100 mrad, the spot-diameter on the surface varies from about 20 m at closest approach to about 100 m at

1000 km distance. On a flat surface with good signal-to-noise ratio, MLA's nominal range precision is approximately ±10 cm (Cavanaugh et al. 2007). However, due to signal-to-noise ratios (SNR), the fact that MLA was not nadir-pointing during data acquisition and due to the limits of knowledge on spacecraft position and pointing, the accuracy of range measurements with respect to Mercury's center of mass was ~16 m (Smith et al. 2010a). The maximum distance up to which range measurements were taken nearly reached 1700 km, several hundred kilometers more than the predicted value of 1200 km. From the flybys, which both were in near-equatorial latitudes, first global equatorial shape determination provided a value of $(a - b)/a = (664 \pm 29) \times 10^{-6}$ m, where a and b are the two polar axes (Smith et al. 2010a). The ellipsoidal shape is consistent with mean slopes of ~0.015° downward to the east, apparent in both flybys. Combined with gravity field data, determinations of long-wavelength topography provided first estimates of Mercury's crustal thicknesses and possible mass anomalies (Smith et al. 2010a).

The two ground-tracks from the flybys complement the main data set that has been obtained after Mercury orbit insertion in March 2011. Because of the highly elliptic polar orbit and an approximate maximum distance for the acquisition of range measurements of up to 1500 km altitude, only the northern hemisphere could be covered by MLA data. The topographic model derived from these data revealed a much smaller dynamic range of elevations (9.85 km) as compared to the Moon (19.9 km) and Mars (30 km) (Zuber et al. 2012). The most prominent features include an extensive lowland at high northern latitudes that hosts the volcanic northern plains and the Caloris basin. Within the northern lowland is a broad topographic rise that experienced uplift after plains emplacement. The interior of the 1500-km-diameter Caloris impact basin has been modified so that part of the basin floor now stands higher than the rim. The elevated portion of the floor of Caloris appears to be part of a quasi-linear rise that extends for approximately half the planetary circumference at mid-latitudes (Zuber et al. 2012).

Figure 3.6 shows a single laser track across an impact crater within the Caloris basin. The tilt of the crater floor is a consequence of the general modification of long-wave topography in that region some time after the formation of the Caloris basin (Zuber et al. 2012).

MLA is operated at 8 Hz. It uses an array of four diffractive telescopes each with an aperture of 14 cm in diameter to detect the return pulse. The different design (as compared, e.g., to MOLA and other altimeters, which use one Cassegrain reflector as receiving device) was chosen because of the extreme thermal environment at Mercury implying significant thermal gradients within the instrument. The signal from the four receiving telescopes is transferred to a single detector APD by fiber optics. Details on the MLA design and its calibration can be found in Cavanaugh et al. (2007).

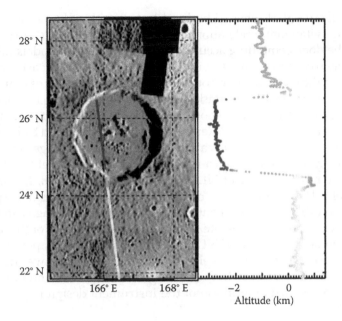

FIGURE 3.6
Single MLA-profile across the 100-km impact crater Atget within the Caloris basin. Note the northward tilt of the crater floor. (From Zuber, M.T., and 23 colleagues. 2012. Topography of the northern hemisphere of mercury from MESSENGER laser altimetry. *Science* 336, 217–220.)

3.3.4 Near-Earth Asteroids

3.3.4.1 NLR on NEAR-Shoemaker

The first laser altimeter flown on an asteroid mission was the NEAR-Shoemaker Laser Rangefinder (NLR) onboard the NEAR-Shoemaker spacecraft that went in orbit around asteroid 433 Eros in 2000. From more than 8 million individual shots, shape models have been constructed with a radial accuracy of about 30 m with respect to the asteroid's center of mass.

The best-fit tri-axial ellipsoid has the following dimensions: $a = 20.591 \pm 0.040$ km, $b = 5.711 \pm 0.040$ km, and $c = 5.332 \pm 0.050$ km (Zuber et al. 2000b). The ellipsoid only poorly fits the real shape of Eros compared with other ellipsoidal fits of asteroids or small moons imaged from close flybys. However, based on NLR data there is no evidence for a dumbbell shape, ruling out the possibility that Eros is a contact binary asteroid with two loosely bound major components. NLR data indicate that asteroid 433 Eros is a consolidated body with a complex shape dominated by collisions. The offset between the asteroid's center of mass and center of figure of a few hundred meters indicates a small deviation from a homogeneous internal structure that is most simply explained by variations in mechanical structure (Zuber et al. 2000b).

For small irregularly shaped bodies like asteroids, it is essential to determine the gravitational acceleration with respect to the bodies' surface slopes. This can be done combining accurate gravity and shape models taking into account the rotation of the asteroid (Miller et al. 2002). Most asteroids rotate very fast (on the order of a few hours) and the centrifugal forces can be of the same order as the small gravitational forces. In case of Eros, regional-scale relief and slope distributions show evidence for control of topography by a competent substrate. Impact crater morphology is influenced by both gravity and structural control. Small-scale topography reveals ridges and grooves that may be generated by impact-related fracturing (Zuber et al. 2000b).

The pulse repetition rate of NLR could be adjusted at 1/8, 1, 2, and 8 Hz, respectively. The small gravity of asteroids implies very small spacecraft velocities in orbit. Therefore, contiguous along-track coverage can easily be achieved with small or moderate shot frequencies. Because of the small distance to the target (in case of NLR ~50 km) small energies per shot can be used to guarantee the required signal-to-noise ratios. NLR was designed to operate at altitudes up to 50 km with range accuracy <6 m and range resolution <1 m. A detailed description of the instrument design is given by Cole et al. (1997).

3.3.4.2 Light Detection and Ranging Instrument on Hayabusa-1

The light detection and ranging instrument (LIDAR) on-board the Japanese Hayabusa-1 spacecraft was operational for three months from September to November 2005 in orbit around asteroid 25143 Itokawa. The instrument detected more than 1.6 million return signals (Mukai et al. 2007) covering most of the surface of the asteroid (Figure 3.7). With dimensions on the order of a few hundred meters only, Itokawa is significantly smaller than asteroid 433 Eros discussed in the previous subsection. The low gravity has challenging implications for navigation and orbit determination. Therefore, measurements of the LIDAR instrument were also important to support the maneuvering during the approach phase and near the asteroid. LIDAR was able to obtain range measurements at distances from 50 m to 50 km.

The accuracy for range measurements at 50 m distance was about 1 m. Operation and data processing of the instrument were severely affected by the failure of two out of three reaction wheels of the Hayabusa spacecraft. This led to a fluctuation of the beam spots on the surface of the asteroid which made the systematic surface survey difficult. However, as shown in Figure 3.7 a good coverage had been obtained for large parts of the asteroid's surface. The typical dimension of the surface spot was a 7 × 12 m ellipse (Mukai et al. 2007). Complementary to the camera data, Itokawa's shape has been derived from LIDAR data yielding a mean density of 1.9 g/cm^3 and a high bulk porosity of 40% (Mukai et al. 2007). The local surface topography is characterized by both large smooth areas and regions covered with meter-sized boulders. The lowlands are filled with regolith with about 2.3 m

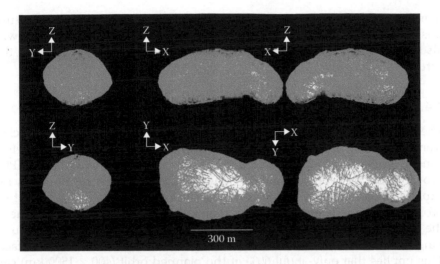

FIGURE 3.7
Shape and LIDAR ground-tracks on the surface of asteroid Itokawa. The spots cover about 87% of the asteroid's surface. The coverage in the polar regions (z-axis corresponds to the rotational axis) is smaller compared to low-latitude regions. (From Mukai, T., and 15 colleagues. 2007. An overview of the LIDAR observations of asteroid 25143 Itokawa. *Adv. Space Res.* 40, 187–192.)

minimum thickness (Barnouin-Jah et al. 2008), which is difficult to explain as a result of impacts only. However, an Itokawa composed of several large masses may have retained this regolith during its formation (Barnouin-Jah et al. 2008). LIDAR data combined with camera data and gravity field determination suggests that Itokawa has a rubble-pile structure made of loosely bound materials.

Laser altimeters have been flown on various missions investigating the terrestrial planets (Mars, Mercury, and the Moon) as well as near-earth asteroids (NEAs) (Itokawa and Eros). The design of the instruments is driven by the scientific objectives (shot-frequency, size of the laser spot etc.), by the trajectory (altitude above the surface), by physical characteristics of the surface (e.g., albedo), and by the given planetary environment (e.g., thermal design at Mercury).

3.4 Future Prospects

Although various missions, especially the lunar missions, have carried laser altimeters onboard, there still remain many objectives regarding planetary evolution that can best be addressed with laser altimetry. In the following, we describe the prospects for laser altimeters to be flown to planetary bodies in the future. We focus on physical aspects relevant for the missions that are in implementation phase or which have been selected as future missions of ESA and NASA. It can be expected that other space agencies will also continue to explore the solar system bodies with laser altimeters in the future.

3.4.1 BELA on BepiColombo

The BepiColombo mission is going to be launched with an Ariane 5 rocket from Kourou in 2016. It consists of two orbiters, the Mercury magnetor-spheric orbiter provided by JAXA and the Mercury Planetary Orbiter (MPO) provided by ESA. BepiColombo is one of ESA's cornerstone missions for exploration of the solar system. After arrival at Mercury in 2022, the two orbiters will be separated. The BepiColombo laser altimeter (BELA) is one of the 11 instruments onboard the MPO. The instrument is designed to acquire range measurements up to a distance of 1055 km from Mercury's surface. This implies that only about 60% of the planned orbit (400 × 1500 km) can be used for range measurements. However, due to Mercury's J_2-value the pericenter at about 400 km altitude of the MPO is shifted in the mission's course from northern to southern latitudes. By taking advantage of this natural perturbation, global coverage will be possible, allowing for altimetry measurements of the Southern Hemisphere for the first time. Interpretation of global gravity and topography will be more reliable with complete coverage of Mercury's surface.

BELA is the first altimeter on a planetary mission for which a full digitization of the transmitted and returned pulse can be used for interpretation of the signal (Thomas et al. 2007). This is important, not only for the exact determination of the peak but also for the instrument performance when the signal-to-noise ratio is small (Gunderson et al. 2006). Furthermore, it can be used for a detailed analysis of the shape of the return pulse at good signal-to-noise levels. The latter bears information on the surface characteristics on the scale of the laser footprint (~40–80 m), which will be investigated in detail with BELA.

BELA can be operated from 1 to 10 Hz and it has a classical single-beam design at 1064 nm wavelength. The receiver is a Cassegrain-type telescope with an aperture of 20 cm. The onboard software is capable of analyzing the return pulse by using polynomial fits to approximate the pulse shapes. In addition, the full digitized pulse can be returned to ground. However, due to constraints in the data-rate for the instruments, this can only be done for few pulses during nominal operations.

The main objective for BELA is to derive a first global network of laser tracks covering the entire surface to derive the topography on global, regional, and local scales. BELA will also assist in determining the amplitude of Mercury's physical librations.

Mercury's rotation state is unique in the solar system. The planet is locked in a 3:2 spin-orbit coupling, which means that the planet undergoes exactly three full rotations while it revolves about the Sun twice. Owing to Mercury's slightly tri-axial figure, the planet is also undergoing small physical librations, that is, small oscillations about its mean rotation. While the main libration period of 88 days is caused by tidal interaction with the Sun, smaller planetary perturbations are expected to contribute. The measurements of amplitudes and phase of the physical librations offer the unique opportunity among all planets to obtain knowledge on the interior state of Mercury ("Peale's experiment"; Peale et al. 2002). In particular, it is possible to verify and to determine the size of Mercury's possibly molten core. If the outer core is molten, the mantle and core are mechanically decoupled, implying that the mantle and core can rotate independently. In particular, the core will not follow the libration of the mantle and the crust. In this case large libration amplitudes will be observed.

For direct determinations of liquid core size from librations, the gravity field coefficients J_2 and C_{22} must be known, from which Mercury's moment of inertia can be derived. Also, the planet must be within the so-called Cassini state 1, in which spin vector, orbit normal and normal of the invariable plane remain coplanar. From Earth-based radar observations (Margot et al. 2007) Mercury's obliquity (rotation axis orientation) was determined at 2.11 arc min, close to the orbit normal, which effectively confirms that Mercury is in the required rotational state. In this case, the moment of inertia, the obliquity and the libration amplitude allow for a determination of the thickness of an outer liquid core. The latter has important implications for the evolution of the planet and for the generation of Mercury's magnetic dipole field.

By matching laser tracks obtained at different times and thus at different librational phases to stereoimaging data, it is possible to determine libration amplitudes. The data-sets obtained by MLA on MESSENGER and the ones to be acquired by BELA are complementary and allow for determination of long-periodic changes in the libration amplitude, which are expected to occur on the time scale of several years. By identifying the long-term effects on the libration signal, the determination of the liquid outer core thickness can be significantly improved.

3.4.2 GALA on the Jupiter Icy Moons Explorer

In 2012 the Jupiter Icy Moons Explorer (JUICE) has been selected as ESA's next L-class mission in the framework of the Cosmic Vision Program. JUICE will explore the Jupiter system including flybys at Europa, Ganymede, and Callisto. JUICE is going to be launched in 2022 and it will arrive at Jupiter after an almost eight years cruise by the end of 2029. After spending several years in Jupiter orbit, investigating the giant planet and its magnetosphere as well as the Galilean Moons, the spacecraft will finally go into orbit around Ganymede, the largest satellite in the solar system. The payload includes the Ganymede laser altimeter (GALA) to investigate the surface and topography of Ganymede in particular of the icy satellites. GALA will also be a key instrument during close flybys at Europa and Callisto. The instrument can be operated when it is closer than about 1000–1300 km altitude (depending on the different albedo values of Europa, Ganymede and Callisto) during flybys and pericenter passages. The main phases for acquiring data at Ganymede are the final circular orbit phases with altitudes around 500 km and finally down to 200 km (for the JUICE mission scenario, see Grasset et al. 2013).

The instrument is a single-beam laser at 1064 nm wavelength. The range-finder module will be capable of a full digitization of the return pulse, similar to the BELA concept. A high shot frequency between 10 and 75 Hz will be challenging for the electronics of the instrument. A high frequency is required to guarantee dense coverage along the tracks when the spacecraft is in the 200 km orbit around Ganymede. Because of the differences in altitude of the orbits, the current design uses two different laser oscillators, optimized for the 500 and 200 km orbit, respectively.

GALA will be the first altimeter to investigate the icy surfaces of three different satellites, Europa, Ganymede, and Callisto. It has to be flexible to adjust to different conditions with respect to different albedo and mean surface slopes and roughness at the different targets. Besides the "classical" scientific objectives of determining the satellites' shapes and in particular the topography and surface roughness of Ganymede on various spatial scales, GALA will aim also at detecting and characterizing the global subsurface ocean at Ganymede. A global subsurface ocean is believed to be present at a depth of about 100 km under the cold ice shell of Ganymede. For a general review on icy satellites, see Hussmann et al. (2007) and references therein. So far, the best experimental evidence for an ocean has been derived from detection of an induced magnetic field superimposed on Ganymede's permanent dipole field. The induced field is interpreted as the response to the rotating Jovian magnetic field. This response requires an excellent electrical conductor relatively close (a few hundred km) below the surface. Ions within a liquid water ocean are best candidates to generate such an induced field. These fields have been detected also at Europa and Callisto during the Galileo mission also suggesting subsurface oceans at these satellites.

Independently from the magnetic signal, the ocean can be detected by the tidal response of the satellite surface to the tide raising potential of Jupiter. The latter is time-variable because of Ganymede's slightly elliptical orbit. During the main tidal cycle of 7.15 days, that is, one revolution of Ganymede around Jupiter, the satellite's gravitational potential is changing and its surface is periodically distorted. The tidal variability of the potential is characterized by the dynamical Love number k_2 to be measured with Doppler tracking from Earth to the spacecraft. The deformation of the surface of Ganymede can be measured with laser altimetry from orbit. The tidal double-amplitudes can reach more than seven meters at Ganymede, which is detectable with altimetry combined with precise orbit determination. The tidal amplitudes depend on longitude and latitude and their detection requires good spatial and temporal coverage of range measurements. From altimetry, the dynamical Love number h_2 can be derived. Linear combinations of h_2 and k_2 are the best way to constrain the ice thickness and the ocean depth (e.g., Wahr et al. 2006). In combination with additional measurements (induced magnetic field and libration amplitude), the extension of the ocean can be well-constrained by the JUICE mission.

To detect the tidal amplitudes, two methods can be chosen. The first one uses cross-over points, that is intersecting laser tracks, where the cross-over points are obtained at different tidal phases, that is different locations of Ganymede in orbit around Jupiter. In principle, the difference between the two range measurements would yield the tidal elevations. However, due to uncertainties in orbital altitude and attitude of the spacecraft, interpolation between the laser spots and other factors, which increase the error budget of the range measurements, the method has to be used in a statistical way including as many cross-over points as possible. In a typical operation scenario, several hundred thousands of cross-over points are analyzed to guarantee sufficient spatial and temporal coverage and to guarantee sufficient reduction of the errors due to the statistical interpretation. The second method would use a global determination of the topography, which is assumed to be time-dependent (at least the low-order coefficients, in particular C_{20} and C_{22}). The inversion would yield the time-variability of the low-order topography from which the Love number h_2 could be derived. With the given mission scenario, the first method turns out to be more promising in determining low error bars for h_2.

In addition to geodetic measurements (shape and topography), GALA will also characterize surface slopes and surface roughness at all three icy satellites. High-resolution along track will allow for geomorphological characterization. For example, the slopes and tilts of ice blocks within the so-called chaos terrain or double-ridges and triple-bands on Europa will give insight in the formation processes of these unique features and the possible relation to water reservoirs near the surface. Profiles across different craters and basins will allow the determination of relaxation processes within the lithosphere which are telltale signs of the satellites' thermal evolutions.

3.4.3 OSIRIS-REx

OSIRIS-REx is a NASA mission targeted for asteroid 101955 Bennu (former 1999 RQ36). The main objective is to return a sample of asteroid material to Earth. The scientific payload also includes the laser altimeter OLA (OSIRIS-REx Laser Altimeter) that will precisely determine the asteroids shape and topography and assist in characterizing the sampling site. OLA is a scanning system that can derive a dense cloud of return pulses with a range accuracy of 5–30 cm (Dickinson et al. 2012). It can use two different oscillators designed to obtain range measurements from 1 to 7.5 km distance and in the approach phase from 500 m to 1 km. The mission is scheduled to be launched in 2016 with an arrival at the target asteroid in 2019.

3.5 Concluding Remarks

Laser altimeters have been included on various missions investigating the geophysics and geology of planetary bodies. So far, only the Moon, the terrestrial planets (Mars and Mercury), and NEAs have been investigated by altimetry. Information obtained from these instruments is essential for interpretation of the gravity signals yielding insight in the processes (globally and locally) that have shaped the planetary surfaces. Besides the topographic measurements, from which crustal thickness and other important features to understand the evolution of terrestrial planets, can be derived, laser altimetry is an excellent tool to obtain absolute reference systems, which are needed for geophysical characterization (e.g., rotational states) as well as for deriving correct cartographic products.

From analysis of the return pulse, physical properties of the planetary surfaces can be derived. On scales of the laser footprint, that is, down to 5 m in the case of LOLA onboard the Lunar Reconnaissance Orbiter, the surface roughness can be measured. In addition, the reflectivity (albedo) at the laser wavelength (1064 nm in most applications) can be derived from the intensity of the transmitted and returned pulses. Such measurements are also possible in permanently shadowed regions (e.g., polar craters) where imaging systems fail to acquire data due to the lack of sunlight.

The examples of the Galilean Moons and Mercury show how laser altimetry can be used to constrain the interior structure and rheology of planetary bodies, in particular the characterization of internal liquid layers. In the case of Mercury, the liquid layer would be the outer liquid core where Mercury's dipole is generated. In the case of the Galilean moons subsurface water oceans expected in the interiors of Europa, Ganymede, and Callisto could be detected. These features are essential to understand the history of planets and satellites and to understand the processes currently at work in their interiors. The subsurface oceans in the outer solar system are also relevant for assessing the astrobiological potential of the icy moons.

The best way to analyze and interpret altimetry data is in combination with Doppler tracking data from the radio science experiments and imaging data. These instruments form a perfect suite to obtain a precise geodetic and geophysical characterization of planetary bodies.

Acknowledgments

The author would like to thank S. Jin, J. Oberst and S. Lau for helpful comments on the manuscript.

References

Abshire, J.B., Sun, X., Afzal, R.S. 2000. Mars orbiter laser altimeter: Receiver model and performance analysis. *J. Appl. Optics* 39, 2449–2460.

Araki, H., Noda, H., Tazawa, S., Ishihara, Y., Goossens, S., Sasaki, S. 2013. Lunar laser topography by LALT on board the KAGUYA lunar explorer—Operational history, new topographic data, peak height analysis of laser echo pulses. *Adv. Space Res.* 52, 262–271.

Araki, H., Tazawa, S., Noda, H., Ishihara, Y., Goossens, S., Sasaki, S., Kawano, N., Kamiya, I., Otake, H., Oberst, J., Shum, C. 2009. Lunar global shape and polar topography derived from Kaguya-LALT Laseraltimetry. *Science* 323, 897–900.

Banerdt, W.B., Neumann, G.A. 1999. The topography (and ephemeris) of Phobos from MOLA ranging. 30th annual Lunar and Planetary Science Conference, March 1999, Houston, Texas. Abstract # 2021.

Barnouin-Jha, O.S., Cheng, A.F., Mukai, T., Abe, S., Hirata, N., Nakamura, R., Gaskell, R.W., Saito, J., Clark, B.E. 2008. Small-scale topography of 25143 Itokawa from the Hayabusa laser altimeter. *Icarus* 198, 108–124.

Cavanaugh, J.F., 18 colleagues. 2007. The Mercury Laser Altimeter Instrument for the MESSENGER mission. *Space Sci. Rev.* 131, 451–480.

Cole, T.D., Boies, M.T., El-Dinary, A.S., Cheng, A., Zuber, M.T., Smith, D.E. 1997. The near-earth asteroid rendezvous laser altimeter. *Space Sci. Rev.* 82, 217–253.

Dickinson, C.S., Daly, M., Barnouin, O., Bierhaus, B., Gaudreau, D., Tripp, J., Ilnicki, M., Hildebrand, A. 2012. An overview of the OSIRIS REx Laser Altimeter (OLA). 43rd Lunar and Planet. Sci. Conf. March, 2012, Woodlands, Texas. Abstract # 1659, id.1447.

Gardner, C.S. 1982. Target signatures for laser altimeters: An analysis. *Appl. Opt.* 21, 3.

Gardner, C.S. 1992. Ranging performance of satellite laser altimeters. *Proc. IEEE* 30, 5, 1061–1072.

Grasset, O., and 17 colleagues. 2013. JUpiter ICy moons Explorer (JUICE): an ESA mission to orbit Ganymede and to characterise the Jupiter system. *Planet. Space Sci.* 78, 1–21.

Gunderson, K., Thomas, N., Rohner, M. 2006. A laser altimeter performance model and its application to BELA. *Geosci. Remote Sens.* 44, 11-2, 3308–3319.

Hussmann, H., Sotin, C., Lunine, J.I. 2007. Interiors and evolution of icy satellites. Schubert, G. (Ed.) *Treatise on Geophysics*, Vol. 10, pp. 509–540, Oxford: Elsevier Ltd.

Jin, S.G., Arivazhagan, S., Araki, H. 2013. New results and questions of lunar exploration from SELENE, Chang'E-1, Chandrayaan-1 and LRO/LCROSS. *Adv. Space Res.* 52, 285–305.

Kamalakar, J.A., and 10 colleagues. 2009. Laser ranging Experiment aboard Chandrayaan-1: Instrumentation and preliminary results. *40th Lunar and Planet. Sci Conf. 2009.* Abstract # 1487.

Kaula, W.M., Schubert, G., Lingenfelter, R.E., Sjogren, W.L., Wollenhaupt, W.R. 1973. Lunar topography from Apollo 15 and 16 laser altimetry. Proceedings of the 4th Lunar Science Conference, *Supplement 4, Geochimica et Cosmochimica Acta 3,* 2811–2819.

Kaula, W.M., Schubert, G., Lingenfelter, R.E., Sjogren, W.L., Wollenhaupt, W.R. 1974. Apollo laser altimetry and inferences as to lunar structure. Proceedings of the 5th Lunar Science Conference, *Supplement 3, Geochimica et Cosmochimica Acta* 3, 3049–3058.

Li, P.J., Hu, X.G., Huang, Y., Wang, G.L., Jiang, D.R., Zhang, X.Z., Cao, J.F., Xin, N. 2012. Orbit determination for Chang'E-2 lunar probe and evaluation of lunar gravity models. *Sci. China-Phys. Mech. Astron.* 55, 514–522.

Margot. J.-L., Peale, S.J., Jurgens, R.F., Slade, M.A., Holin, I.V. 2007. Large longitude libration of Mercury reveals a molten core. *Science* 316, 710–714.

Mazarico, E., Rowlands, D.D., Neumann, G.A., Smith, D.E., Torrence, M.H., Lemoine, F.G., Zuber, M.T. 2012. Orbit determination of the Lunar Reconnaissance Orbiter. *J. Geod.* 86, 193–207.

Mazarico, E., Neumann, G.A., Rowlands, D.D., Smith, D.E. 2010. Geodetic constraints from multi-beam laser altimeter crossovers. *J. Geod.* 84, 343–354.

Miller, J.K., Konopliv, A.S., Antreasian, P.G., Bordi, J.J., Chesley, S., Helfrich, C.E., Owen, W.M., Wang, T.C., Williams, B.G., Yeomans, D.K., Scheeres, D.J. 2002. Determination of shape, gravity and rotational state of asteroid 433 Eros. *Icarus* 155, 3–17.

Mukai, T. and 15 colleagues. 2007. An overview of the LIDAR observations of asteroid 25143 Itokawa. *Adv. Space Res.* 40, 187–192.

Namiki, N. and 18 colleagues. 2009. Farside gravity field of the Moon from four-way Doppler measurements of SELENE (Kaguya) *Science* 323, 900–905.

Neumann, G.A., Smith, D.E., Zuber, M.T. 2003b. Two years of clouds detected by the Mars Orbiter Laser Altimeter. *J. Geophys. Res.* 108, 5023.

Noda, H., Araki, H., Goossens, S., Ishihara, Y., Matsumoto, K., Tazawa, S., Kawano, N., Sasaki, S. 2009. Illumination conditions at the lunar polar regions by KAGUYA(SELENE) laser altimeter. *Geophys. Res. Lett.* 35, CiteID L24203.

Peale, S.J., Phillips, R.J., Salomon, S.C., Smith, D.E., Zuber, M.T. 2002. A procedure for determining the nature of Mercury's core. *Met. Planet. Sci.* 37, 1269–1283.

Roberson, F.I., Kaula, W.M. 1972. Apollo 15 laser altimeter. *Apollo 15: Preliminary Science Report.* NASA SP-289, NASA, Washington, D.C., 25–50.

Rowlands, D.D., Pavlis, D.E., Lemoine, F.G., Neumann, G.A., Luthke, S.B. 1999. The use of laser altimetry in the orbit and attitude determination of Mars Global Surveyor. *Geophys. Res. Lett.* 26, 1191–1194.

Rowlands, D.D., Lemoine, F.G., Chinn, D.S., Luthke, S.B. 2009. A simulation study of multi-beam altimetry for lunar reconnaissance orbiter and other planetary missions. *J. Geod.* 83, 709–721.

Rummel, R. 2005. Gravity and topography of moon and planets. *Earth Moon Planets* 94, 103–111.

Shum, C.K., Zhang, B.H., Schutz, B.E., Tapley, B.D. 1990. Altimeter crossover methods for precision orbit determination and the mapping of geophysical parameters. *J. Astronaut. Sci.* 38, 355–368.

Sjogren, W.L., Wollenhaupt, W.R. 1973. Lunar shape via the Apollo laser altimeter. *Science* 179, 275–278.

Smith, D.E., and 16 colleagues. 2010a. The equatorial shape and gravity field of Mercury from MESSENGER flybys 1 and 2. *Icarus* 209, 88–100.

Smith, D.E., and 18 colleagues 1999. The global topography of Mars and implications for surface evolution. *Science* 284, 1495–1503.

Smith, D.E., and 19 colleagues 2010b. Initial observations from the Lunar Orbiter Laser Altimeter (LOLA). *Geophys. Res. Lett.* 37, L18204.

Smith, D.E., and 23 colleagues 2001a. Mars Orbiter Laser Altimeter: Experiment summary after the first year of global mapping of Mars. *J. Geophys. Res.*106, 23, 689–722.

Smith, D.E., and 30 colleagues 2010c. The Lunar Orbiter Laser Altimeter investigation on the Lunar Reconnaissance Orbiter mission. *Space Sci. Rev.* 150, 209–241.

Smith, D.E., Zuber, M.T., Neumann, G.A. 2001b. Seasonal variations of snow depth on mars. *Science* 294, 2141–2146.

Smith, D.E., Zuber, M.T., Neumann, G.A., Lemoine, F.G. 1997. Topography of the moon from the Clementine LIDAR. *J. Geophys. Res.* 102, 1591–1611.

Sridharan, R., Pratim Das, T., Ahmed, S.M., Supriya, G., Bhardwaj, A., Kamalakar, J.A. 2013. Spatial heterogeneity in the radiogenic activity of the lunar interior: Inferences from CHACE and LLRI on Chandrayaan-1. *Adv. Space Res.* 51, 168–178.

Thomas, N., and 26 colleagues. 2007. The BepiColombo Laser Altimeter (BELA): Concept and baseline design. *Planet. Space Sci.* 55, 1398–1413.

Wahr, J.M., Zuber, M.T., Smith, D.E., Lunine, J.I. 2006. Tides on Europa, and the thickness of Europa's icy shell. *J. Geophys. Res.* 111, E12005.

Wollenhaupt, W.R., Sjogren, W.L. 1972. Apollo 16 laser altimeter. *Apollo 16: Preliminary Science Report*. NASA SP-315, NASA, Washington, D.C., 30-1-30-5.

Wollenhaupt, W.R., Sjogren, W.L., Lingenfelter, R.E., Schubert, G., Kaula, W.M. 1973. Apollo 17 laser altimeter. *Apollo 17: Preliminary Science Report*. NASA SP-330, NASA, Washington, D.C., 33-41-33-44.

Zuber, M.T., and 11 colleagues. 2000b. The shape of 433 Eros from the NEAR-Shoemaker laser rangefinder. *Science* 289, 2097–2101.

Zuber, M.T., and 14 colleagues. 2000a. Internal structure and early thermal evolution of Mars from mars global surveyor topography and gravity. *Science* 287, 1788–1793.

Zuber, M.T., and 23 colleagues. 2012. Topography of the northern hemisphere of Mercury from MESSENGER laser altimetry. *Science* 336, 217–220.

Zuber, M.T., Smith, D.E., Lemoine, F.G., Neumann, G.A. 1994. The shape and internal structure of the Moon from the Clementine mission. *Science* 266, 1839–1843.

4

*Photogrammetric Processing of Chang'E-1 and Chang'E-2 STEREO Imagery for Lunar Topographic Mapping**

Kaichang Di, Yiliang Liu, Bin Liu, and Man Peng

CONTENTS

4.1 Introduction 77
4.2 Rigorous Geometric Modelling of CE-1 and CE-2 CCD Images...........79
    4.2.1 Interior Orientation...79
    4.2.2 Exterior Orientation...82
    4.2.3 Space Intersection and Back-Projection 83
4.3 Sensor Model Refinements 83
    4.3.1 Refinement on Exterior Orientation...83
    4.3.2 Refinement on IO...85
    4.3.3 Refinement on Both Exterior and IOs 85
4.4 Experimental Results...87
    4.4.1 Intratrack Adjustment of CE-1 CCD Images 87
    4.4.2 Intratrack Adjustment of CE-2 CCD Images 89
    4.4.3 Intertrack Adjustment of CE-2 CCD Images............................91
4.5 Conclusion and Discussion 94
Acknowledgments...95
References...95

4.1 Introduction

The Chinese Lunar Exploration Program (CLEP), usually known as the Chang'E program is a program of robotic and human missions to the Moon undertaken by the China National Space Administration (CNSA). This program has three phases: orbital mission, soft lander and rover, and automated

---

* An early version of this paper was presented at the 22nd ISPRS Congress, Melbourne, Australia, August 25 to September 1, 2012.

sample return (Ouyang and Li 2010). Chang'E-1 (CE-1) and Chang'E-2 (CE-2) are the orbiters in Phases I and II, respectively.

The CE-1 orbiter is the first lunar probe of China launched on October 24, 2007. It carries a three-line push-broom CCD camera, which has a ground resolution of 120 m and a swath width of 60 km at 200 km orbit altitude. It is implemented on an area array CCD sensor and uses only the 11th, 512th, and 1013th lines to generate the forward-, nadir-, and backward-looking images simultaneously in the flight direction (Li et al. 2010; Peng et al. 2010).

The CE-2 orbiter, launched on October 1, 2010, is the follow-up orbiter of CE-1. As a mission highlight, a local image map at Sinus Iridum, the pre-selected landing site of Chang'E-3 lunar rover produced using CE-2 CCD images was released on November 8, 2010 (NAOC 2010). CE-2 was broadly similar to CE-1 mission, but had important differences. CE-2 flew at two kinds of orbit altitude, 100×100 km circular orbit at which the CCD camera can reach a resolution of 7 m and 100×15 km elliptical orbit with a resolution of 1.5 m around the perilune. The former kind of orbit provides high-resolution global image coverage of the Moon for various scientific researches, while the latter mainly provides more detailed information for Chang'E-3 lunar landing and surface operation.

The CE-2 CCD stereo camera adopts a stereo imaging solution with the single lens and two angles of view in the same track and a self-push-broom imaging mode with high sensitivity time delay integration (TDI) CCD. Comparing with the three-line array mode of CE-1 CCD camera, CE-2 adopts a two-line push-broom camera to reduce the pressure of mass data, and meanwhile, assures stereo imaging condition. It can acquire forward- and backward-looking images simultaneously to consist of stereo pairs (Zhao et al. 2011a, 2011b, 2011c).

At present, a global image mosaic of the Moon using CE-1 CCD images and laser altimeter (LAM) data has been produced (Li et al. 2010). A global image map of the Moon with a resolution of 7 m has also been released by the CLEP on February 6, 2012 (Xinhua News Agency 2012). But there's no CE-2 topographic products (e.g., digital elevation model (DEM) and digital ortho map (DOM)) released so far. In our previous research, we have developed a rigorous sensor model for CE-1 stereo imagery (Peng et al. 2010) and methods of automatic DEM and DOM generation, and coregistration of CE-1 CCD images and LAM data (Di et al. 2010, 2012).

In this chapter, we elaborate the rigorous sensor models of CE-1 and CE-2 CCD cameras for photogrammetric processing of the lunar images acquired by the two cameras. First, interior orientation (IO) model of CE-1 and CE-2 CCD cameras including parameters and structure of the camera are introduced in detail. Second, the formulas of exterior orientation model are given to constitute a rigorous geometric model for both CE-1 and CE-2 cameras. After analyzing the distribution patterns of back-projection residuals, we propose different methods to refine the sensor model: (a) refining exterior orientation parameters by correcting the attitude angle bias for CE-1 intratrack adjustment, (b) refining the IO model by calibration of the relative position of the two linear CCD arrays for CE-2 intratrack adjustment, (c) refining the exterior orientation parameters and the IO model simultaneously for CE-2 intertrack adjustment. Finally, experimental results with CE-1 and CE-2 images are presented to demonstrate the effectiveness of the proposed methods.

4.2 Rigorous Geometric Modelling of CE-1 and CE-2 CCD Images

4.2.1 Interior Orientation

The CE-1 CCD camera is a three-line push broom camera, which is implemented on an area array CCD sensor. The CCD array has 1024 × 1024 pixels, with each pixel being 14 × 14 μm in the chip. The forward-, nadir-, and backward-looking images of the lunar surface are generated by reading the 11th, 512th, and 1013th rows that are perpendicular to the flight direction (Figure 4.1) (Li et al. 2010). The convergence angle between the adjacent views is 16.7°. At a 200-km altitude, the image spatial resolution is 120 m and the swath width is about 60 km. The focal length of the CCD camera is 23.33 mm. The actual imaging area is 1024 lines by 512 samples. The focal plane frame is shown in Figure 4.2.

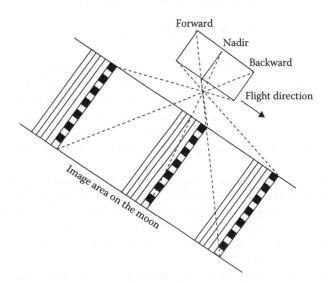

FIGURE 4.1
CE-1 stereo camera imaging configuration.

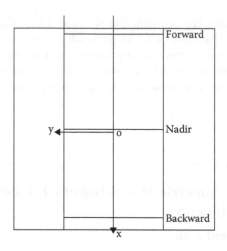

FIGURE 4.2
Focal plane of CE-1 CCD camera.

According to the configuration of the CE-1 CCD camera and its focal plane frame, we have developed the IO model of CE-1 CCD camera to calculate focal plane coordinates from pixel coordinates using Equation 4.1 (Peng et al. 2010).

$$x = (x_p - ccd\_line) \cdot pixsize - x_0$$

$$y = (y_p - col) \cdot pixsize - y_0 \tag{4.1}$$

where *pixsize* stands for the pixel size of the CCD array; *col* is the pixel position in column direction; (x_p, y_p) are the center position (511.5, 255.5) of the actual imaging area (1024 lines by 512 samples); (x_0, y_0) represents principal point position in the focal plane frame; (x, y) are the focal plane coordinates of forward-, nadir-, or backward-looking images. *ccd_line* is 11, 512, and 1013 for forward-, nadir-, and backward-looking images, respectively.

Unlike the CE-1 CCD camera implemented on an area array CCD sensor, the CCD camera (shown in Figure 4.3) carried by CE-2 is a two-line push-broom sensor assembled on a focal plane separately. The two CCD arrays share the same optical axis with a focal length of 144.3 mm. Each CCD line array has 6144 pixels. The primary technical parameters are listed in Table 4.1 (Zhao et al. 2011c).

According to the focal plane configuration of CE-2 CCD camera shown in Figure 4.4, we can derive the following IO equation to calculate focal plane coordinates from pixel coordinates.

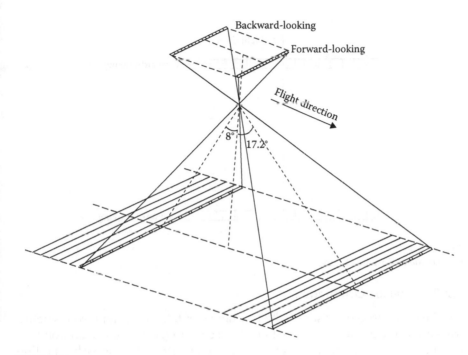

FIGURE 4.3
CE-2 stereo camera imaging configuration.

TABLE 4.1

Technical Parameters of C

Parameters	Orbit	
	100 km Circular Orbit	**100 × 15 km Elliptical Orbit**
Image swath	≥43 km	≥6 km
Spatial resolution	≤10 m	≤1.5 m
Look angle	Forward-looking +8°, backward-looking −17.2°	
Focal length	144.3 mm	

$$x = x_0 - \tan(\theta) \cdot f$$

$$y = y_0 - (col - s_0) \cdot pixsize \qquad (4.2)$$

where θ represents the viewing angle, which is 8° for forward-looking images and −17.2° for backward-looking images; f is the focal length; *col* is the column number of an image point; s_0 is the center of CCD, the value of which is 3071.5; *pixsize* stands for the pixel size of the CCD array, which is 10.1 µm (Zhao et al. 2011b); (x_0, y_0) represent the principal point position in the focal plane frame; (x, y) are the focal plane coordinates of forward- or backward-looking images.

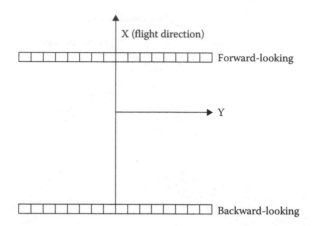

FIGURE 4.4
Focal plane of CE-2 CCD camera.

4.2.2 Exterior Orientation

For the push-broom sensor, as the frequency of telemetry data from satellite tracking is much lower than that of CCD scanning, exterior orientation (EO) parameters are interpolated from original telemetry data for each scan line. Lagrange polynomials and third-order polynomials have been used for the interpolation of EO parameters in previous research (Peng et al. 2010; Liu and Di 2011). Given the IO model and EO parameters after interpolation, we developed the rigorous sensor model of CE-1 and CE-2 CCD imagery in the form of a collinearity equation:

$$\begin{bmatrix} X - X_s \\ Y - Y_s \\ Z - Z_s \end{bmatrix} = \lambda \mathbf{R}_{ol} \mathbf{R}_{bo} \mathbf{R}_{ib} \begin{bmatrix} x \\ y \\ -f \end{bmatrix} = \lambda \mathbf{R} \begin{bmatrix} x \\ y \\ -f \end{bmatrix} \qquad (4.3)$$

where (x, y) are focal plane coordinates of an image point; f is the focal length; (X, Y, Z) and (X_s, Y_s, Z_s) are the three-dimensional (3D) ground coordinate and the camera center position in lunar body-fixed (LBF) coordinate system; $\mathbf{R}_{ib}$ is the rotation matrix from image space coordinate system (ISCS) to spacecraft body coordinate system (BCS); $\mathbf{R}_{bo}$ is the rotation matrix from BCS to orbit coordinate systems (OCS); $\mathbf{R}_{ol}$ is the rotation matrix from OCS to LBF, λ is a scale factor, $\mathbf{R}$ represents the overall rotation matrix from the ISCS to the LBF. In Equation 4.3, (X_s, Y_s, Z_s) for each scan line are obtained from spacecraft ephemeris and interpolation; $\mathbf{R}_{ib}$ and $\mathbf{R}_{bo}$ are obtained from telemetry data. In this research, we use third-order polynomial model for interpolation of the EO parameters, with the image scan time t as independent variable (as shown in Equation 4.4).

$$X_s(t) = a_0 + a_1 t + a_2 t^2 + a_3 t^3$$

$$Y_s(t) = b_0 + b_1 t + b_2 t^2 + b_3 t^3$$

$$Z_s(t) = c_0 + c_1 t + c_2 t^2 + c_3 t^3 \qquad (1.1)$$

$$\varphi_o(t) = d_0 + d_1 t + d_2 t^2 + d_3 t^3$$

$$\omega_o(t) = e_0 + e_1 t + e_2 t^2 + e_3 t^3$$

$$\kappa_o(t) = f_0 + f_1 t + f_2 t^2 + f_3 t^3$$

4.2.3 Space Intersection and Back-Projection

Stereopairs can be formed by CE-1 and CE-2 images of different looking angles. Based on the rigorous sensor model, the 3D coordinate of a ground point in LBF can be calculated by space intersection from the image coordinates of the conjugate points in stereo images, and the image coordinate can be calculated from 3D coordinate by back-projection. Ideally, using the 3D coordinate from space intersection, the back-projected image positions should be the same as the measured image points, which are used in space intersection. However, due to the orbit uncertainty and IO uncertainty, the back-projected image points are different from the measured point. The differences are called back-projection residuals. Some regular patterns usually appear in these residuals, which are extremely useful for analyzing and finding out the error sources and eliminating the inconsistencies.

4.3 Sensor Model Refinements

In order to reduce the inconsistencies (back-projection residuals) of stereo images from single track or adjacent tracks and improve mapping precision, refinements are performed on exterior orientation or IO process or both according to the camera structure and the distribution patterns of the residuals.

4.3.1 Refinement on Exterior Orientation

The CE-1 camera has a stable internal structure since it is implemented on one area array sensor. Thus, the back-projection residuals of CE-1 images are mainly caused by errors of EO parameters. For satellite images with high altitude, the position parameters and the attitude parameters have a strong correlation. So correcting attitude angle bias can also compensate errors caused by position parameters. For each image, Equation 4.3 can be simplified as Equation 4.5.

$$\frac{\mathbf{u}_2}{|\mathbf{u}_2|} = \mathbf{R}\,\frac{\mathbf{u}_1}{|\mathbf{u}_1|}$$

$$\mathbf{u}_1 = \begin{bmatrix} x & y & -f \end{bmatrix}^T$$

$$|\mathbf{u}_1| = \sqrt{x^2 + y^2 + (-f)^2} \tag{4.5}$$

$$\mathbf{u}_2 = \begin{bmatrix} X - X_s & Y - Y_s & Z - Z_s \end{bmatrix}^T$$

$$|\mathbf{u}_2| = \sqrt{(X - X_s)^2 + (Y - Ys)^2 + (Z - Z_s)^2}$$

where R is the rotation matrix from image space to LBF. Then we get the partial derivatives $\partial \mathbf{R}/\partial \varphi$, $\partial \mathbf{R}/\partial \omega$, $\partial \mathbf{R}/\partial \kappa$ in the form of 3×3 matrix. The observation equation for attitude angle bias correction can be represented as Equation 4.6 (Yuan and Yu 2008).

$$\mathbf{v} = \mathbf{A}\mathbf{x} - \mathbf{L} \tag{4.6}$$

$$\mathbf{A} = [\frac{\partial \mathbf{R}}{\partial \varphi}\mathbf{u}_1{}' \quad \frac{\partial \mathbf{R}}{\partial \omega}\mathbf{u}_1{}' \quad \frac{\partial \mathbf{R}}{\partial \kappa}\mathbf{u}_1{}']$$

$$\mathbf{L} = -\mathbf{R}\mathbf{u}_1{}' + \mathbf{u}_2{}'$$

$$\mathbf{x} = [d\varphi'\ d\omega'\ d\kappa']^T$$

$$\mathbf{u}_i{}' = \frac{\mathbf{u}_i}{|\mathbf{u}_i|} \quad (i = 1, 2)$$

Given a set of initial values of attitude angle biases as $d\varphi = 0$, $d\omega = 0$, and $d\kappa = 0$, rotation matrix **R** can be calculated using Euler angles $\varphi + d\varphi$, $\omega + d\omega$, and $\kappa + d\kappa$. In each iteration, the correction values are added to the bias values as shown in Equation 4.7.

$$d\varphi \leftarrow d\varphi + d\varphi'$$

$$d\omega \leftarrow d\omega + d\omega' \tag{4.7}$$

$$d\kappa \leftarrow d\kappa + d\kappa'$$

Iteration stops when the correction values are less than a predefined threshold. And finally the attitude angle biases can be figured out.

4.3.2 Refinement on IO

For the CE-2 CCD camera, the two linear CCD arrays are assembled separately on the focal plane, which have a relatively unstable relationship and may cause significant errors of the sensor model during operations in orbit. As a result, back-projection residuals usually appear to have a regular pattern in image space. These residuals can be used to refine the IO and/or exterior orientation so that to improve the mapping precision.

For the single track of CE-2 CCD images, back-projection residuals between forward- and backward-looking images mainly exist in column direction and generally have a linear relationship with the column number. Thus, refinement of IO by correction of only y coordinate in a single track adjustment can effectively reduce the residuals. For one point k with measured coordinate (x_k, y_k), the corrected coordinate y_k' can also be represented as:

$$y_k' = y_k + r_k \cdot pixsize \tag{4.8}$$

where r_k is the residual of the point, which can be fitted using a least-squares line (LSL) according to the residuals distribution. With col_k as the column number of point k, a line function can be represented as Equation 4.9. By minimizing the sum of the squares of r_k, the two parameters a and b can be estimated.

$$r_k = a(col_k - s_0) + b \tag{4.9}$$

4.3.3 Refinement on Both Exterior and IOs

For a more complicated condition, for example, multitrack adjustment, inconsistencies on both interior and exterior orientation should be considered. A self-calibration bundle adjustment method has been developed to refine the spatial relationship between the two CCD camera line arrays as well as the EO parameters.

We modified the IO model by adding four additional parameters for forward- and backward-looking images.

$$x' = (x - x\_offset)/x\_scale$$
$$y' = (y - y\_offset)/y\_scale \tag{4.10}$$

where (x, y) and (x', y') are the measured coordinates and the corrected coordinates for one point; x_offset and x_scale represent the translation and rotation in the x direction, respectively; similarly, y_offset and y_scale represent the translation and rotation in the y direction, respectively.

According to the general form of collinearity equation in 4.3 and the modified IO model in 4.10, the self-calibration bundle adjustment model can be represented as

$$x = -x\_scale \cdot f \cdot \overline{X} / \overline{Z} + x\_offset$$
$$y = -y\_scale \cdot f \cdot \overline{Y} / \overline{Z} + y\_offset$$

(4.11)

$$\overline{X} = a_1(X_A - X_s) + b_1(Y_A - Y_s) + c_1(Z_A - Z_s)$$
$$\overline{Y} = a_2(X_A - X_s) + b_2(Y_A - Y_s) + c_2(Z_A - Z_s)$$
$$\overline{Z} = a_3(X_A - X_s) + b_3(Y_A - Y_s) + c_3(Z_A - Z_s) .$$

Using Taylor's formula, the linearized observation equations can be derived as

$$v_x = \frac{\partial x}{\partial X_S} dX_S + \frac{\partial x}{\partial Y_S} dY_S + \frac{\partial x}{\partial Z_S} dZ_S + \frac{\partial x}{\partial \varphi_o} d\varphi_o + \frac{\partial x}{\partial \omega_o} d\omega_o$$

$$+ \frac{\partial x}{\partial \kappa_o} d\kappa_o + \frac{\partial x}{\partial X} dX + \frac{\partial x}{\partial Y} dY + \frac{\partial x}{\partial Z} dZ$$

$$+ \frac{\partial x}{\partial x\_scale} dx\_scale + \frac{\partial x}{\partial x\_offset} dx\_offset + (x) - x$$

$$v_y = \frac{\partial y}{\partial X_S} dX_S + \frac{\partial y}{\partial Y_S} dY_S + \frac{\partial y}{\partial Z_S} dZ_S + \frac{\partial y}{\partial \varphi_o} d\varphi_o + \frac{\partial y}{\partial \omega_o} d\omega_o$$

$$+ \frac{\partial y}{\partial \kappa_o} d\kappa_o + \frac{\partial y}{\partial X} dX + \frac{\partial y}{\partial Y} dY + \frac{\partial y}{\partial Z} dZ$$

$$+ \frac{\partial y}{\partial y\_scale} dy\_scale + \frac{\partial y}{\partial y\_offset} dy\_offset + (y) - y$$

(4.12)

The matrix form can be represented the same as Equation 4.6, and additional IO parameters are solved with other unknowns (polynomial coefficients of the EO parameters, and corrections of ground points) together in the self-calibration bundle adjustment model. Given that no ground control is available during the adjustment, the coefficient matrix of the observation equations is rank-deficient. Thus, several lines of original EO parameters at certain intervals are selected as pseudo observations to improve the stability of the observation equations (Li et al. 2011).

4.4 Experimental Results

4.4.1 Intratrack Adjustment of CE-1 CCD Images

Considering that Sinus Iridum is the preselected landing site of the Chang'E-3 lunar rover mission, in our experiment, we choose three tracks (No.0561, No.0562, and No.0563) of CE-1 CCD images located at Sinus Iridum. Figure 4.5 shows the partial images (1000 lines, after grayscale stretching) from three view angles in No.0562 track.

In this experiment, we use 1000 lines of the forward-, nadir-, and backward-looking CCD images from these three tracks. Conjugate points of stereo images are obtained through feature point extraction by SIFT operator (Lowe 1999) and feature point matching by cross-correlation. Statistics of back-projection residuals of more than 15,000 matched feature points for each image is shown in Table 4.2. The attitude angle biases shown in Table 4.3 for each image of these three tracks are added to the EO parameters, then ground points are back-projected to image space again and back-projection residuals are recalculated in Table 4.4.

Figure 4.6 shows the residuals distributed in column direction before and after adjustment (correction of attitude angle bias) for forward-, nadir-, backward-looking images of No.0562 track. It can be easily observed that after attitude angle bias correction, mean back-projection residuals of CE 1 CCD images are reduced to better than 1/100 pixel. After this refinement, DEM

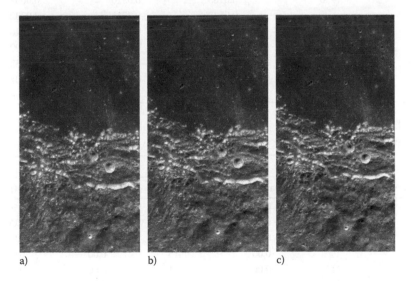

a)   b)   c)

FIGURE 4.5
Partial CCD images taken by CE-1 from No.0562 track (120 m resolution), a), b) and c) are the forward-, nadir- and backward-looking images, respectively.

TABLE 4.2

Back-Projection Residuals of CE-1 CCD Images before Correction of Attitude Angle Biases (F, N and B are Forward-, Nadir-, and Backward-Looking for Short, the Same Below

Track No.	Look Angle	Column Direction		Row Direction	
		Mean (pixel)	Mean (pixel)	Mean (pixel)	RMS (pixel)
0561	F	−1.07	0.12	0.00	0.00
	N	−0.05	0.20	0.00	0.00
	B	1.13	0.18	0.00	0.00
0562	F	−1.51	0.13	0.00	0.00
	N	0.06	0.21	0.00	0.00
	B	1.45	0.16	0.00	0.00
0563	F	−1.40	0.16	0.00	0.00
	N	0.02	0.26	0.00	0.00
	B	1.38	0.19	0.00	0.00

TABLE 4.3

Correction Values of Attitude Angle Biases for CE-1 CCD Images

Track No.	Look Angle	φ (degree)	ω (degree)	κ (degree)
0561	F	−0.0099	−0.0324	0.0137
	N	−0.0062	0.0001	−0.0008
	B	0.0163	0.0320	0.0057
0562	F	−0.0532	−0.0311	0.0332
	N	−0.0254	0.0119	0.0046
	B	0.0149	0.0432	0.0090
0563	F	−0.0162	−0.0416	0.0183
	N	−0.0035	0.0022	−0.0004
	B	0.0202	0.0390	0.0056

TABLE 4.4

Back-Projection Residuals of CE-1 CCD Images after Attitude Angle Bias Correction

Track No.	Look Angle	Column Direction		Row Direction	
		Mean (pixel)	RMS (pixel)	Mean (pixel)	RMS (pixel)
0561	F	0.00	0.12	0.00	0.00
	N	−0.00	0.20	0.00	0.00
	B	0.00	0.19	0.00	0.00
0562	F	−0.00	0.13	0.00	0.00
	N	−0.00	0.21	0.00	0.00
	B	−0.00	0.16	0.00	0.00
0563	F	0.00	0.16	0.00	0.00
	N	0.00	0.26	0.00	0.00
	B	0.00	0.18	0.00	0.00

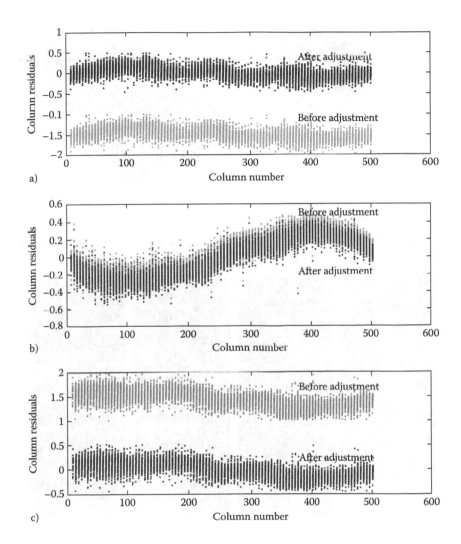

FIGURE 4.6
The residuals distributions in column direction before and after adjustment (taking No.0562 track for example), a), b) and c) are the residuals of the forward-, nadir-looking and backward-looking images, respectively.

and DOM generated are automatically generated using CE-1 images and are shown in Figure 4.7.

4.4.2 Intratrack Adjustment of CE-2 CCD Images

By now, all the images of the CE-2 CCD camera with a resolution of 7 m can be applied from the China Lunar Exploration and Aerospace Engineering Center. We also choose three tracks (No. 0579, No. 0580, and No. 0581) of CE-2 CCD images located at Sinus Iridum. Figure 4.8 shows partial forward- and

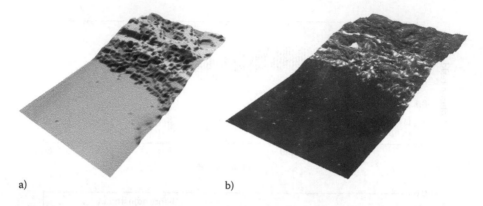

a)   b)

FIGURE 4.7
Partial DEM and DOM generated using CE-1 CCD images (taking No. 0562 track for example),
a) and b) are the DEM and DOM, respectively.

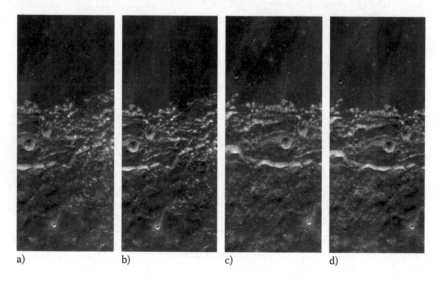

a)   b)   c)   d)

FIGURE 4.8
Partial forward- and backward-looking lunar images taken by CE-2 CCD camera from No. 0580
and No. 0581 track (7 m resolution), a) and b) are forward- and backward-looking image of
No. 0580 track, c) and d) are forward- and backward-looking image of No. 0581 track.

backward-looking images (15,000 lines, after grayscale stretching) of No. 0580
and No. 0581 tracks.

In this experiment, we use 60,000 lines by 6,144 samples forward- and backward-looking images for each track. After feature point extraction, matching and gross error elimination (using RANSAC procedure), around 60,000
evenly distributed conjugate points for each stereo pair are obtained and
used for space intersection and back-projection. Table 4.5 shows the residuals
between the feature points and the back-projected points correspondently

in image space, which indicate that the residuals mainly exist in the column direction.

Compared with Table 4.2, it is obvious that the back-projection residuals of CE-2 images are much larger than those of CE-1 images. As analyzed early, this is mainly because the two linear arrays of the CE-2 camera are separately assembled on the focal plane, while the three line CE-1 camera is realized by one area array sensor. We fitted LSLs of column residuals versus column number for forward- and backward-looking images of these three tracks. The coefficients *a* and *b* of them are listed in Table 4.6.

Back-projection residuals are calculated again after the IO model refinement. Table 4.7 shows the results, which indicate that residuals in the column direction are reduced significantly from over 20 pixels to 1/100 pixel level.

The residuals distributed in the column direction before and after adjustment (IO model refinement) as well as the LSLs are shown in Figure 4.9 which visually demonstrates the effectiveness of this method.

DEM and DOM are generated using the corrected IO model and partial of them are shown in Figure 4.10.

4.4.3 Intertrack Adjustment of CE-2 CCD Images

Intertrack adjustment experiment is performed using forward- and backward-looking images (15,000 lines by 6144 samples) from No. 0580 and No. 0581 track. Hundreds of tie points (including intratrack tie points and

TABLE 4.5

Back-Projection Residuals of 60,000 Lines of CE-2 CCD Images before Refinement

		Column Direction		Row Direction	
Track No.	Look Angle	Mean (pixel)	RMS (pixel)	Mean (pixel)	RMS (pixel)
0579	F	−22.77	2.22	0.00	0.00
	B	22.73	2.21	0.00	0.00
0580	F	−23.46	1.93	0.00	0.00
	B	23.42	1.92	0.00	0.00
0581	F	−22.34	1.81	0.00	0.00
	B	22.30	1.81	0.00	0.00

TABLE 4.6

Coefficients *a* and *b* of the Fitted Lines of Column Residuals for Forward- and Backward-Looking CE-2 Images

	Coefficient *a*		Coefficient *b*	
Track No.	F	B	F	B
0579	−0.0011020	0.0011000	−23.02	22.97
0580	−0.0008601	0.0008583	−23.46	23.42
0581	−0.0009592	0.0009575	−22.19	22.15

TABLE 4.7

Back-Projection Residuals of 60,000 Lines of CE-2 CCD Images after Intratrack Adjustment

Track No.	Look Angle	Column Direction		Row Direction	
		Mean (pixel)	Mean (pixel)	Mean (pixel)	RMS (pixel)
0579	F	−0.02	1.22	0.00	0.00
	B	0.02	1.22	0.00	0.00
0580	F	−0.02	1.22	0.00	0.00
	B	0.02	1.22	0.00	0.00
0581	F	−0.02	0.81	0.00	0.00
	B	0.02	0.80	0.00	0.00

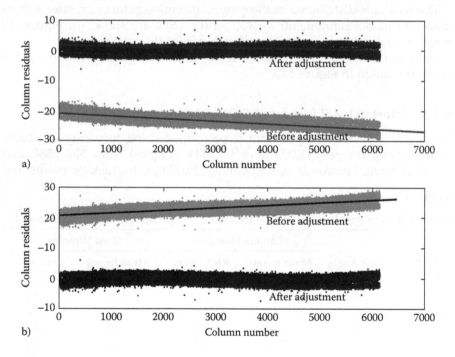

FIGURE 4.9

Back-projection residuals in column direction before and after adjustment as well as the LQL (taking No.0580 track for example). a) and b) are the residuals and LSL for forward- and backward-looking image, respectively.

intertrack tie points) evenly distributed in the overlapping area are selected and used to generate the corresponding ground points as initial adjustment values. Back-projection residuals before and after adjustment are shown in Table 4.8. The results show that the intertrack self-calibration bundle adjustment can effectively reduce the back-projection residuals from over 20 pixels to an order of magnitude of 1/100 pixel. Changes in EO parameters and ground points are shown in Table 4.9. For the intertrack adjustment,

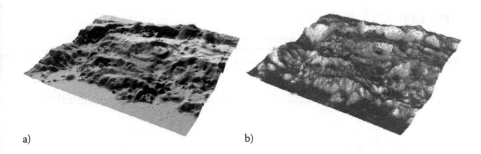

a)   b)

FIGURE 4.10

Partial DEM and DOM generated using CE-1 CCD images (taking No. 0581 track for example).

TABLE 4.8

Back-Projection Residuals before and after Intertrack Adjustment of CE-2 Images

	Track No.	Look Angle	Column Direction (pixel)		Row Direction (pixel)	
			Average	RMSE	Average	RMSE
	0580	F	−0.61	0.44	20.92	0.51
		B	9.69	0.61	16.72	1.51
Before adjustment	0581	F	−9.53	0.39	−16.84	0.36
		B	1.04	0.79	−20.99	1.42
	0580	F	0.04	0.36	0.01	0.13
		B	−0.03	0.40	0.02	0.59
After adjustment	0581	F	−0.06	0.37	−0.01	0.30
		B	0.04	0.45	−0.04	0.56

TABLE 4.9

Changes of EO Parameters and Ground Points after Intertrack Adjustment

	Change of EO Parameters		Change of Ground Points		
Track No.	(X_s, Y_s, Z_s) (m)	$(\varphi_o, \omega_o, \kappa_o)$ (″)	X (m)	Y (m)	Z (m)
0580	47.07	1.43	4.08	4.07	8.85
0581	41.71	1.27			

additional IO parameters for different tracks (see Table 4.10) are considered the same. From the DEMs generated before and after the intertrack adjustment (shown in Figure 4.11a and b), we can see that the inconsistencies (e.g., artifacts in the intertrack overlapping area) between two adjacent tracks are effectively eliminated using bundle adjustment.

TABLE 4.10

Additional IO Parameters for Intertrack Adjustment

Look Angle	x_scale	x_offset	y_scale	y_offset
F	0.99999963	0.00000471	1.00026279	−0.00005342
B	0.99999762	−0.00002736	0.99973929	0.00004101

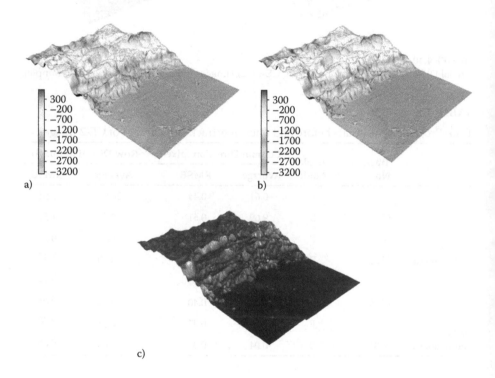

a)   b)

c)

FIGURE 4.11

DEM and DOM of CE-2 CCD images, a) and b) are DEMs generated using CE-2 data before and after adjustment, respectively, and c) is the DOM overlaid on the DEM after adjustment.

4.5 Conclusion and Discussion

In this research, we have developed the rigorous sensor models of CE-1 and CE-2 CCD cameras for photogrammetric processing of the images. By analyzing back-projection residuals, we proposed several methods to refine the sensor models so as to reduce the residuals. For CE-1 images, the EOP parameters are refined by correcting the attitude angle biases; for CE-2 single track images, the IO model is refined by calibration of the relative position of the two linear CCD arrays; for multitrack CE-2 images, both the IO model and EO parameters are refined through an intertrack self-calibration bundle adjustment. Experimental results indicate that these methods are very effective for CE-1 and CE-2 images, respectively; the mean back-projection residuals for both CE-1 and CE-2 CCD images can be significantly reduced to an order of magnitude of 1/100 pixel either in one single track or between adjacent tracks. Consequently, the mapping products DEM and DOM have better precision after the sensor model refinement.

Acknowledgments

Funding of this research by National Key Basic Research and Development Program of China (2012CB719902) and National Natural Science Foundation of China (41171355, 40871202) is acknowledged. We thank the Lunar and Deep Space Exploration Science Applications Center of the National Astronomical Observatories (NAOC) and Beijing Aerospace Control Center for providing the CE-1 and CE-2 images and telemetry data.

References

Di, K., Yue, Z., Peng, M., et al. 2010. Co-registration of CHANG'E-1 stereo images and laser altimeter data for 3D mapping of lunar surface. ASPRS/CaGIS 2010 Specialty Conference, Orlando, Florida, USA, December 15–19.

Di, K., Hu, W., Liu, Y., Peng, M. 2012. Co-registration of Chang'E-1 stereo images and laser altimeter data with crossover adjustment and image sensor model refinement. *Adv. Space Res.* 50(12), 1615–1628.

Li, C., Liu, J., Ren, X., et al. 2010. The global image of the Moon obtained by the Chang'E-1 data processing and lunar cartography. *Sci. China Earth Sci.* 40(3), 294–306.

Li, R., Hwangbo, J., Chen Y., Di, K. 2011. Rigorous photogrammetric processing of HiRISE stereo imagery for Mars topographic mapping. *IEEE Trans. Geosci. Remote Sens.* 49(7), 2558–2572.

Liu, Y., Di, K. 2011. Evaluation of rational function model for geometric modeling of Chang'E-1 CCD images. In: *ISPRS Workshop, Geospatial Data Infrastructure: Form Data Acquisition and Updating to Smarter Services*, Guilin, China, 121–125 (CD-ROM).

Lowe, D.G. 1999. Object recognition from local scale-invariant features, *Proc. of the 1999 International Conference on Computer Vision*, 20–25 September, Corfu, Greece, 1150–1157.

National Astronomical Observatories, Chinese Academy of Science (NAOC), 2010. Release of the first Chang'E-2 image of Sinus Iridum. http://moon.bao.ac.cn/templates/T_yestem_articelcontent/index.aspx?nodeid=13&page=ContentPage&contentid=184 (last accessed 29 Mar. 2012)

Ouyang, Z., Li, C. 2010. The primary science results from the Chang'E-1 probe. *Sci. China Ser. D-Earth Sci.* 53(11), 1565–1581.

Peng, M., Yue, Z., Liu, Y., et al. 2010. Research on lunar and Mars orbital stereo image mapping. Proc. of SPIE Remote Sensing of Environment: The 17th Remote Sensing Conference of China, 27–31 August, Hangzhou, China.

Xinhua News Agency, 2012. China released a high resolution global map of the Moon. Xinhua Net. http://news.xinhuanet.com/politics/2012-02/06/c_111491784_3.htm?prolongation=1 (last accessed 29 Mar. 2012)

Yuan, X., Yu, J. 2008. Calibration of constant angular error for high resolution remotely sensed imagery. *Acta Geodaet. Et Acrtogr. Sinica* 37(1), 36–41.

Zhao, B., Wen, D., Yang, J. 2011a. Two bore-sight stereo mapping with single lens, TDI CCD pushing model imaging and compensations of the speed-to-height rate — Chang'E-2 CCD camera. *Acta Opt. Sinica* 31(9), 126–133.

Zhao, B., Yang, J., Wen, D., 2011b. Chang'E-2 satellite CCD stereo camera design and verification. *Spacecr. Eng.* 20(1), 14–21.

Zhao, B., Yang, J., Wen, D., et al. 2011c. Overall scheme and on-orbit images of Chang'E-2 lunar satellite CCD stereo camera. *Sci. China Ser. E-Tech Sci.* 54(9), 2237–2242.

5

Integration and Coregistration of Multisource Lunar Topographic Data Sets for Synergistic Use

Bo Wu, Jian Guo, and Han Hu

CONTENTS

5.1 Introduction 97
5.2 Lunar Topographic Data Sets from Chang'E-1, SELENE, and
LRO Missions 99
5.3 A Multifeature-Based Surface Matching Method for
Coregistration of Multisource Lunar DTMs 103
5.4 Coregistration of Multiple Lunar DTMs in the Apollo 15
Landing Area 103
    5.4.1 Coregistration of Chang'E-1 and LRO DTMs in the
Apollo 15 Landing Area 103
    5.4.2 Coregistration of SELENE and LRO DTMs in the
Apollo 15 Landing Area 106
5.5 Coregistration of Multiple Lunar DTMs in the Sinus Iridum Area 108
    5.5.1 Coregistration of Chang'E-1 and SELENE DTMs in the
Sinus Iridum Area 108
    5.5.2 Coregistration of SELENE and LRO DTMs in the Sinus
Iridum Area 112
5.6 Synergistic Use of Multisource Lunar Topographic Data Sets 113
5.7 Summary 115
References 117

5.1 Introduction

Lunar topographic information is essential for lunar scientific investigations and lunar exploration missions. For example, high-resolution topographic data are critical for understanding the ring structures, mare fill, ejecta, and other crustal features of impact basins. They have major implications for determining the origin and evolution of the Moon. The Moon's crustal features have largely been established from a combination of satellite-altime-try-derived topography analyses and satellite observations of the lunar grav-ity field (Potts and von Frese 2005). The topographic information of lunar impact craters can provide fundamental insights into lunar crust properties, the role of volatiles, and the relative surface age and physics of the craters (Garvin et al. 1999), and their morphological characterizations are signifi-cant sources for scientific research. Surface slope measurements are reliable indicators of the importance of gravity-driven processes to a surface. Lunar topographic data also play a critical role in landing-site selection, precision landing, and ground science experiments in lander, vehicle/robot, astronaut, and outpost explorations.

Starting in the 1960s, various lunar topographic products have been created in the Apollo missions (Livingston 1980; Mellberg 1997) and the Clementine mission (Smith et al. 1997; Rosiek et al. 1999) to fulfill geoscientist demand. Over the past several years, recent lunar exploration missions, such as the Chinese Chang'E-1 (Ouyang et al. 2010) and Chang'E-2 (Zou et al. 2012), the Japanese SELenological and ENgineering Explorer (SELENE)/Kaguya (Kato et al. 2008), and NASA's Lunar Reconnaissance Orbiter (LRO) (Chin et al. 2007) have collected a vast amount of lunar topographic data. Laser altimeters and cameras onboard the Chang'E-1, SELENE, and LRO are the primary sensors for collecting lunar topographic information. They have different configurations providing elevation measurements with differ-ent characteristics. Various lunar digital terrain models (DTMs) have been generated using the data, for example, from the Chang'E-1 Laser AltiMeter (LAM) (Li et al. 2010), SELENE Laser ALTimeter (LALT) (Araki et al. 2009; Noda et al. 2009), and LRO's lunar orbiter laser altimeter (LOLA) (Smith et al. 2010; Mazarico et al. 2011). They provide great value to scientists who need DTMs for various lunar scientific investigations and explorations. However, among these lunar DTMs derived from different mission data, there are usually inconsistencies, such as translational shifts, angular rotations, or scale variations due to differences in sensor configurations, data acquisition periods, and production techniques (Wu et al. 2011). To obtain maximum value for science and exploration, the multiresolution and multiscale DTMs derived from different sensors must be coregistered in a common reference frame (Kirk et al. 2012). Only such an effort will ensure (a) the study of off-sets, trends, and error analysis in various lunar topographic data sets, (b) the proper calibration and registration of the data sets, (c) the full comparative and synergistic use of the data sets, and (d) the generation of consistent and precise lunar topographic products.

This chapter first presents a discussion about multisource topographic data sets, using the Chang'E-1, SELENE, and LRO LAM data as examples. Then, a novel surface matching method is described for the coregistration of multisource topographic data sets. DTMs derived from the Chang'E-1, SELENE, and LRO LAM data in the Apollo 15 landing area and the Sinus Iridum area were examined, and the inconsistencies among them were quantitatively analyzed. Based on the surface matching results, the multi-source topographic data sets can be coregistered together, from which DTMs with greater quality (e.g., spatial resolution) can be generated. Examples are given in this chapter.

5.2   Lunar Topographic Data Sets from Chang'E-1, SELENE, and LRO Missions

The laser altimeter data collected from the Chang'E-1, SELENE, and LRO missions are used here as examples to discuss the multisource lunar topographic data sets. Laser altimeter is used to measure the altitude of lunar surface. In more specific terms, the laser altimeter transmits a beam of high-power narrow laser pulse to the lunar surface, and receives its returned signals by an optical telescope. By measuring out and home time delay of the laser, the distance between the satellite and the lunar surface can be calculated (Smith et al. 2010).

The LAM onboard Chang'E-1 is one of the major payloads of the Chinese Chang'E-1 probe. Chang'E-1 runs on a 200-km height circle orbit. The LAM operates at 1 Hz and is able to measure the distance to lunar surface with a distance resolution of 1 m and distance deviation of 5 m. The spacing resolution along the flight direction is approximately 1.4 km. The LAM has successfully collected approximately 9.12 million measurements, over 1000 orbits, covering the entire surface of the Moon (Li et al. 2010).

The SELENE LALT is one of the main scientific payloads on board the Japanese SELENE satellite. The SELENE operates on a 100 km height circle orbit, which is lower than the one of Chang'E-1. SELENE LALT data gained measurements with a height resolution of 5 m at a sampling interval smaller than 2 km. The SELENE LALT collected more than 10 million high-quality range measurements covering the entire lunar surface (Araki et al. 2009). A global lunar topographic map with a spatial resolution of 0.5 degree was derived from the SELENE LAM data (Araki et al. 2009).

The LOLA is one of the important payloads onboard the NASA LRO robotic spacecraft. LOLA is a five-spot x-pattern pulse detection altimeter orbiting the Moon on a low 50 km polar mapping orbit. LOLA operates at 28 Hz (Smith et al. 2010). The five-spot pattern provides five adjacent profiles for each track. The spacing resolution in the along-track direction is 10–12 m within the combined measurements in the five adjacent profiles. The average distance between LOLA tracks is in the order of 1–2 km at the equator and decreases at higher latitudes. LOLA measurements are generally very dense (varying from a few meters to tens of meters) at polar sites. LOLA digital elevation models (LDEM) are built by binning all valid measurements into the map grid cells and are generated at multiple resolutions. The LDEM_1024 from the LOLA release eight in the planetary data system has the best resolution of 1024 pixel per degree, which is equivalent to about 30 m/pixel in latitude. However, it should be noted that not every pixel in the LDEM_1024 has a measurement, depending on the latitudes of the locations. The accuracy of the topographic measurements and models derived from LOLA ranging data is very high. As reported by Mazarico et al. (2011), LOLA measurements have a precision about 10 m in the along-track and cross-track directions and 1.5 m in the radial direction, respectively, which are estimated as the root-mean-square deviation of the LRO orbit solutions, after cross-over analysis and depending on the orbit solution (Mazarico et al. 2011). The absolute accuracy of the orbit solutions, and therefore the error of the true position of the terrain point, is estimated to be of the same order of magnitude. The LOLA data set is considered to be the most accurate topographic data set of the Moon to date. It provides much denser range measurements compared with other similar instruments from other missions.

In order to have a systematic investigation of the laser altimeter data sets from Chang'E-1, SELENE, and LRO missions, two typical experimental areas were chosen. The first one is the Apollo 15 landing site, which is located at the foot of the Apennine Mountain range (3.66°E, 26.08°N). This selected experimental area covers a region of 0.5° to 8°E and 21° to 32°N, including diverse terrain features such as the Apennine Mountain, Autolycus Crater, and Hadley Rille. The maximum elevation difference is about 8 km, from the peak of the Apennine Mountain to the bottom of the Autolycus Crater. Figure 5.1 shows the distributions of the laser altimeter points from Chang'E-1 LAM, LRO LOLA, and SELENE LALT in the study area. There were 9,683, 1,132,676, and 11,595 points from Chang'E-1 LAM, LRO LOLA, and SELENE LALT, respectively. DTMs with the same spatial resolution, 0.02° grid spacing (about 600 m in latitude and 400 m in longitude), were interpolated from the LAM points, respectively. The points were overlaid on their corresponding DTMs as illustrated in Figure 5.1a, c, and e. Figure 5.1b, d, and f show the three-dimensional (3D) view of the DTMs.

The second experimental area is the Sinus Iridum area. Sinus Iridum (Latin for "Bay of Rainbows") is a plain area that forms a northwestern extension to the Mare Imbrium. The Chinese first lunar lander/rover Chang'E-3 is planned to land in the Sinus Iridum area in 2013. The Sinus Iridum area covers a region of 26° to 38°W and 40° to 50°N, including a large flat area surrounded by mountains. The maximum elevation difference was about 5 km. Figure 5.2 shows the distributions of the laser altimeter points from Chang'E-1 LAM, SELENE LALT, and LRO LOLA in this study area. There were 15,581, 18,379, and 1,428,761 points from Chang'E-1 LAM, SELENE LALT, and LRO LOLA, respectively. DTMs with the same spatial resolution (0.02° grid spacing) were interpolated from the LAM points, respectively. The points were overlaid on their corresponding DTMs as illustrated in Figure 5.2a, c, and e. Figure 5.2b, d, and f show the 3D view of the DTMs.

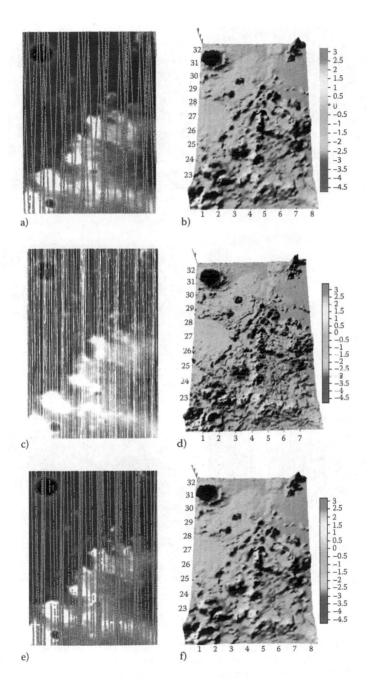

FIGURE 5.1
Chang'E-1 LAM, LRO LOLA, and SELENE LALT data and DTMs in the Apollo 15 landing
area. a) Chang'E-1 LAM data (9,863 points) overlaid on the DTM, b) 3D view of the Chang'E-1
DTM, c) LRO LOLA data (1,132,676 points) overlaid on the DTM, d) 3D view of the LRO DTM,
e) SELENE LALT data (11,595 points) and f) 3D view of the SELENE DTM.

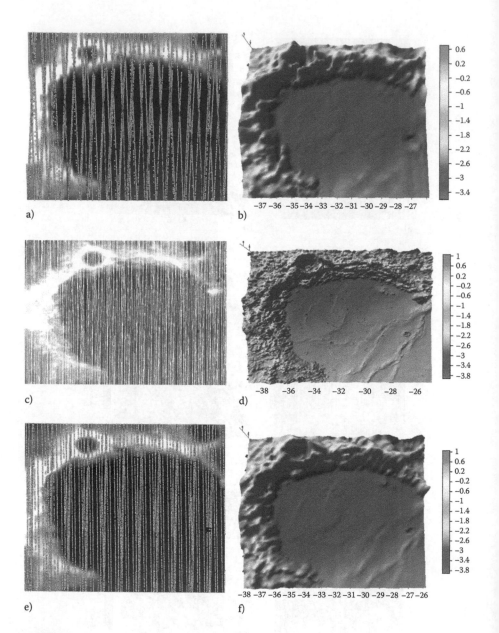

FIGURE 5.2
Chang'E-1 LAM, SELENE LALT, and LRO LOLA data and DTMs in the Sinus Iridum area.
a) Chang'E-1 LAM data (15,581 points) overlaid on the DTM, b) 3D view of the Chang'E-1
DTM, c) LRO LOLA data (1,428,761 points) overlaid on the DTM, d) 3D view of the LRO DTM,
e) SELENE LALT data (18,379 points) overlaid on the DTM and f) 3D view of the SELENE DTM.

5.3 A Multifeature-Based Surface Matching Method for Coregistration of Multisource Lunar DTMs

A multifeature-based surface matching method is developed for the comparison and coregistration of multiple lunar DTMs, which incorporates feature points, lines, and surface patches in surface matching to guarantee robust surface correspondence. From this surface matching method, seven transformation parameters (one scale factor, three rotations, and three translations) can be determined, from which the multiple DTMs could be compared and coregistered.

The multifeature-based surface matching method is described as follows. For two lunar DTMs derived from different sources, one is treated as a reference DTM (e.g., DTM 1) and the other as a matching DTM (e.g., DTM 2). Feature points, lines, and surface patches are identified on both DTMs for surface matching. The feature points are normally the centers of craters or other terrain feature points. The feature lines are the ridge or valley lines detected from the DTMs. To obtain feature surface patches, the DTMs are segmented through triangulations, and the triangles that share surface normals are merged to form local surface patches. The feature points, lines, and surface patches are then used as inputs in the multifeature-based surface matching model. The seven unknown transformation parameters are calculated through a least squares minimization of difference between the matching surface and the reference surface based on the surface matching model. The seven transformation parameters indicate the differences between the two DTMs and from which the matching DTM can be coregistered to the referencing DTM. Figure 5.3 shows the overview of the approach. Details about this multifeature-based surface matching method can be found in Wu et al. (2013).

It should be noted that if the spatial resolutions of the two DTMs are different, then the DTM with a higher resolution will be resampled to a new DTM with the same resolution. Surface matching is carried out between the two equal-resolution DTMs, while the obtained transformation parameters are applied to the original DTMs for further comparison and coregistration.

5.4 Coregistration of Multiple Lunar DTMs in the Apollo 15 Landing Area

5.4.1 Coregistration of Chang'E-1 and LRO DTMs in the Apollo 15 Landing Area

The multifeature-based surface matching method was used to compare and coregister the Chang'E-1 DTM to the LRO DTM. Feature points, lines, and surface patches were detected from both of the DTMs. Finally, three point pairs (two crater centers and one mountain peak) and six line pairs

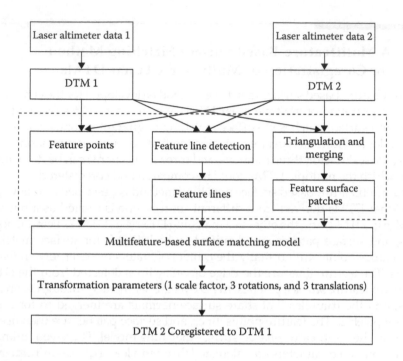

FIGURE 5.3
Framework of the surface matching method.

were identified from the two DTMs as illustrated in Figure 5.4a and b. Three pairs of surface patches were also identified from the two DTMs as shown in Figure 5.4c and d. They were then used as inputs in the surface matching model for surface matching.

After surface matching using the developed method, seven transformation parameters including one scale factor, three translations, and three rotations were obtained, from which the two DTMs can be compared. They are listed in Table 5.1. The results show that for the DTMs derived from the Chang'E-1 LAM and LRO LOLA data in the Apollo 15 landing area, there were about 25 m offset between these two data sets in the horizontal direction, and the LRO data were about 218 m higher than the Chang'E-1 data. The deviations in rotations between these two data sets were small, and the scales were very similar.

The obtained transformation parameters were used to transform the Chang'E-1 LAM point set to a new frame, and a new DTM was interpolated from the new point set. This coregistration process brought the new Chang'E-1 DTM into alignment with the LRO DTM. To evaluate the coregistration method performances, four reference lines were selected for profile comparison analysis as shown in Figure 5.5, of which lines 1 and 2 are along the north–south direction and lines 3 and 4 are along the east–west direction.

Three profiles were derived for each reference line, including a profile derived from the LRO DTM, a profile derived from the original Chang'E-1

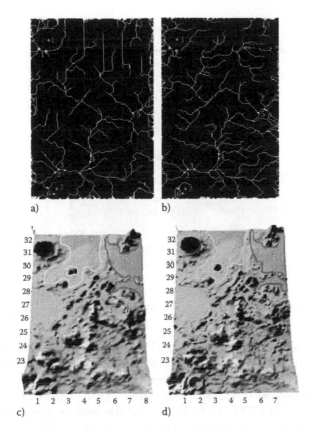

a) b) c) d)

FIGURE 5.4
Feature points, lines, and surface patches determined on the Chang'E-1 and LRO DTMs in the Apollo 15 landing area. a) The selected feature points (marked with purple) and line segments (cyan) on the Chang'E-1 DTM, b) the corresponding feature points and line segments on the LRO DTM, c) the selected surface patches on the Chang'E-1 DTM and d) the corresponding surface patches on the LRO DTM.

TABLE 5.1

Transformation Parameters between the Chang'E-1 and LRO DTMs in the Apollo 15 Landing Area

Transformation Parameters	Scale	ΔX (longitude)	ΔY (latitude)	ΔZ (altitude)	Δφ (roll)	Δω (pitch)	Δκ (yaw)
Values	0.9978	−1.92 m	24.44 m	−217.63 m	0.014°	0.011°	−0.001°

DTM, and a profile derived from the Chang'E-1 DTM after coregistration. Figure 5.6a–d shows the results from the multifeature-based coregistration.

Figure 5.6 shows that the general trends among these profiles are consistent for the four reference lines in this study area. The profiles derived from the Chang'E-1 DTMs show relatively smooth topography compared

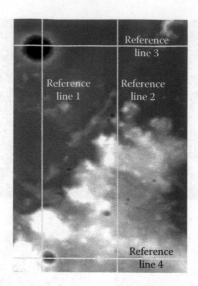

FIGURE 5.5
Four reference lines selected for comparison analysis in the Apollo 15 landing area.

with the profiles derived from the LRO DTM. This is because the Chang'E-1 DTMs were interpolated from the relatively sparse Chang'E-1 LAM points and may not be sufficient to represent the actual topography in this area. Figure 5.6 also shows obvious offsets between the profiles derived from the original Chang'E-1 DTMs (cyan lines) and the profiles from the LRO DTM (blue lines). After coregistration using the proposed multifeature-based surface matching method, the profiles derived from the coregistered Chang'E-1 DTM (red lines) are quite consistent with the LRO profiles for all the four reference lines as shown in Figure 5.6.

The absolute elevation differences of the profiles were calculated and the statistics including the average, maximum, and minimum were obtained. They are shown in Table 5.2. The results indicate that the multifeature-based coregistration method performed well to remove the inconsistency between the original Chang'E-1 and LRO topographic models.

5.4.2 Coregistration of SELENE and LRO DTMs in the Apollo 15 Landing Area

Similar processes were used to compare and coregister the DTM derived from the SELENE LALT data with the DTM derived from the LRO LOLA data in the Apollo 15 landing area. Table 5.3 lists the seven transformation parameters of the SELENE and LRO DTMs obtained using the multifeature-based surface matching method. There is about a 40 m offset between the SELENE and LRO DTMs in the horizontal direction, and the SELENE data are generally about 164 m higher than the LRO data. The rotation

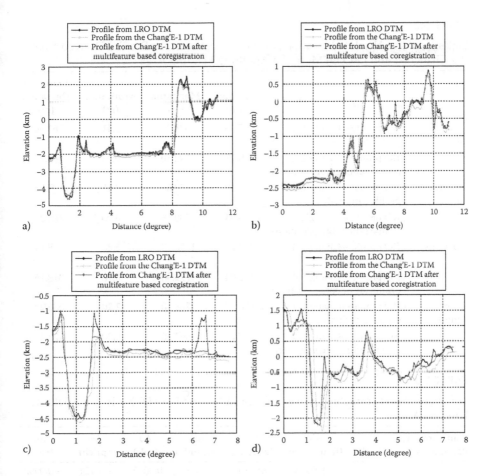

FIGURE 5.6
Profiles derived from the LRO and Chang'E-1 DTMs using multifeature based coregistration in the Apollo 15 landing area. a) Profiles for reference line 1, b) profiles for line 2, c) profiles for line 3, and d) profiles for line 4.

deviations between the two data sets are small, and their scales are very similar.

Figure 5.7 shows the profiles for the same four reference lines illustrated in Figure 5.5, including profiles derived from the LRO DTM, the original SELENE DTM, and the SELENE DTM after multifeature-based coregistration. Table 5.4 lists the statistics of the absolute elevation differences of these profiles.

Figure 5.7 and Table 5.4 show that the SELENE and LRO profiles along the north–south direction are quite consistent. The averages of the absolute elevation differences between the profiles derived from the original SELENE DTM and LRO DTM are 31.98 m for reference line 1 and 40.22 m for line 2. For the profiles along the east–west direction (reference lines 3 and 4), the differences

TABLE 5.2

Statistics of the Absolute Elevation Differences between the Profiles Derived from the Chang'E-1 and LRO DTMs in the Apollo 15 Landing Area

		Average (m)	Maximum (m)	Minimum (m)	Standard Deviation (m)
Reference line 1	LRO profile—Chang'E-1 profile (original)	97.48	838.38	3.41	261.86
	LRO profile—Chang'E-1 profile (after coregistration)	9.27	792.40	0.25	189.75
Reference line 2	LRO profile—Chang'E-1 profile (original)	73.19	958.86	0.11	197.33
	LRO profile—Chang'E-1 profile (after coregistration)	8.84	700.15	0.00	188.22
Reference line 3	LRO profile—Chang'E-1 profile (original)	223.75	1756.30	6.30	338.14
	LRO profile—Chang'E-1 profile (after coregistration)	102.04	1724.90	0.10	318.15
Reference line 4	LRO profile—Chang'E-1 profile (original)	138.47	2785.30	2.20	536.22
	LRO profile—Chang'E-1 profile (after coregistration)	57.24	1063.30	0.30	172.40

TABLE 5.3

Transformation Parameters between the SELENE and LRO DTMs in the Apollo 15 Landing Area

Transformation Parameters	Scale	ΔX (longitude)	ΔY (latitude)	ΔZ (altitude)	$\Delta\phi$ (roll)	$\Delta\omega$ (pitch)	$\Delta\kappa$ (yaw)
Values	0.9995	−39.08 m	9.89 m	164 m	0.0025°	−0.0121°	−0.0002°

are much larger, showing averages of 189.29 m for reference line 3 and 91.57 m for line 4. In comparing the profiles at reference line 3, it can be seen that the original SELENE data are clearly higher than the LRO data in the upper-right part of the study area. After applying multifeature-based coregistration, the differences are reduced significantly for all four reference lines.

5.5 Coregistration of Multiple Lunar DTMs in the Sinus Iridum Area

5.5.1 Coregistration of Chang'E-1 and SELENE DTMs in the Sinus Iridum Area

In the Sinus Iridum area, the Chang'E-1 DTM and SELENE DTM were examined first. Three point pairs (two crater centers and one feature point) and

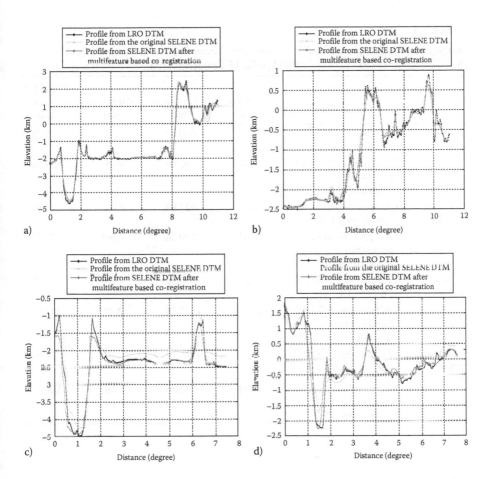

FIGURE 5.7
Profiles derived from the LRO DTM, the original SELENE DTM, and the SELENE DTM after multifeature based coregistration in the Apollo 15 landing area. a) Profiles for reference line 1, b) profiles for line 2, c) profiles for line 3, and d) profiles for line 4.

six line pairs were identified from the two DTMs as illustrated in Figure 5.8a and b. Four pairs of surface patches were also identified from the two DTMs as shown in Figure 5.8c and d. They were then used as inputs in the surface matching model.

After surface matching using the multifeature-based method, seven transformation parameters were obtained. They are listed in Table 5.5. The results show that for the DTMs derived from the Chang'E-1 LAM and SELENE LALT data in the Sinus Iridum area, there were about 470 m offset between the two data sets in the horizontal direction, and the SELENE data were about 230 m higher than the Chang'E-1 data.

TABLE 5.4

Statistics of the Absolute Elevation Differences between the Profiles Derived from the SELENE and LRO DTMs in the Apollo 15 Landing Area

		Average (m)	Maximum (m)	Minimum (m)	Standard Deviation (m)
Reference line 1	LRO profile—SELENE profile (original)	31.98	865.26	0.04	193.01
	LRO profile—SELENE profile (after coregistration)	7.67	652.89	0.01	138.38
Reference line 2	LRO profile—SELENE profile (original)	40.22	818.99	0.04	136.99
	LRO profile—SELENE profile (after coregistration)	13.64	730.46	0.001	131.77
Reference line 3	LRO profile—SELENE profile (original)	189.29	858.40	2.50	231.75
	LRO profile—SELENE profile (after coregistration)	48.35	673.30	0.00	142.97
Reference line 4	LRO profile—SELENE profile (original)	91.57	1023.70	2.70	270.85
	LRO profile—SELENE profile (after coregistration)	2.55	848.40	0.50	169.67

The obtained transformation parameters were used to co-register the Chang'E-1 DTM to the SELENE DTM. To evaluate the performances of the coregistration methods, four reference lines were selected for profile comparison analysis as shown in Figure 5.9, of which lines 1 and 2 are along the north–south direction and lines 3 and 4 are along the east–west direction.

Three profiles were derived for each reference line, including a profile derived from the SELENE DTM, a profile interpolated from the original Chang'E-1 DTM, and a profile interpolated from the Chang'E-1 DTM after co-registering to the SELENE DTM. Figure 5.10a–dshow the results from the multifeature-based surface matching method. Figure 5.10 shows that the general trends among these profiles are consistent for all the four reference lines in this study area. There are obvious offsets between the profiles derived from the original Chang'E-1 DTMs (cyan lines) and the profiles from the SELENE DTMs (blue lines). After coregistration using the multifeature-based method, the profiles derived from the coregistered Chang'E-1 DTM (red lines) are closely attached to the SELENE profiles for the four reference lines as shown in Figure 5.10.

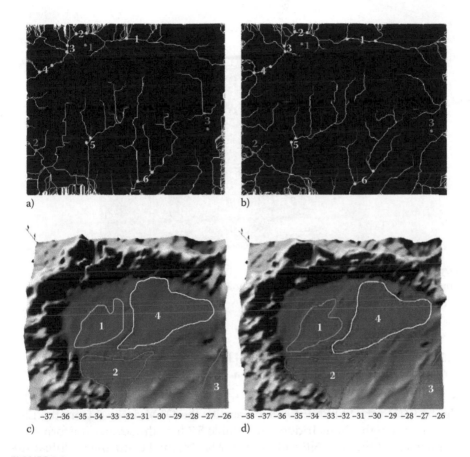

FIGURE 5.8

Feature points, lines, and surface patches determined on the Chang'E-1 and SELENE DTMs in the Sinus Iridum area. a) The selected feature points (marked with purple) and line segments (cyan) on the Chang'E-1 DTM, b) the corresponding feature points and line segments on the SELENE DTM, c) the selected surface patches on the Chang'E-1 DTM and d) the corresponding surface patches on the SELENE DTM.

TABLE 5.5

Transformation Parameters between the Chang'E-1 and SELENE DTMs in the Sinus Iridum Area

Transformation Parameters	Scale	ΔX (longitude)	ΔY (latitude)	ΔZ (altitude)	$\Delta\phi$ (roll)	$\Delta\omega$ (pitch)	$\Delta\kappa$ (yaw)
Values	1.0024	−301.18 m	−357.18 m	−230.91 m	0.0073	0.0028	−0.0072

The absolute elevation differences of the profiles were calculated and the statistics are given in Table 5.6. The results show that differences between original Chang'E-1 and SELENE topographic models were reduced significantly after using the multifeature-based surface matching method.

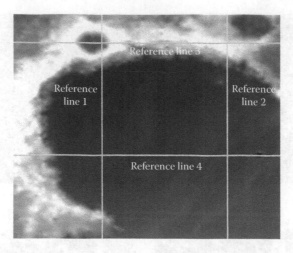

FIGURE 5.9
Four reference lines selected for comparison analysis in the Sinus Iridum area.

5.5.2 Coregistration of SELENE and LRO DTMs in the Sinus Iridum Area

Similar processes were performed to compare and coregister the DTM derived from the SELENE LALT data with the DTM derived from the LRO LOLA data in the Sinus Iridum area. Table 5.7 lists the seven transformation parameters of the SELENE and LRO DTMs obtained using the multifeature-based surface matching method. The results show that there are about 56 m offset between the SELENE and LRO DTMs in the horizontal direction, and the elevations of the two data sets are fairly consistent in the study area. The rotation deviations and scale differences between the two data sets are all very small.

Figure 5.11 shows the profiles for the same four reference lines illustrated in Figure 5.9, including profiles derived from the LRO DTM, the original SELENE DTM, and the SELENE DTM after multifeature-based surface matching. Table 5.8 lists the statistics of the absolute elevation differences of these profiles. Figure 5.11 and Table 5.8 show that the profiles derived from the original SELENE DTM and LRO DTM are quite consistent. The averages of the absolute elevation differences between the profiles for all four reference lines are less than 10 m. The differences are further reduced for the profiles derived from the SELENE DTM after multifeature-based surface matching. The obvious discrepancies in the right part of the profiles in Figure 5.11d can be attributed to better resolution of the LRO LOLA data with respect to the SELENE LALT data when generating the DTMs.

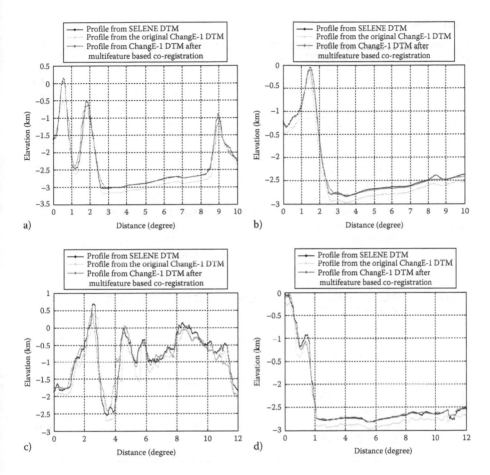

FIGURE 5.10
Profiles derived from the SELENE and Chang'E-1 DTMs using multifeature based coregistration in the Sinus Iridum area. a) Profiles for reference line 1, b) profiles for line 2, c) profiles for line 3, d) profiles for line 4.

5.6 Synergistic Use of Multisource Lunar Topographic Data Sets

As mentioned previously, one of the benefits of multiple DTM surface matching is the full synergistic use of multiple data sets and generation of consistent and better lunar topographic products. After the SELENE and Chang'E-1 DTMs were coregistered to the LRO DTM using the multifeature-based surface matching method, the derived transformation parameters were used to bring the SELENE LALT and Chang'E-1 LAM data sets into alignment with

TABLE 5.6

Statistics of the Absolute Elevation Differences between the Profiles Derived from the Chang'E-1 and SELENE DTMs in the Sinus Iridum Area

		Average (m)	Maximum (m)	Minimum (m)	Standard Deviation (m)
Reference line 1	SELENE profile—Chang'E-1 profile (original)	176.80	734.17	1.41	233.35
	SELENE profile—Chang'E-1 profile (after coregistration)	15.18	521.28	0.07	124.74
Reference line 2	SELENE profile—Chang'E-1 profile (original)	154.78	520.44	2.02	74.17
	SELENE profile—Chang'E-1 profile (after coregistration)	15.72	276.47	0.91	68.16
Reference line 3	SELENE profile—Chang'E-1 profile (original)	192.86	1244.5	0.60	300.26
	SELENE profile—Chang'E-1 profile (after coregistration)	29.76	1118.9	0.50	230.40
Reference line 4	SELENE profile—Chang'E-1 profile (original)	149.55	697.50	0.80	123.97
	SELENE profile—Chang'E-1 profile (after coregistration)	17.78	402.80	0.00	9.48

TABLE 5.7

Transformation Parameters between the SELENE and LRO DTMs in the Sinus Iridum Area

Transformation Parameters	Scale	ΔX (longitude)	ΔY (latitude)	ΔZ (altitude)	$\Delta \varphi$ (roll)	$\Delta \omega$ (pitch)	$\Delta \kappa$ (yaw)
Values	0.9989	−22.87 m	51.42 m	7.74 m	−0.0003°	0.00008°	0.00014°

the LRO LOLA. The DTMs with 0.01° grid spacing resolutions (about 300 m in latitude and 200 m in longitude) were generated using all of the LAM points in the Apollo 15 landing area and the Sinus Iridum area.

Figure 5.12 shows a side-by-side visualization of the original DTMs derived from the LRO LOLA data and the new DTMs, indicating that the latter provides finer and more detailed topographic information. The yellow boxes marked in Figure 5.12b and d show obvious examples.

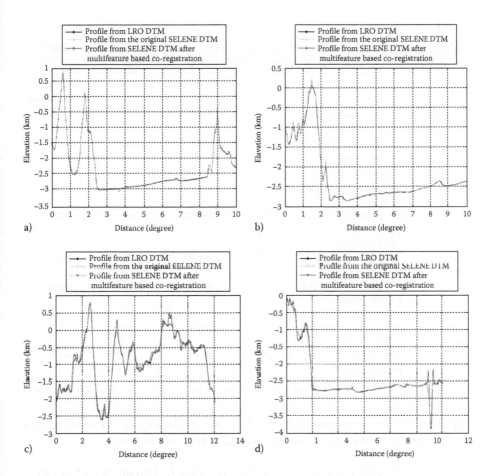

FIGURE 5.11
Profiles derived from the LRO DTM, the original SELENE DTM, and the SELENE DTM after multifeature based coregistration in the Sinus Iridum area. a) Profiles for reference line 1, b) profiles for line 2, c) profiles for line 3, and d) profiles for line 4.

5.7 Summary

This chapter has presented a comparative study of multisource lunar topographic data sets using the Chang'E-1, SELENE, and LRO LAM data as examples. A multifeature-based surface matching method was developed and applied to the comparison and coregistration of multiple lunar DTMs derived from Chang'E-1, SELENE, and LRO LAM data. On the basis of the

TABLE 5.8

Statistics of the Absolute Elevation Differences between the Profiles Derived from the SELENE and LRO DTMs in the Sinus Iridum Area

		Average (m)	Maximum (m)	Minimum (m)	Standard Deviation (m)
Reference line 1	LRO profile—SELENE profile (original)	9.82	248.1	0.11	40.29
	LRO profile—SELENE profile (after coregistration)	9.66	170.58	0.08	35.19
Reference line 2	LRO profile—SELENE profile (original)	5.41	199.87	0.11	41.34
	LRO profile—SELENE profile (after coregistration)	5.09	185.81	0.01	34.70
Reference line 3	LRO profile—SELENE profile (original)	8.38	441.20	0.20	109.28
	LRO profile—SELENE profile (after coregistration)	1.65	413.50	0.20	93.90
Reference line 4	LRO profile—SELENE profile (original)	1.44	670.3	0.00	98.28
	LRO profile—SELENE profile (after coregistration)	0.47	560.4	0.00	79.79

experimental analysis that used the data sets in the Apollo 15 landing area and the Sinus Iridum area leads to the following conclusions:

1. In the Apollo 15 landing area, the LRO LOLA data are about 218 m higher than the Chang'E-1 LAM data, and there are about 25 m offset between the two data sets in the horizontal direction. For the SELENE LALT and LRO LOLA data sets, the SELENE data are generally about 164 m higher than the LRO data, and there are about 40 m offset between them in the horizontal direction. In the Sinus Iridum area, the SELENE LALT data are about 230 m higher than the Chang'E-1 LAM data, and there are about 470 m offset between them in the horizontal direction. There are about 56 m offset between the SELENE LALT and LRO LOLA data sets in the horizontal direction, and their elevations are fairly consistent in this study area.

2. After applying the surface matching method to the multiple DTMs, the differences among them could be significantly reduced in both the Apollo 15 landing area and the Sinus Iridum area.

3. The coregistration of multisource lunar topographic data sets enables the generation of consistent and better lunar topographic products for the synergistic use of them.

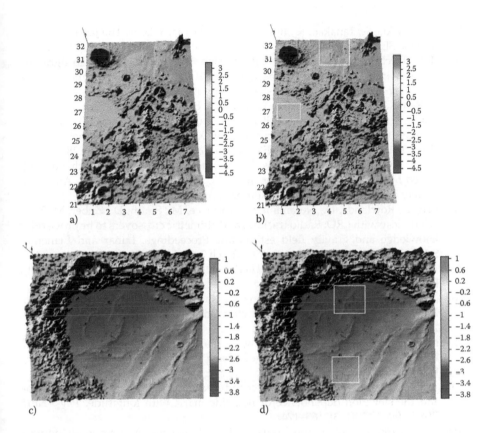

FIGURE 5.12
Side by side visualization of the DTMs. a) The original LOLA DTM in the Apollo 15 landing area, b) the DTM derived by using all the aligned LAM points in the Apollo 15 landing area, c) the original LOLA DTM in the Sinus Iridum area, and d) the DTM derived by using all the aligned LAM points in the Sinus Iridum area.

References

Araki, H., Tazawa, S., Noda, H., Ishihara, Y., Goossens, S., Sasaki, S., Kawano, N., Kamiya, I., Otake, H., Oberst, J., Shum, C. 2009. Lunar global shape and polar topography derived from Kaguya-LALT laser altimetry. *Science* 323(5916), 897–900.

Chin, G., Brylow, S., Foote, M., Garvin, J., Kasper, J., Keller, J., Litvak, M., Mitrofanov, I., Paige, D., Raney, K., Robinson, M., Sanin, A., Smith, D., Spence, H., Spudis, P., Stern, S.A., Zuber, M. 2007. Lunar reconnaissance orbiter overview: The instrument suite and mission. *Space Sci. Rev.* 129(4), 391–419.

Garvin, J.B., Sakimoto, S.E.H., Schnetzler, C., Frawley, J.J. 1999. Global geometric properties of Martian impact craters: A preliminary assessment using Mars Orbiter Laser Altimeter (MOLA). The Fifth International Conference on Mars, July 18-23, Pasadena, CA.

Kato, M., Sasakia, S., Tanakaa, K., Iijimaa, Y., Takizawaa, Y. 2008. The Japanese lunar mission SELENE: Science goals and present status. *Adv. Space Res.* 42(2), 294–300.

Kirk, R.L., Archinal, B.A., Gaddis, L.R., Rosiek, M.R. 2012. Lunar cartography: Progress in the 2000s and prospects for the 2010s. *Int. Arch. Photogram. Remote Sens. Spat. Infor. Sci.*, vol. XXXIX-B4, pp. 489–494.

Li, C.L., Ren, X., Liu, J., Zou, X., Mou, L., Wang, J., Shu, R., Zou, Y., Zhang, H., Lv, C., Liu, J., Zuo, W., Su, Y., Wen, W., Bian, W., Wang, M., Xu, C., Kong, D., Wang, X., Wang, F., Geng, L., Zhang, Z., Zheng, L., Zhu, X., Li, J. 2010. Laser altimetry data of chang'E-1 and the global lunar DEM model. *Sci. China Earth Sci.* 40(3), 281–293.

Livingston, R.G. 1980. Aerial cameras, In *Manual of Photogrammetry* (4th Edition), edited by C.C. Slama, C. Theurer, and S.W. Henriksen, pp. 187–278, American Society of Photogrammetry and Remote Sensing, Falls Church, VA.

Mazarico, E., Rowlands, D., Neumann, G., Torrence, M., Smith, D., Zuber, M. 2011. Selenodesy with LRO: Radio tracking and altimetric crossovers to improve orbit knowledge and gravity field estimation. Proceedings, Lunar and Planetary Science Conference, Woodlands, TX, March 7–11, LPS XLII, 2215.

Mellberg, W.F. 1997. *Moon Missions: Mankind's First Voyages to Another World*. Plymouth Press, Ltd., Plymouth, MI.

Noda, H., Araki, H., Tazawa, S., Goossens, S., Ishihara, Y.; The Kaguya LALT Team. 2009. Kaguya (SELENE) Laser altimeter: One year in orbit. *Geophys. Res. Abstr.* 11, EGU2009-3841-1.

Ouyang, Z.Y., Li, C., Zou, Y., Zhang, H., Lv, C., Liu, J., Liu, J., Zuo, W., Su, Y., Wen, W., 2010. Preliminary scientific results of Chang'E-1 lunar orbiter. *Sci. China* 53(11), 1565–1581.

Potts, L.V., von Frese, R.R.B. 2005. Impact-induced mass flow effects on lunar shape and the elevation dependence of nearside maria with longitude. *Phys. Earth Planet. Inter.* 153(1–3), 165–174.

Rosiek, M., Kirk, R., Howington-Kraus, E. 1999. Lunar topographic maps derived from Clementine imagery, In *Lunar Planetary Science XXX*. Abstract #1853, LPI, Houston, TX, March 15–19.

Smith, D.E., Zuber, M.T., Neumann, G.A., Lemoine, F.G. 1997. Topography of the moon from the Clementine LiDAR. *J. Geophys. Res. Planet.* 102(E1), 1591–1611.

Smith, D.E., Zuber, M.T., Jackson, G.B., Cavanaugh, J.F., Neumann, G.A., Riris, H., Sun, X., Zellar, R.S., Coltharp, C., Connelly, J., Katz, R.B., Kleyner, I., Liiva, P., Matuszeski, A., Mazarico, E.M., McGarry, J.F., Novo-Gradac, A., Ott, M.N., Peters, C., Ramos-Izquierdo, L.A., Ramsey, L., Rowlands, D.D., Schmidt, S., Scott, V.S., Shaw, G.B., Smith, J.C., Swinski, J., Torrence, M.H., Unger, G., Yu, A.W., Zagwodzki, T.W. 2010. The lunar orbiter laser altimeter investigation on the lunar reconnaissance orbiter mission. *Space Sci. Rev.* 150(1–4), 209–241.

Wu, B., Guo, J., Zhang, Y., King, B., Li, Z., Chen, Y. 2011. Integration of Chang'E-1 imagery and laser altimeter data for precision lunar topographic modeling. *IEEE Trans. Geosci. Remote Sens.* 49(12), 4889–4903.

Wu, B., Guo, J., Hu, H., Li, Z., Chen, Y., 2013. Coregistration of lunar topographic models derived from Chang'E-1, SELENE, and LRO laser altimeter data based on a novel surface matching method. *Earth Planet. Sci. Lett.* 364(2013), 68–84.

Zou, X.D., Liu, J.J., Mou, L.L., Ren, X., Li, K., Zhao, J.J., Liu, Y.X., Li, C.L. 2012. Topographic analysis of the proposed landing area of Sinus Iridum. *EPSC Abstr.* 7, EPSC2012-151-1.

6

Estimates of the Major Elemental Abundances with Chang'E-1 Interference Imaging Spectrometer Data

Yunzhao Wu

CONTENTS

6.1 Overview of Remote Measurements of Lunar Major Elements 120
    6.1.1 Remote Measurements of FeO and TiO$_2$ 120
        6.1.1.1 Remote Measurements of FeO 121
        6.1.1.2 Remote Measurements of TiO$_2$ 121
    6.1.2 Remote Measurements of Nontransition Elements 122
6.2 Calibration of IIM Data 123
    6.2.1 Overview of IIM Data 123
    6.2.2 Laboratory Calibration of IIM Data 126
    6.2.3 In-Flight Calibration of IIM Data 126
        6.2.3.1 Photometric Model 126
        6.2.3.2 Data Selection 127
        6.2.3.3 Correction for Inhomogeneity of Spatial Response 128
        6.2.3.4 Correction for Systematic Artifacts 129
        6.2.3.5 Cross-Calibration of IIM NIR Bands 130
6.3 Prediction of FeO and TiO$_2$ 133
    6.3.1 Inversion of FeO Content 133
    6.3.2 Inversion of TiO$_2$ Content 138
    6.3.3 Inversion of Nontransition Elements 139
6.4 Abundance and Distribution of Major Elements on the Moon 141
    6.4.1 Global Elemental Abundances 141
    6.4.2 Global Mg# 150
References 152

6.1   Overview of Remote Measurements of Lunar Major Elements

Usually, lunar major elements contain Fe, Ti, Mg, Al, Ca, Si, and O. They play very important roles in understanding the origin and evolution of the Moon. It is well known that Si is the essential element in classifying rock types into ultramafic, mafic, intermediate, and felsic rock. Aluminum is commonly used to constrain the origin and evolution of the Moon. For example, the bulk Al content of the Moon can control the point at which plagioclase comes onto the liquidus of the lunar magma ocean (LMO), which in turn affects the timing of primary crust formation (Taylor 1987; Prettyman et al. 2006). Fe and Ti allow us to distinguish among the known types of lunar rocks, and Ti forms the basis for classifying the basalts that make up the lunar maria (Taylor et al. 1991). The abundance and distribution of Mg on the Moon is of interest to many lunar scientists because, in conjunction with mineral data, it can be used to indicate mode of formation (i.e., cumulate vs. lava flow) and in crystallized rock, it indicates the evolution of magma formation.

Lunar samples from the U.S. Apollo and Soviet Luna missions provide the most direct information about the Moon's elements. However, now we have known that the lunar samples from distinct areas cannot represent the elemental abundances of the whole Moon. Up to now, orbital remote sensing technology is the only way that can allow scientists to measure the concentrations of major elements for the entire Moon. Direct elemental remote sensing mostly exploits energy-dependent variations in neutron, x-ray, or gamma-ray flux. Global maps of SiO_2, Al_2O_3, CaO, MgO, FeO, and TiO_2 have been derived from gamma ray spectroscopy (GRS) data acquired by Lunar Prospector (e.g., Elphic et al. 2002; Lawrence et al. 2002; Prettyman et al. 2006). Unfortunately, the usefulness of these compositional maps is significantly limited due to the low spatial resolution inherent to the instrument (generally at the tens to hundreds of kilometers scale). In this chapter, we mostly focus on the assessment of elemental abundances with the spectral reflectance remote sensing within the visible and near infrared wavelength (VNIR), the spatial resolution of which is much higher than that of high-energy techniques.

6.1.1   Remote Measurements of FeO and TiO_2

Among the seven major elements, Fe and Ti are transition elements that have unfilled d orbitals. Electronic transitions between Fe or Ti and their surrounding ligands influence the albedo and shapes of the spectral continuum. This provides the opportunity to determine their abundances with optical spectroscopy (Burns 1993). The two elements have been quantitatively estimated, even prior to the orbital remote-sensing missions using data from Earth-based telescopes (e.g., Charette et al. 1974, 1977; Johnson et al. 1991). The FeO and TiO_2 concentrations derived from Clementine spectral reflectance data have been investigated by many researchers (e.g., Lucey et al. 1995, 2000; Blewett et al. 1997; Gillis et al. 2004, 2006). These high spatial resolution FeO and TiO_2 maps have been used widely, such as for inferring ages and stratigraphy of lunar mare basalts (Bugiolacchi et al. 2006; Hiesinger et al. 2010), estimating the thickness of mare basalts (Thomson et al. 2009), etc.

6.1.1.1   Remote Measurements of FeO

Iron is the first element to be remotely assessed in terrestrial planets and their Moons because it can contribute to remote-sensed spectra due to its high cosmic abundance. Furthermore, reflectance spectroscopy is so sensitive to both crystalline and amorphous Fe that it is possible to detect at very little quantity. Lunar Fe can exist in two forms: ferrous iron (Fe^{2+}) within the minerals and nanophase metallic iron ($npFe^0$). Ferrous iron has two overlapping absorptions in the VNIR region: an intense allowed absorption in the UV and a weak forbidden absorption in the NIR at ~1000 nm (Lucey et al. 1998). The nanophase metallic iron, which is produced in lunar soil during soil maturation caused by space weathering processes, can reduce both of its albedo and spectral contrast, and introduce red spectral slope. The spectral changes occur due to the space weathering even with the same iron concentration. Therefore, it is crucial to decouple the composite spectral effects of crystallized iron and $npFe^0$ to estimate the bulk Fe concentration of the lunar surface. Lucey et al. (1995) noted that the spectral effects of maturity and FeO content observed in spectra of lunar samples and in low spatial resolution Clementine data appeared to be orthogonal. They developed an algorithm for the derivation of FeO abundance from VIS reflectance and NIR/VIS ratio of lunar soils, which is insensitive to the maturity level of the soil. Based on this algorithm and Clementine spectral reflectance data, the first near-global map showing the abundance and distribution of iron on the Moon was derived. Later, a series of algorithms were developed for improving the predicting accuracy of lunar Fe contents (Lucey et al. 1998, 2000; Blewett et al. 1997; Lawrence et al. 2002; Gillis et al. 2004).

6.1.1.2   Remote Measurements of TiO_2

Titanium concentration is the most useful discriminator for classifying lunar mare basalts because of its substantial variation (from <1 wt.% to >14 wt.%). Ti is present in lunar materials primarily in the opaque mineral ilmenite. Ilmenite is dark and spectrally neutral, so that it reduces the spectral contrast and causes the generally red lunar material to become less red (i.e., more blue). Titanium can also exist in silicates, particularly silicate glasses. The presence of Ti in a glass will produce strong Fe–Ti charge-transfer bands, which appear at ultraviolet light and extend into the visible region. For the charge-transfer absorption, the strength depends on both the abundance of Fe and of Ti (Bell et al. 1976). Strom was perhaps the first to propose that the marked differences observed between ultraviolet and visible images of the lunar maria were due to variations in Ti abundance (Whitaker 1972). Based on the observation that Fe-bearing silicates have strong absorption bands in the UV due to the charge transfer and that ilmenite often exhibits a reflectivity upturn below about 450 nm, Charette et al. (1974) constructed an empirical relationship between UV ratio and TiO_2 for mature mare soils. The "Charette relation" was subsequently modified and improved by many researches (e.g., Johnson et al. 1977, 1991; Melendrez et al. 1994; Pieters 1993; Robinson et al. 2007).

The Charette relationship is limited to mature mare areas because the UV/VIS ratio is also influenced by soil Fe content. Mare basalts have relatively narrow ranges of Fe, but at low abundances of Ti, the UV/VIS ratio is dominated by variations in Fe. Johnson et al. (1991) concluded that the reliability of the Charette relationship is poor at TiO_2 contents less than about 3 wt.%. Lucey et al. (1996, 1998, 2000) developed a method for the determination of TiO_2, which is applicable to much lower Ti contents and is insensitive to Fe, thus allowing Ti-mapping to be extended to the lunar highlands and low-Ti maria. They defined titanium sensitive parameter by coordinate transformation in UV/VIS versus VIS space. The spectral parameter can simultaneously be correlated with titanium and insensitive to maturity. Noting that data points for Apollo 11, Luna 16, and Luna 24 data deviate significantly from the others, and also noting that large regions of nearside maria have similar spectral characteristics to these points, Gillis et al. (2003) modified the single regression method by using a dual-regression procedure for two different sets of data points. Titanium concentrations derived from the dual-regression method are more consistent with observed epithermal-to-thermal neutron-flux ratios than are previous Clementine-based derivations of TiO_2 for basaltic regions.

6.1.2 Remote Measurements of Nontransition Elements

Elemental assessment with optical spectroscopy is not only limited to the chromophore elements such as Fe and Ti. Many researches predicted nonchromophore elements with optical spectroscopy for both point spectra and image data. Jaumann (1991) predicted SiO_2, Al_2O_3, MgO, CaO, MnO as well as FeO and TiO_2 with the laboratory reflectance spectra by combining multivariate analysis and principal component analysis (PCA). These spectral-chemical parameters were further applied to telescopic data and the concentrations estimated agreed well with the true values estimated from soil samples. Pieters et al. (2002) also predicted lunar soil chemistry from laboratory reflectance spectra with this approach, which can be named principal component regression (PCR). Li (2006) and Zhang et al. (2009) predicted many major elements with the partial least squares regression (PLSR) and the combination of PCA and support vector machine (SVM), respectively. The approach used by Li (2006) and Zhang et al. (2009), that is, PLSR and SVM, improved the prediction accuracy over the PCR method. All the researches by Pieters et al. (2002), Li (2006), and Zhang et al. (2009) used the spectral and chemical data measured by Lunar Soil Characterization Consortium (LSCC).

The successful predictions discussed above with laboratory spectra confirmed that nonchromophore elements can be estimated by optical spectroscopy and that the techniques used have the potential to be applied to remote sensing image data. Fischer and Pieters (1995) mapped the concentration of Al on the Moon by the Galileo solid state imaging system (SSI) based on the positive linear correlation between the major Al-bearing phase, plagioclase, and albedo. Shkuratov et al. (2003) predicted mineral and chemical compositions with a series of linear combinations of optical parameters derived from 24 mare LSCC samples and applied the weight coefficients to Clementine images. Shkuratov et al. (2005) produced global maps of several major elements (Fe, Ti, Ca, Mg, Al, O) with Clementine spectral reflectance data by using Lunar Prospector gamma ray spectroscopy (LP-GRS) data as "ground truth" to establish relationships linking optical data and chemical information.

The prediction of nonchromophore elements such as Al, Ca, Mg, and Si with optical spectroscopy is not pure mathematic statistic, on the contrary, the physical mechanism exists, which makes it possible for the prediction. Although nonchromophore elements do not have diagnostic absorption features, they can affect the Moon's reflectance values. For example, the more aluminous and calcic materials exhibit higher reflectance, while the more magnesian materials exhibit lower reflectance. This is because Al and Ca are carried predominantly in plagioclase, which has high reflectance, while Mg is carried predominantly in pyroxene, which has low reflectance. Moreover, nonchromophore elements are often naturally correlated with chromophore elements, such as Fe and Ti, within the lunar regolith. For example, Al and Fe in lunar soils are inversely correlated with each other (Heiken et al. 1991).

6.2   Calibration of IIM Data

6.2.1   Overview of IIM Data

The IIM is a Sagnac-based spatially modulated Fourier transform imaging spectrometer that uses an interference pattern to derive a spectrum. It is the first planetary application of this technology. The performance characteristics and the optical layout of IIM are shown in Table 6.1 and Figure 6.1, respectively. IIM contains four major optical subsystems: an Objective Lens, which images the scene onto a slit; a Sagnac interferometer, by which the rays are split, and slightly sheared; a Fourier lens collimates the light and a cylindrical lens images the energy onto the detector (Zhao et al. 2009). A cut-off filter with out-of-band rejection less than 1% is placed in front of the objective lens to limit the instrument response to 480.9–946.8 nm by preventing photons of the out-of-band wavelength entering the optical system. Within the wavelength range of IIM, that is, 480.9–946.8 nm, IIM has 32 channels with

TABLE 6.1

Performance Characteristics of IIM

Swath width	25.6 km
Imaging coverage	Between 75°S and 75°N
Digital level	12 bit
Pixel no.	128*128 (256*256 after 2*2 combination)
Spectral range	480.9–946.8 nm
Spectral resolution	330 cm^{-1}
Space resolution	200 m (orbit altitude = 200km)
Spectral channels	32 bands
Modulated transferring function	0.51 (snapshot)
SNR	375 (snapshot)
F/#	F/2.4
Instantaneous-field-of-view	1 milliradian
Slit width	0.6 milliradian
Data compression ratio	No
Apodization	No
Electronic gain	1 (typical used), 1.5 and 2
Exposure time	140 ms (typical used) and 70 ms
Frame rate	7.1 FPS (typical used) and 14.2 FPS
Data rate	1363 kbps
Read noise	No measurement
Interferogram sampling mode	Double sided sampling
Fringe visibility	64%@543.5nm
Out-of-band rejection	<1%

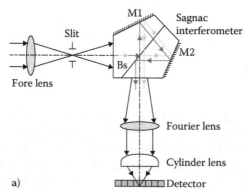

a)

b)

FIGURE 6.1

a) Schematic of the IIM. The lunar surface is imaged onto a slit aperture using the Fore Lens. Light emerging from the slit is sheared by the Sagnac interferometer and collimated by the Fourier lens. The cylindrical lens reimages the spatial axis of the slit onto the detector array while preserving the spectral interference pattern. b) The IIM instrument, CCD camera to the left, Sagnac interferometer in the center, and the fore lens is the cylindrical optical assembly on the right.

a theoretical spectral resolution of 330 cm^{-1} according to the Sparrow's criterion (Zhao et al. 2009). The spectral resolution and wavelength position in the laboratory test with the gas laser and semiconductor laser show that the actual resolution is about 355 cm^{-1} and maximal shift of 2.48 nm@831.2 nm for the wavelength position (Zhao et al. 2010). The measured value of the fringe visibility is 64%@543.5 nm, and the ratio of the side lobe to the maximum peak is about 21%. The Nyquist minimum of at least two channels per fringe is maintained throughout the range of spectral response of the optics and detector.

IIM data were received, processed, archived, and distributed by the Ground Segment for Data, Science and Application (GSDSA) of China's Lunar Exploration Program, which was located in the National Astronomical Observatories, Chinese Academy of Sciences. Two flight modes were designed for CE-1 according to the solar beta angle (the angle between the orbital plane and the vector to the Sun), which relates to the energy supply to the satellite. When the solar beta angle is between −45° and 45° the solar energy is relatively large, the spacecraft is in the normal flight, and the IIM instrument can image the Moon. When the solar beta angle is outside this range the solar energy is relatively weak and to acquire enough solar energy the spacecraft adjusted its attitude and hence the IIM instrument could not image the Moon. Ten orbits of IIM data are available to the public from the first optical period obtained between November 26, 2007, and November 29, 2007. Most of the IIM data (565 orbits) were measured during the second optical period and the data collection period of the released data was between May 15, 2008 and July 28, 2008. The beta angle of IIM was relatively large during the first optical period (minimum value 18°), whereas the second optical period was better optimized for spectral imaging, being centered on a 0° beta angle on June 17, 2008, when the position of the subsolar point was

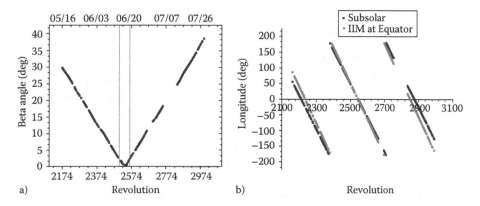

FIGURE 6.2
a) The relation between the β angle and the IIM revolution Φ. The orbits within the range of the two red lines are used to calculate the in-flight flat-field image. b) The longitude of the subsolar point and sub-IIM point of each revolution at equator.

~12.73°E, ~1.33°N. Figure 6.2 shows the beta angle of each revolution and the corresponding subsolar point.

6.2.2 Laboratory Calibration of IIM Data

The IIM instrument shares electronics with the CCD stereo cameras. To ensure global mapping of the Moon with the CCD stereo camera, no in-flight radiometric calibration for IIM was performed. However, the radiometric properties of IIM interference intensity data were carefully characterized using the preflight data obtained in laboratory experiments. By adjusting the slit width (0.06 mm) the maximum saturation radiance is set at 54.3 W/ m^2.sr.(480–960 nm). The dark current, linearity, and relative calibration were performed at the interferogram stage. The dark current was measured for each detector 25 times, and their arithmetic mean was subtracted to remove the dark current. The linearity and relative calibration was measured with a well-calibrated integrating sphere. Seventeen light source sets were used to examine the linearity. IIM is highly linear with the largest residual of 2.5% for one lamp, and less than 1% for all other lamps. However, the linearity with interference intensity of the in-flight data beyond the scope measured in the laboratory is unknown. So for the photometric correction, it should be very careful with the data selection to ensure that the data are linear. For the absolute spectral calibration, an XTH-2000 solar simulator and an ASD Fieldspace@ Pro UV/VNIR were used as the standard sources. The maximum uncertainty for the absolute calibration is 6% with an average of 4%. The Nyquist minimum of at least two channels per fringe is maintained throughout the range of spectral response of the optics and detector. Due to a long integration time (140 ms) the frame transfer offset can be neglected. Further details regarding the CE-1 preflight calibration can be found in Zhao et al. (2010).

6.2.3 In-Flight Calibration of IIM Data

6.2.3.1 Photometric Model

The radiance data of IIM were converted to radiance factor (RADF) with the correction of the Sun–Moon distance for each pixel with the following formula:

$$\frac{I}{F} = \frac{\pi L(\lambda, i, e, \alpha) \cdot d^2}{E_0(\lambda)} \tag{6.1}$$

where λ, i, e, and α are wavelength, incidence angle, emission angle, and phase angle. L is the radiance of IIM data. E_0 is the solar irradiance at 1 AU from the Sun. d is the Sun–Moon distance in kilometers at the observation time divided by the standard Sun–Moon distance (149,597,870 km). The solar irradiance from ATLAS 3 (Thuillier et al. 2004) was adopted. The solar irradiance of ATLAS 3 was resampled according to the wavelength and full-width-at-half-max (FWHM) of IIM. An empirical photometric function based on the Lommel–Seeliger model was employed to correct the solar photometric effects:

$$\frac{I}{F} = \frac{\mu_0}{\mu_0 + \mu} \cdot f(\alpha) \tag{6.2}$$

$$f(a) = b_0 e^{-b_1 \alpha} + a_0 + a_1 a + a_2 a^2 + a_3 a^3 + a_4 a^4 \tag{6.3}$$

In this formula, the phase function, $f(\alpha)$, is separated from the incidence and emission angles. To the first order it is appropriate to use it on the Moon, which has a relatively low albedo and hence singularly scattered light dominates the reflectance. The parameters b_0, b_1, and a_0–a_4 are adjusted to make the phase function fit the data.

6.2.3.2 Data Selection

IIM observed radiance depends on both the observation geometry and the composition of the lunar surface. Multiple observations of the same location at various geometries should be performed to determine the parameters in the photometric function. However, it is difficult to observe any particular area on the lunar surface at various geometries with low-orbit spacecraft observations. Moreover, IIM did not obtain observations at large phase angles, and most of the IIM observations are nadir or near-nadir pointing (with an emission angle between 0° and 4°). Thus, in selecting images to use in the study, it is important to include images from as wide a range of viewing and illumination geometries as possible. Besides, the purpose of this study is to produce a photometric function representative of the entire lunar surface and apply it to the global IIM data. For these reasons, we began our investigation with the global data set. The stability of the IIM instrument response during the mission was checked to ensure that the data used are stable. The average spectra of each revolution was calculated and it was found that there was an obvious difference in the radiance spectra between orbits 2216 and 2217. Consequently, only the data between orbits 2217 and 3000, which represent 92% of the total orbits, were used. During this period the response of IIM was stable.

To reduce the effects caused by the variations in compositions rather than the observation geometry, a new data selection method by using the FeO as the classification criterion was developed. Even at much different solar phase angles, high-Ti basalts and low-Ti basalts can have the same albedo due to the variation of their compositions. Similarly, the fresh rays that are very bright should be separated from the mature highlands materials because they have a different photometric behavior. The global FeO map derived by Wu et al. (2012) was used to separate materials. By careful checking, the maria are classified into two classes according to their FeO contents: Class 3 (7%<FeO<14%) and Class 4 (14%<FeO<21%). Class 2 (2.5%<FeO<7%) is for highland and Class 1 (0.5%<FeO<2.5%) is for fresh bright rays. The pixels outside this range, that is, with FeO less than 0.5% and greater than 21%, were neglected because they were mostly from inclined facets.

To minimize the influence of topography on the matching, the shadowed region was masked during the data extraction. The data extracted with the above method were then processed with a running median filter of 0.1° phase angle resolution for statistically reducing the effects of inclined facets as well as random noise. Moreover, IIM image has some bad pixels, which are singularly and discretely distributed in the IIM image. These bad pixels have values either much larger or much smaller than their immediate neighbors and should be removed. The spectral shape of the bad pixels is much different from the normal pixels. Therefore, multiple bands were used to detect bad pixels with spectral angle mapping or correlation coefficients between adjacent pixels. Subsequently, the bad pixels were replaced by the average value of their immediate neighbors. The linearity of the sensor response is another important factor to consider. The observations of some orbits with small beta angle in low latitude highland areas are saturated. Although some pixels are not saturated, the flux received is well beyond the linearity of the detector. These pixels were not selected for derivation of photometric properties.

6.2.3.3 Correction for Inhomogeneity of Spatial Response

To correct for the inhomogeneity of spatial response, laboratory measurements were acquired across the field of view (FOV) from a uniform source. The in-flight flat field correction was also suggested after the laboratory flat-field correction (e.g., Kodama et al. 2010). However, the usual in-flight flat-field correction often removes the cross-track photometric signal. We provided a method for the correction of heterogeneity of pixel-to-pixel response. The cross-track FOV of IIM is 7.33°, which corresponds to a very small emission angle at nadir, with the largest emission angle of approximately 4°. Therefore, considering the photometric function used in this study (Equations 6.2 and 6.3), if both the incidence angle and phase angle exhibit very small variations across the full cross-track swath (e.g., less than 1°), then the albedo vary along the cross-track can be regarded as due to the nonuniformity of the detector. For this reason, the orbits with beta angles less than 2° were selected. They are symmetrical in distribution with the center line of the zero beta angle. The correction procedures include the following steps:

1. Each pixel within a column is normalized by the average of the whole column.
2. The residual flat-field image is produced by stacking all lines with a phase angle in cross-track swath less than 0.5°. Generally, the relative

response is high at and near the center of the detector and smaller towards both edges. The 918 nm band has the poorest cross-track spectral uniformity with the worst from −0.238 to 0.143.

3. The values of the residual flat-field image data are normalized by dividing the average of each line. It should be noted that, to ensure the precision, the average of each line is calculated with the first 113 samples only because some pixels of these small beta angle orbits on the last 15 samples on the right are saturated.

4. The correction factor is derived by dividing the values of (3), and the correction for the inhomogeneity of spatial response is done by multiplying the correction factor. Most of the optical remote sensing data have wider ranges of emission angles than IIM. For example, Moon Mineralogy Mapper (M^3) has a FOV of 24° (Green et al. 2011), and the Lunar Reconnaissance Orbiter Camera (LROC) Wide Angle Camera (WAC) has a 90° FOV in black and white mode and 60° FOV in 7-color mode (320 nm to 689 nm) (Robinson et al. 2010). So the incidence angles and phase angles vary much more across the full cross-track swath for those instruments than the angles for the IIM. To find a more universal method, we made a second experiment by using the orbits that have larger ranges of incidence angles and phase angles across the cross-track swath. Those orbits have beta angles within 25°–30° and are symmetrical with respect to zero beta angle. The correction for the inhomogeneity of spatial response is done by the same method as above. The result is very similar to the first method. Therefore, the method can also be applied to those instruments with larger FOV than the IIM to decouple the cross-track photometric effects and the inhomogeneity of sensor response. This is because the orbits used for deriving the correction factor are systematically distributed with respect to zero beta angle, the photometric effects can be counteracted by stacking all the lines. The residual pixel-by-pixel variation is only caused by the detector response variations.

6.2.3.4 Correction for Systematic Artifacts

IIM spectra have systematic artifacts. Before applying photometric correction, the artifacts are removed with the multiplication of a correction factor. The factor is the ratio of the smoothed average spectra of the entire data set (excluding problematic pixels) divided by the raw average spectra of the same data set. To find a method of smoothing the IIM average spectra, we averaged the highlands, maria and all the spectra from the RELAB (Reflectance Experiment Laboratory) spectral library (http://www.planetary.brown.edu/pds/LSCCsoil.html). Figure 6.3 shows that all three average spectra (in blue, green, and red) behave similarly, that is, their spectral shapes match a second order polynomial in the visible bands, and a fourth order polynomial in the

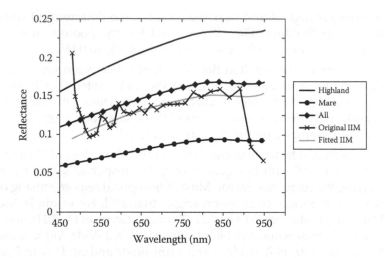

FIGURE 6.3
The raw average spectra of the entire IIM data set (problematic pixels removed) and the smoothed IIM spectra. The average of the mare, highland, and all the spectra from the RELAB data are also shown. It can be seen that the smoothed averaged IIM spectra parallels the average spectra of all the RELAB data set.

near infrared bands. The smoothed average spectra of the entire IIM data set was required to be fitted separately with a second order polynomial in the visible bands and fourth order polynomial in the near infrared bands with the restriction that the fitted curve paralleling to the average spectra of all RELAB spectra. To achieve this, the average spectra of the IIM data set are fitted with a second order polynomial for 531–818 nm and fourth order polynomial for 721–818 nm, and the average spectra of all RELAB spectra are fitted with a second order polynomial for 520–820 nm and fourth order polynomial for 750–950 nm. The smoothed spectra of IIM for 721–946 nm was calculated with the polynomial coefficients from the fourth order polynomial of RELAB data with the difference between the fitted IIM with fourth order polynomial for 721–818 nm and RELAB with fourth order polynomial for 750–950 nm as the constant of the fourth order polynomial. The results show that the final smoothed average spectra of the entire IIM data set parallels to the average spectra of all the RELAB spectra (Figure 6.3) and the resulting R^2 of the fitted polynomial is 0.999 for the visible bands and 0.998 for the near infrared bands.

6.2.3.5 Cross-Calibration of IIM NIR Bands

In order to validate the IIM data processed above, reflectance spectra of IIM were compared with Earth-based telescopic spectra (Pieters and Pratt, 2000). Figure 6.4 shows the IIM spectra with 25 × 25 blocks of 200-m pixels (5 × 5 km) averaged for Aristarchus Plateau3, MS2, Aristarchus South Rim, and Hadley A. It should be noted that the signal-to-noise ratio (SNR) of the shortest and

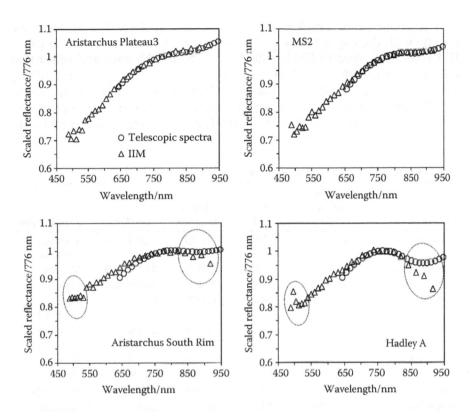

FIGURE 6.4
IIM and Earth-based telescopic spectra of four representative areas including mature (Aristarchus Plateau3 and MS2) and fresh (Aristarchus South Rim and Hadley A) lunar surfaces. To emphasize the spectral properties, these spectra are scaled to unity at 776 nm.

longest bands of IIM (Bands 1 and 32) is very low and they were not shown in Figure 6.4, that is, 30 IIM points in each plot. As shown in Figure 6.4, the remaining first six bands (Bands 2–7) are problematic for unknown reasons. These bands should be removed before they are used. For the mature areas where the near-infrared absorption peak of the mafic minerals is weak, reflectance spectra of the IIM and Earth-based telescopes match well in their mutual spectral range (e.g., Aristarchus Plateau3 and MS2 shown in Figure 6.4). However, for the fresh areas where the near-infrared absorption peak of the mafic minerals is relatively strong, reflectance of the near-infrared wavelength of the IIM is lower than that of Earth-based telescopic spectra (e.g., Aristarchus South Rim and Hadley A shown in Figure 6.4). The longer the wavelength, the lower of the near-infrared reflectance of the IIM is relatively to that of Earth-based telescopes. The near-infrared is particularly important for the FeO prediction because the absorption center of the lunar mafic minerals is near 1000 nm. Therefore, to improve its usability the IIM reflectance spectra must be recalibrated.

In response to the problem mentioned above, all Earth-based telescopic spectra with the highest data quality were used for the cross-calibration of IIM reflectance spectra (Table 6.2). At each band, the gain and offset that minimized the differences between $Telescopic_\lambda/Telescopic_{776}$ and IIM_λ/IIM_{776} was found using a least squares method. The gain and offset was applied to the IIM data to perform the cross-calibration. Table 6.3 shows the gains and offsets of their mutual bands of IIM and telescopic spectra. The improvement

TABLE 6.2

Summary of Sites for Cross-Calibration

Site Name	Telescopic Spectral ID	IIM Revolution	IIIM Location	Lat./Long.	Feature Type
Sulpicius Gallus 2	HA1055	2230	72,6701	20.4N, 9.7E	Mantling material
Apennine Front	HA0819	2856	53,7620	23.1N, 1.3E	Mountain
Sinus Aestuum 3	HA0857	2550	35,9877	10.8N, 3.2W	Mare
Black Spot 2	HA1094	2864	84,9923	8.3N, 8.3W	Mantling material
Aristarchus Plateau1	HA0979	2902	6,7030	27.1N, 52.8W	Mantling material
Vitruvius Floor	HA1088	2521	69,8546	17.7N, 31.2E	Mare
Ms2	HB0916	2220	72,6929	18.7N, 21.4E	Mare
Apollo 16	H90366	2225	71,11159	9.0S, 15.1E	Highlands
Aristarchus Plateau2	HA0130	2902	47,6948	28.4N, 48.9W	Mantling material
Aristarchus Plateau3	HA1032	2899	40,6984	27.5N, 52.3W	Mantling material
Hadley A	HA0811	2852	108,7344	25.0N, 6.6E	Crater
Aristarchus South Rim	HA1040	2898	60,7693	23.3N, 47.7W	Crater feature (WALL)
Mare Serentatis 2 AVG.	HC0033	2220	86,6915	28.7N, 21.4E	Mare
Littrow NR	HA1082	2521	67,7840	22.3N, 31.2E	Mountain
Apollo 16 sample 62231.1 (smooth)	HC0028	2225	125,11161	9S, 15.1E	Highlands
Apollo 14A	HC1192	2252	70,10191	2.8S, 16.8W	Highlands
Apollo 14B	HC1194	2252	101,10250	3.2S, 16.6W	Highlands
Copernicus Mare 5	HC0758	2569	39,8357	22N, 25.2W	Mare
Copernicus Ray 6	HD0352	2562	61,7800	24N, 16.8W	Crater feature (Ray)
Aratus	HA0809	2853	3,7661	23.6N, 4.5W	Crater (Highland)
Aristarchus East Wall	HA1038	2897	12,7511	23.8N, 46.8W	Crater feature (WALL)
Aristarchus Peak	HA0971	2898	90,7598	23.7N, 47.5W	Central peaks

TABLE 6.3

Cross-Calibration Parameters of IIM and Statistic Information before and after the Cross-Calibration[b]

Band(nm)	Gain	Offset	RMSE(Before)	RMSE(After)	R[a]
673.28	0.651388	0.316103	0.0132471	0.00701614	0.909077
688.582	0.724891	0.248255	0.0147996	0.0065062	0.906617
704.595	0.690325	0.292375	0.00903555	0.0060937	0.88352
721.371	0.634521	0.351801	0.00801728	0.00503132	0.875551
738.966	0.549169	0.441591	0.00673573	0.00401339	0.84518
757.44	0.662994	0.333749	0.00294673	0.00234847	0.8068
797.305	0.8821	0.112839	0.00730744	0.00295562	0.867823
818.854	0.914933	0.075334	0.0125673	0.00499281	0.897621
841.6	0.919126	0.078665	0.00785591	0.00695857	0.933322
865.646	0.81787	0.192553	0.0140016	0.00805246	0.944487
891.106	0.916068	0.109767	0.028779	0.0107007	0.941614
918.109	0.688302	0.364343	0.0684196	0.0176575	0.88011

[a] Correlation coefficients between IIM and telescopic spectra.
[b] Only bands longer than 776 nm were cross-calibrated.

can be seen from the root mean square error (RMSE), which is decreased significantly for the longer bands (e.g., 918 and 891 nm). Figure 6.5 shows the comparison of the reflectance spectra between Earth-based telescopic and the cross-calibrated IIM. It can be seen that for whatever the mature or fresh areas reflectance spectra of the cross-calibrated IIM and Earth-based telescopes match well in their mutual spectral range. The reflectance of the fresh areas at the longer bands such as 891 and 918 nm were improved much.

6.3   Prediction of FeO and TiO_2

Iron and titanium are two principal elements that dominate the spectral features of the Moon so they are also the two most commonly estimated elements by reflectance spectra. We first discuss their predictions with IIM data. Based on the successful prediction of the two spectrally dominant elements, we will further discuss the prediction for all the major elements with IIM data.

6.3.1   Inversion of FeO Content

Quantitative assessment of the lunar compositions with the reflectance spectra needs the ground-truth data for the calibration. Blewett et al. (1997) and Lucey et al. (2000) used 40 and 47 samples for the calibration with

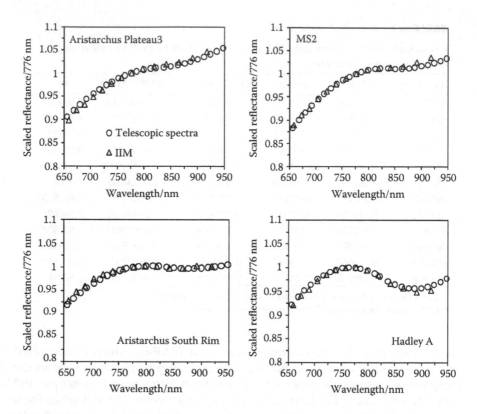

FIGURE 6.5
Cross-calibration IIM and Earth-based telescopic spectra. To emphasize the spectral properties, these spectra are scaled to unity at 776 nm.

Clementine data, respectively. IIM data did not cover the Apollo 15 area. Moreover, the spatial resolution of IIM is 200 m, which is lower than that of Clementine spectral data. Some individual sampling locations for Apollo 16 and Apollo 17 cannot be reliably resolved in the IIM images. Though the latitude and longitude of each station has been shown in Table 6.1 in Lucey et al. (2000), we manually extracted the reflectance spectra of each station from the IIM images by comparison with published traverse maps and with the help of Apollo Pan photography. For some individual station, the reliability for the identification of the station from IIM image is very low, we would rather abandon it in order not to reduce the degree of confidence. In order to reduce the noise of IIM data, the pixels of each sampling station were averaged. Boxes used for averaging pixels of each sampling station are variable as their case might be. After careful selection, 17 sites with high confidence were selected from the IIM images (Table 6.4 except NFH).

The Lucey method for the inversion of FeO relies on a plot of the NIR/VIS ratio versus the VIS reflectance. The NIR band used by IIM data is 918 nm (the second longest band of the IIM data) because the SNR of the longest

TABLE 6.4

Major Elements Abundances for Lunar Sample-Return Stations and Northern Farside Highlands (NFH) Used for the Prediction

Site	Pixels Averaged	FeO (wt.%)	TiO$_2$ (wt.%)	MgO (wt.%)	CaO (wt.%)	Al$_2$O$_3$ (wt.%)	SiO$_2$ (wt.%)	Sample	References
A11	3 × 3	15.8	7.5	7.81	12.01	13.45	41.86	10002.10010.10084 12001.12003.12023.120	Rhodes and Blanchard (1981), Heiken et al. (1991)
A12	4 × 4	15.4	3.1	9.66	10.58	13.86	45.62	30.12032.12033.12034.1 2037.12041.12042.1204 4.12057.12070 14003.14148.14149.141	Heiken et al. (1991), Frondel et al. (1971)
A14 LM-Cone	4 × 4	10.4	1.67	9.29	11.12	17.57	47.94	56.14049.14163.14240.1 4259.14421 61141.61161.61241.612 81.61501.62241.62281.6	Heiken et al. (1991), Rose et al. (1972)
A16S1-9	10 × 10	5.5	0.61	6.04	15.51	26.67	45.07	4421.64501.65501.6570 1.65901.66041.66081.68 121.68501.68821.68841. 69921.69941.69961	Korotev (1981)
A16S11	3 × 3	4.2	0.4	4.3	16.5	28.9	45.1	61141.61161.61241.612 81.61501	Korotev (1981)
A16S13	4 × 4	4.8	0.5	5.4	15.8	27.6	45.1	63321.63341.63501	Korotev (1981)

(Continued)

TABLE 6.4 (Continued)

Major Elements Abundances for Lunar Sample-return Stations and Northern Farside Highlands (NFH) Used for the Prediction

Site	Pixels Averaged	FeO (wt.%)	TiO$_2$ (wt.%)	MgO (wt.%)	CaO (wt.%)	Al$_2$O$_3$ (wt.%)	SiO$_2$ (wt.%)	Sample	References
A17LM	2 × 2	16.6	8.5	9.8	11.04	12.07	40.73	70019.70161.70181.700 11	Rhodes et al. (1974), Rose et al. (1974)
A17S1	2 × 2	17.8	9.6	9.62	10.75	10.87	39.93	71501.71041.71061.711 31.71151	Rhodes et al. (1974), Korotev and Kremser (1992)
A17S3	2 × 2	8.7	1.8	10.25	12.89	20.29	44.94	73221.73241.73261.732 81	Rose et al. (1974)
A17S5	2 × 2	17.7	9.9	9.51	10.85	10.97	39.86	75061.75081	Rhodes et al. (1974)
A17S6-7	3 × 3	11.15	3.65	10.54	12.05	17.67	43.3	76241.76261.76281.763	Rhodes et al. (1974)
A17S8	2 × 2	12.3	4.3	9.91	11.77	15.73	42.67	21.76501.77531	Rhodes et al. (1974)
A17LRV7-8	3 × 3	15.9	6.7	10.06	11	13.1	41.85	78501	Korotev and Kremser (1992)
A17LRV12	2 × 2	17.4	10	9.36	10.7	11.15	39.9	75111.75121	Korotev and Kremser (1992)
Luna16	5 × 5	16.7	3.3	8.8	12.5	15.3	41.7	70311.70321	Heiken et al. (1991)
Luna20	5 × 5	7.5	0.5	9.8	15.1	22.3	45.1		Heiken et al. (1991)
Luna24	5 × 5	19.6	1	9.4	12.3	12.5	43.9		Heiken et al. (1991)
NFH	10 × 10	3.8	0.16	4.28	18.47	29.4	43.9	NWA 482	Korotev et al. (2003)

band (946 nm) is very low. The key to the iron parameter, the location of the "hypermature" end-member, was selected according to maximizing the correlation between θ_{Fe} and the FeO content of the lunar soils and minimizing the residual effects of maturity. However, many possible values of the "hypermature" end-member gave similarly satisfied results. Ultimately, the same origin as that used by Lucey et al. (2000) was used and the same origin could have some implications, though for different data and wavelengths. The Fe parameter algorithm and regression equation are:

$$\theta_{Fe} = -\arctan\left[\frac{(R_{918} / R_{757}) - 1.19}{R_{757} - 0.08}\right] \tag{6.4}$$

In contrast to Lucey et al. (1995) and subsequent papers, we used a power function for the regression instead of the linear function, which was often used in previous research. Lawrence et al. (2002) showed that a linear relationship between θ_{Fe} and the FeO content is not correct for the full range of lunar iron contents. Figure 6.6 shows the plot of station FeO against the spectral Fe parameter. It can be seen that a power function results in a higher correlation than that of the linear function and avoids negative values that can be caused by the linear function. The standard deviation for the accuracy of FeO determination with IIM is 1.29 wt.% FeO. The mapping of the spectral Fe parameter to FeO follows:

$$wt.\%FeO = 8.878 \times \theta_{Fe}^{1.8732} \tag{6.5}$$

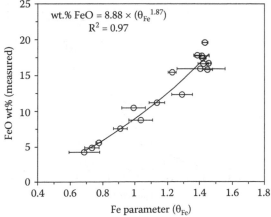

FIGURE 6.6
Plot of FeO content measured for returned lunar soils versus the spectral iron parameter derived from remote measurements by IIM. The equation of the best fit and correlation coefficient is indicated. The error bars are shown with 2× standard error encompassing more than 95% of the sample.

6.3.2 Inversion of TiO$_2$ Content

The Lucey method for the inversion of TiO$_2$ relies on a plot of the UV/VIS ratio versus the VIS reflectance (Lucey et al. 1998, 2000). Because the available ground-based telescopic data do not cover the shortest IIM wavelengths, the uncross-calibrated IIM data were used for the inversion of TiO$_2$ content. As stated above and shown in Figure 6.4, the first seven bands are problematic. Moreover, to ensure the quality of the data are as good as possible for the application the remaining first two bands (Bands 8–9) were also removed due to their low SNR. Hence, the UV band in the titanium-sensitive parameter was replaced with 561 nm (Band 10). The location of the optimized origin was determined in a manner similar to that of the iron origin. The Ti parameter algorithm and regression equation are:

$$\theta_{Ti} = -\arctan\left[\frac{R_{561} / R_{757} - 0.71}{R_{757} - 0.07}\right] \tag{6.6}$$

$$TiO_2 = 2.6275 \times \theta_{Ti}^{4.2964} \tag{6.7}$$

Figure 6.7 shows a plot of station TiO$_2$ against the spectral Ti parameter. It can be seen that the points for Apollo 11, Luna 16, and Luna 24 are significantly deviated from the trend of the other sampling sites and stations. The same discrepancies have been noted by previous researchers using Clementine data (e.g., Lucey et al. 2000; Gillis et al. 2003) so IIM demonstrates

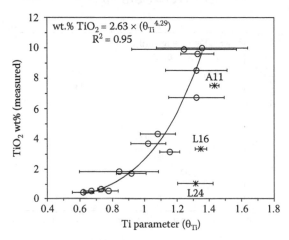

FIGURE 6.7

Plot of TiO$_2$ content measured for returned lunar soils versus the spectral titanium parameter derived from remote measurements by IIM. The equation of the best fit and correlation coefficient is indicated. The Apollo 11, Luna 16, and Luna 24 samples lie well off this solid trend line and they were excluded. The error bars are shown with 2× standard error encompassing more than 95% of the sample.

that this characteristic is a feature of lunar reflectance and not a calibration artifact in the Clementine data.

Though the wavelength of IIM data used for predicting TiO_2 (561 nm) is significantly longer than that of Clementine band (415 nm), the relationships between the spectral titanium parameter and the measured TiO_2 for IIM data and Clementine are strikingly similar (Figure 6.7). The Ti parameter of IIM is highly correlated with TiO_2 content with $R^2 = 0.95$. The standard deviation value for the fit is 1.21 wt.% TiO_2, similar to that of the Clementine data (Lucey et al. 2000). This correlation supports the hypothesis that the spectral influence of titanium is due to the low reflectance of ilmenite, not a UV absorption process (Gillis et al. 2003). This can also be seen from similar maps derived from IIM and Clementine and the good correlation between FeO and TiO_2.

6.3.3 Inversion of Nontransition Elements

The partial least squares regression (PLSR) method was used to predict all the elemental abundances. PLSR is a bilinear model developed by Herman Wold under the name of nonlinear iterative partial least squares (NIPALS) (Wold, 1966a, 1966b). The general underlying model of multivariate PLSR is

$$X = TP^T + E \tag{6.8}$$

$$Y = UQ^T + F, \tag{6.9}$$

where X is an n × m matrix of predictors, Y is an n × p matrix of responses; T and U are n × l matrices that are, respectively, projections of X (the X score, component or factor matrix) and projections of Y (the Y scores); P and Q are, respectively, m × l and p × l orthogonal loading matrices; and matrices E and F are the error terms. The decompositions of X and Y are made so as to maximize the covariance of T and U.

There are two types of PLSR algorithms. PLS1 deals with only one response at a time, while PLS2 can handle several responses at a time. The PLS1 was used to predict each element at one time. PLSR does not require all the components to be retained in the regression model. Incorporating too many components often complicates interpretation of the model and increases the risk of overfitting. A very important issue in building the PLSR model is the choice of the optimal number of components. The most common methods use a test set or cross-validation. Because a limited number of samples were available for lunar research, a leave-one-out cross-validation procedure was adopted to determine the optimal number of components. With leave-one-out cross-validation, one sample was left out of the global data set and the model was calculated on the remaining data points. The value for the left-out sample was then predicted, and the prediction residual was computed.

The process was repeated with another sample of the data set, and so on, until every sample had been left out once; then all prediction residuals were combined to compute the root-mean-square error of prediction (RMSEP). The optimum number of terms was taken as the number resulting in the minimum RMSEP. For a comparison between the calibration and validation, at the calibration stage, a measurement of the average difference between predicted and measured response values, that is, RMSEC, was also calculated.

The prediction accuracy of chemical composition with optical spectroscopy by means of empirical methods depends on the input data. The samples of the calibration data set should uniformly distribute and cover the possible abundance range of the material that is going to be analyzed. Among all the Apollo and Luna landing sites, the Apollo 16 landing site is the only one landing in the highlands. However, it is not typical of the highlands crust with high FeO while low Al_2O_3 (Lucey et al. 1995). The previous IIM results also show that the highlands of the south hemisphere (e.g., Apollo 16 site) have higher FeO than the northern far-side highland, which has the lowest concentration of FeO of the Moon (Wu et al. 2012). For mapping compositions of the entire Moon, expansion of the ground truth data to include a broader range of soil compositions, especially to highland soils that have high Al and Ca and low Fe and Ti, is strongly needed. Several studies have demonstrated that the compositions of feldspar-rich lunar meteorites can be used as proxies for the composition of the far-side lunar highlands (e.g., Korotev et al. 2003; Gillis et al. 2004). To compensate for the lack of the highland samples and extend the abundance range of Al_2O_3 and CaO of the calibration data set, the feldspathic lunar meteorite was used as a proxy for northern lunar far-side highlands. Although the mean composition of several lunar meteorites provides a better estimate of the far-side highlands than does one meteorite alone, the meteorite with the highest Al_2O_3 and CaO and the lowest FeO and TiO_2 is the optimal choice in order to cover the largest abundance range of the calibration data set. Based on this criterion, the meteorite of northwest Africa (NWA) 482 was selected to serve as anchors at the low-concentration ends of Fe and Mg and at the high-concentration ends of Al and Ca (Table 6.4). In a comprehensive analysis of the samples shown in Table 6.4, Apollo 16 and NWA 482 provide constraints for the feldspathic highland compositions, Apollo 12 and 14 provide constraints for the low-Ti mare compositions, and Apollo 11 and 17 provide constraints for high-Fe, high-Ti basalt compositions. Before establishing the regression model, an exploratory analysis was carried out to detect outliers. The outliers can decrease the prediction accuracy; hence, they should be omitted. After trial and error, the remaining data set, which give the highest prediction accuracy, was used for building the model.

Table 6.5 shows the prediction results of the optimal PLSR models, which were determined by the least RMSEP at the validation stage. In addition to the RMSEC and RMSEP, the prediction accuracy is also described by the correlation coefficient (R), the regression slope, and the offset of the predicted vs. measured values. On the whole, the PLSR method gives very good

TABLE 6.5

Prediction Results of the Optimal PLSR Models for the Six Major Elements

	FeO (wt.%)	TiO$_2$ (wt.%)	MgO (wt.%)	CaO (wt.%)	Al$_2$O$_3$ (wt.%)	SiO$_2$ (wt.%)
Calibration						
R	0.95	0.83	0.88	0.97	0.96	0.87
RMSEC	1.58	2.00	1.00	0.65	1.68	0.75
Slope	0.90	0.68	0.78	0.93	0.92	0.76
Offset	1.29	1.43	1.86	0.86	1.36	10.21
Validation						
R	0.94	0.77	0.85	0.96	0.95	0.82
RMSEP	1.76	2.26	1.14	0.74	1.92	0.91
Slope	0.89	0.66	0.78	0.92	0.91	0.73
Offset	1.34	1.48	1.86	1.03	1.63	11.52

predictions for all the six elements with the correlation coefficients ranging from the minimum of 0.77 for TiO$_2$ to the maximum of 0.96 for CaO. From the correlation coefficients, it can be seen that the prediction accuracy for FeO, Al$_2$O$_3$, and CaO is higher than MgO, TiO$_2$, and SiO$_2$. This can also be verified from the regression slope and offset (MgO, TiO$_2$, and SiO$_2$ have lower slope and higher offset than those of FeO, Al$_2$O$_3$, and CaO).

However, as shown in Table 6.5, the regression slope is less than one and the offset greater than zero for all the elements, which caused an overprediction for the lower elemental abundances and underprediction for the higher elemental abundances. So for the final mapping of the global elements the models were readjusted to make the offset zero and slope unity by subtracting the offset and dividing by the slope of Table 6.5.

6.4 Abundance and Distribution of Major Elements on the Moon

6.4.1 Global Elemental Abundances

Figure 6.8 shows the global distribution of the six major elements derived with the modified PLSR model and IIM data at 200 m/pixel spatial sampling at the equator. For each plot, the lower limit of the legend (color bar) encompasses all the values below this limit and the higher limit of the legend encompasses all the values above this limit. The legend is limited because the two ends (very negative and very positive) are meaningless because the linear model suffered from topographic effects and very bright ejecta.

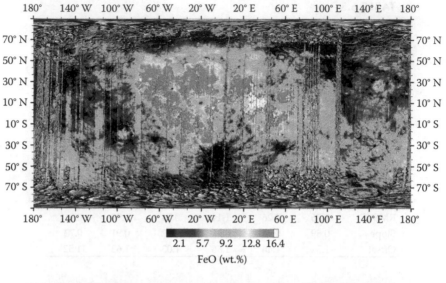

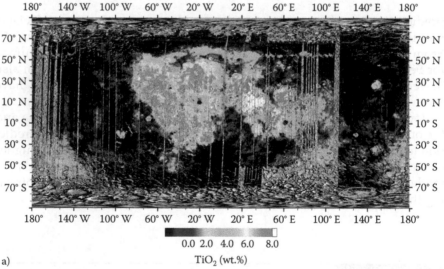

FIGURE 6.8
Equidistant cylindrical maps of major elemental abundances determined by IIM data. For each plot, the lower limit of the legend (color bar) encompasses all the values below this limit and the higher limit of the legend encompasses all the values above this limit. (*Continued*)

The 200-m resolution global maps of major elements derived from IIM data enable us to explore their statistical characteristics. The histogram reveals a unimodal distribution of TiO_2 abundances in the lunar maria (Figure 6.9). A unimodal distribution and the IIM-derived abundances in the maria are consistent with previous results (e.g., Giguere et al. 2000; Gillis et al. 2003), which

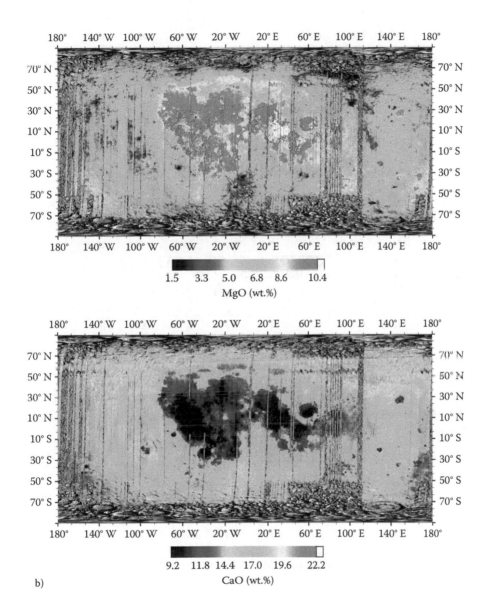

b)

FIGURE 6.8 (*Continued*)

demonstrates an apparent sampling bias in the mare basalt collection because TiO_2 abundances in the samples have a bimodal distribution (Giguere et al. 2000). Except for Ti, all the other elements exhibit bimodal distributions of their histograms of the entire Moon, which corresponds to the mare and highlands. The lower modal Fe abundance of ~5.57 FeO wt.% corresponds to the highland area and is close to the abundance of ~5.7 wt.% given by Gillis et al. (2004) and ~5.5 FeO wt.% derived by Lawrence et al. (2002) but higher

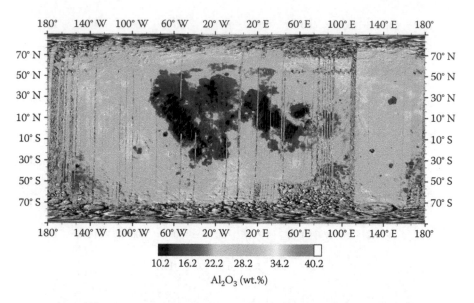

Al_2O_3 (wt.%)

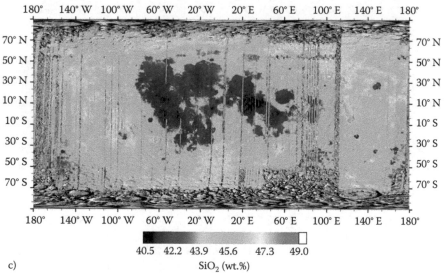

c)

SiO_2 (wt.%)

FIGURE 6.8 (Continued)

than the modal abundance of ~4.5 wt.% given by Lucey et al. (1998). For mare basalts, the FeO abundances derived from IIM are a little lower than that derived from LP-GRS by Lawrence et al. (2002) and Prettyman et al. (2006). Figure 6.9 shows that the IIM-derived FeO upper limit is ~21 wt.% while GRS FeO can be as high as ~25 wt.%. By checking all the Apollo and Luna samples, the most basalt-rich soils only reach ~20 wt.% FeO (Heiken et al. 1991). As argued by Gillis et al. (2004), not only do the extremely high FeO values in the

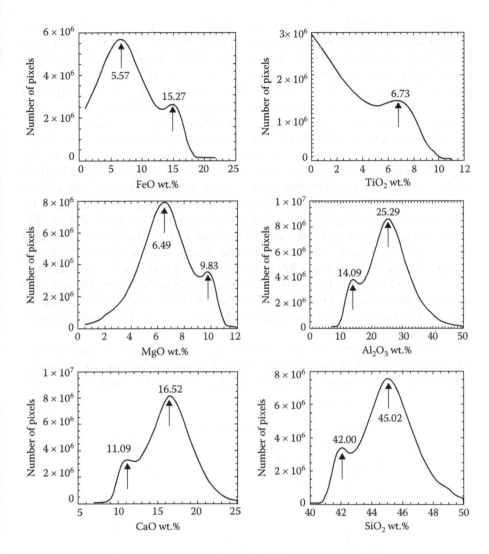

FIGURE 6.9
Histograms of elemental abundances for the entire Moon derived from the modified PLSR model.

LP-GRS data set exceed those expected on the basis of sample data, but petrologic modeling and basaltic phase equilibria do not support the existence of extensive regions of mare basalts with FeO concentrations above ~22 wt.%. Some local areas show differences between IIM-derived and LP-GRS elemental abundances, however, the average values of the major elements for the whole Moon is approximately consistent between the two data sets. As can be seen in Figure 6.10, the correlation coefficient between IIM measured and LP-GRS measured average abundance is very high ($R^2 = 0.99$). Moreover, the slope of the linear relationship is close to one and the offset is close to zero.

Although the PLSR model has been modified by means of the normal-ization of the regression slope and offset, systematic errors still appear to exist. For example, as discussed above the Al_2O_3 abundances in the maria seem higher and FeO lower compared to previous results. There are two possible explanations. One is due to the limited samples. For the empirical model, a wide range of known compositions in the calibration data set is necessary, especially applied to the entire Moon. Moreover, the exact posi-tion of the meteorite of NWA 482 is unknowable and hence the correspond-ing IIM spectra may have caused inaccuracies. The second reason is due to the limited band range. The spectral range used in this study is limited within 561 to 918 nm. For this issue, we further investigated the effects of different spectral ranges on the prediction accuracy using LSCC data. As can be seen in Figure 6.11 for whatever spectral sections, the predic-tion accuracy with a spectral range of 500 nm all have low slope and large offset. Although the R^2 derived from the spectral range of 1500–2000 nm is close to that derived from 500–2500 nm, its slope and offset depart from those derived from the later. The wide spectral range between 500 and 2500 nm has a slope close to one and offset close to zero. These results suggest that an instrument with a wider spectral range, such as M^3 would produce more accurate predictions.

Figure 6.12 shows the average values of the six major elements for four rep-resentative areas: mare basalts (include all the maria, dark mantle deposits, craters filled with basalts or pyroclastic deposits), NFH, southern nearside

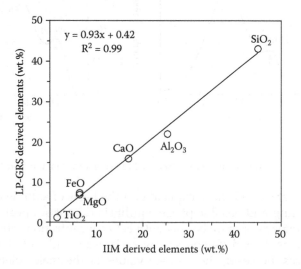

FIGURE 6.10
IIM-measured vs. GRS-measured average abundances of all major elements for the entire Moon. (Prettyman, et al. (2006). Elemental composition of the lunar surface. Analysis of gamma ray spectroscopy data from Lunar Prospector. *J. Geophys. Res.* 111(E12), E12007.)

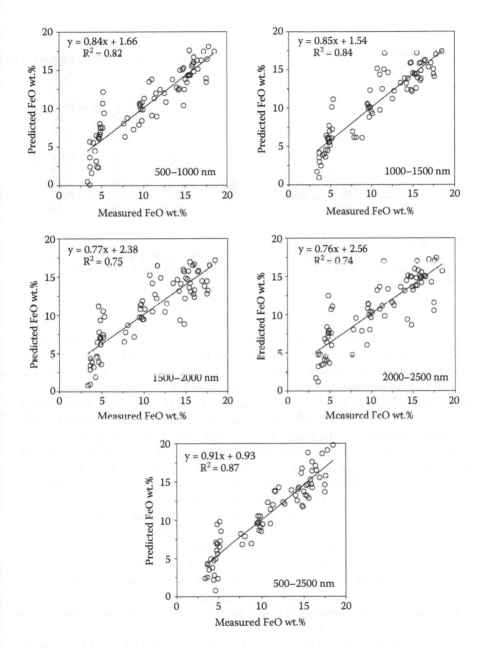

FIGURE 6.11

Plots of measured vs. predicted concentrations of FeO for five spectral ranges. In order to compare between the different spectral ranges the coefficients of determination (R^2) and the regression slope and offset are shown.

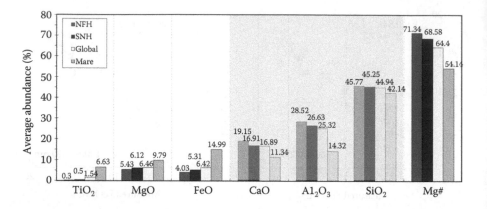

FIGURE 6.12

Average abundances of the lunar major elements and Mg# derived from the modified PLSR model and IIM data. *Abbreviations*: NFH, North far-side highland; SNH, South nearside highland.

highlands (SNH), and the entire Moon. The pixels, which are affected by the topography and overexposure, were eliminated by mask during calculating the average abundance. Because the TiO_2 is very low for the highland areas for the calculation of the wt% TiO_2 average for the entire Moon, the 0.3% and 0.5% was used (Wu et al. 2012) for NFH and SNH, respectively. As can be seen in Figure 6.12, the order of the average abundances of the six major elements of the highlands are $SiO_2 > Al_2O_3 > CaO > MgO > FeO > TiO_2$, while their orders of the mare are $SiO_2 > FeO > Al_2O_3 > CaO > MgO > TiO_2$. For both mare and highlands, the SiO_2 average abundance is the highest and the TiO_2 average abundance is the lowest. For SiO_2 the difference between highlands and mare is the smallest among the six elements. The average Al_2O_3 of the highlands is much higher than FeO, however, the average abundances of Al_2O_3 and FeO for mare basalts are almost the same. In the mare areas, FeO is higher than MgO while in the highland areas MgO is higher than FeO. But for the entire Moon, FeO and MgO is almost the same. With the global elemental maps derived from IIM data, the average abundance of the lunar major oxides, lunar major elements, and the molar percentage of the lunar major elements for any terrane and the entire Moon can be estimated (Figures 6.13–6.15).

The difference between the two highlands areas, SNH and NFH, is worth discussing. The LMO hypothesis, in which the Moon's crust formed through the flotation of plagioclase in a global fractionation event (e.g., Wood et al. 1970), implies that the lunar crust is uniform. However, the one-way analysis of variance (one-way ANOVA) followed Wu et al. (2012) for testing the significance difference between the two groups shows that the highlands between nearside and far side are significantly different for all the elements

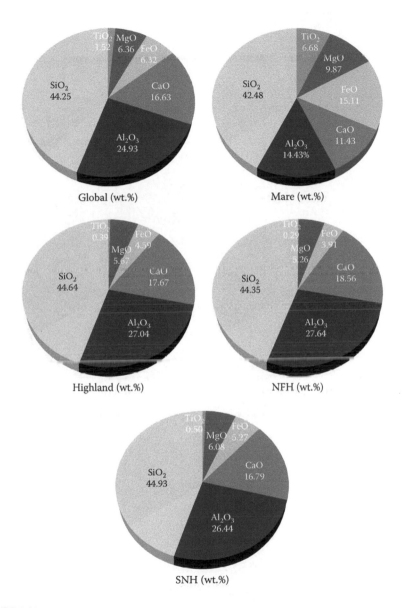

FIGURE 6.13
Pie chart showing the average abundances of the lunar major oxides.

($p < 0.01$). The highlands of the south hemisphere (SNH) have higher FeO, MgO, and TiO_2 but lower CaO, Al_2O_3, and SiO_2 than the northern far-side highland (NFH). The results of this study indicate that the lunar highland crust is inhomogenous on the global scale and supports the viewpoint of the crustal heterogeneity derived from the mineralogy (Pieters 1993; Tompkins and Pieters 1999).

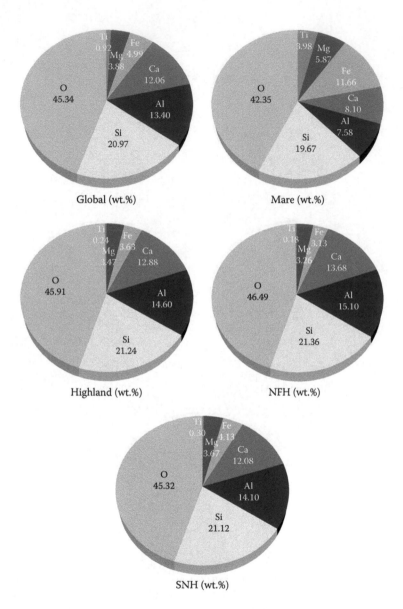

FIGURE 6.14
Pie chart showing the average abundances of the lunar major elements.

6.4.2 Global Mg#

With the major elements derived in this study the molar or atom ratio of Mg/(Mg+Fe), which is a very useful quantity for studying how composition changes in a magma as it crystallizes, can be acquired at a spatial resolution

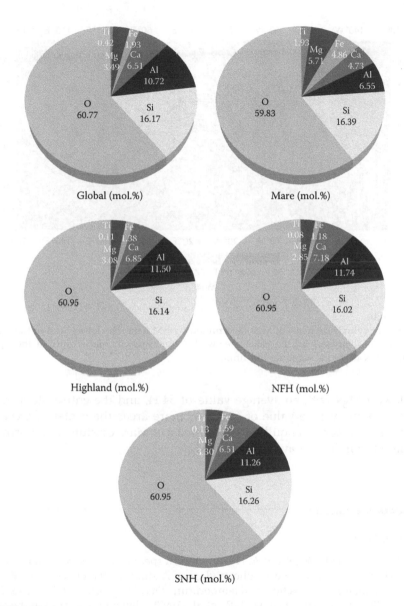

FIGURE 6.15
Pie chart showing the molar percentage of the lunar major elements.

of 200 m. Figure 6.16 shows the 200-m resolution global map of Mg# determined by IIM data. It can be seen that Mg# of the highlands is higher than that of the mare, which is consistent with the fact that in the mare areas FeO is higher than MgO while in the highland areas MgO is higher than FeO. NFH have the highest Mg# with an average value of 71.34, mare basalts have

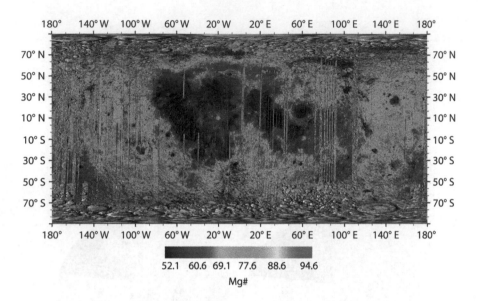

FIGURE 6.16
Equidistant cylindrical maps of Mg# determined by IIM data. The lower limit of the legend (color bar) encompasses all the values below this limit and the higher limit of the legend encompasses all the values above this limit.

the lowest Mg# with an average value of 54.14, and the entire Moon has a Mg# with an average value of 64.4. For mare areas the center of Oceanus Procellarum, Mare Tranquillitatis, southeast of Mare Crisium, and northeast of Mare Fecunditatis have the lowest Mg#.

References

Bell, P.M., Mao, H.K., Weeks, R.A. (1976). Optical spectra and electron paramagnetic resonance of lunar and synthetic glasses: A study of the effects of controlled atmosphere, composition, and temperature. *Proc. Lunar Sci. Conf. 7th*, 2543–2559.

Blewett, D.T., Lucey, P.G., Hawke, B.R., et al. (1997). Clementine images of the lunar sample-return stations: Refinement of FeO and TiO_2 mapping techniques. *J. Geophys. Res.* 102(E7), 16319–16326.

Bugiolacchi, R., Spudi, P.D., Guest, J.E. (2006). Stratigraphy and composition of lava flows in Mare Nubium and Mare Cognitum. *Meteorit. Planet. Sci.* 41(2), 285–304.

Burns, R.G. (1993). *Mineralogical Application of Crystal Field Theory*. Cambridge University Press, Cambridge.

Charette, M.P., McCord, T.B., Pieters, C.M., et al. (1974). Application of remote spectral reflectance measurements to lunar geology classification and determination of titanium content of lunar soils. *J. Geophys. Res.* 79(11), 1605–1613.

Charette, M.P., Taylor, S.R., Adams, J.B., et al. (1977). The detection of soils of Fra Mauro basalt and anorthositic gabbro composition in the lunar highlands by remote spectral reflectance techniques. *Proc. Lunar Sci. Conf. 8th*, 1049–1061.

Elphic, R.C., Lawrence, D.J., Feldman, W.C., et al. (2002). Lunar Prospector neutron spectrometer constraints on TiO_2. *J. Geophys. Res.* 107(E4), 5024.

Fischer, E.M., Pieters, C.M. (1995). Lunar surface aluminum and iron concentration from Galileo solid state imaging data and the mixing of mare and highland materials. *J. Geophys. Res.* 100(E11), 23279–23290.

Frondel, C., Klein, C. Jr., Ito J. (1971). Mineralogical and chemical data on Apollo 12 lunar fines. *Proc. Lunar Sci. Conf.* 2, 719–726.

Giguere, T.A., Taylor, G.J., Hawke, B.R., et al. (2000). The titanium contents of lunar mare basalts. *Meteorit. Planet. Sci.* 35(1), 193–200.

Gillis, J.J., Jolliff, B.L., Elphic, R.C. (2003). A revised algorithm for calculating TiO2 from Clementine UVVIS data: A synthesis of rock, soil, and remotely sensed TiO2 concentrations. *J. Geophys. Res.* 108(E2), 5009.

Gillis, J.J., Jolliff, B.L., Korotev, R.L. (2004). Lunar surface geochemistry: Global concentrations of Th, K, and FeO as derived from lunar prospector and Clementine data. *Geochim. Cosmochim. Acta* 68(18), 3791–3805.

Gillis, J.J., Lucey, P.G., Hawke, B.R. (2006). Testing the relation between UV-vis color and TiO2 content of the lunar maria. *Geochim. Cosmochim. Acta* 70(24), 6079–6102.

Green, R.O., Pieters, C., Mouroulis, P., et al. (2011). The Moon Mineralogy Mapper (M3) imaging spectrometer for lunar science: Instrument description, calibration, on orbit measurements, science data calibration and on orbit validation. *J. Geophys. Res.* 116(E10), E00G19.

Heiken, G., Vaniman, D.T., French, B.M. (1991). *Lunar Sourcebook: A User's Guide to the Moon.* Cambridge University Press, Cambridge.

Hiesinger, H., Head, J.W. III, Wolf, U., et al. (2010). Ages and stratigraphy of lunar mare basalts in Mare Frigoris and other nearside maria based on crater size frequency distribution measurements. *J. Geophys. Res.* 115(E3), E03003.

Jaumann, R. (1991). Spectral-chemical analysis of lunar surface materials. *J. Geophys. Res.* 96(E5), 22793–22807.

Johnson, J.R., Larson, S.M., Mosher, J.A. (1977). A TiO2 abundance map for the northern maria. *Proc. Lunar. Sci. Conf. 8th*, 1, 1029–1036.

Johnson, J.R., Larson, S.M., Singer, R.B. (1991). Remote sensing of potential lunar resources: 1. Near-side compositional properties. *J. Geophys. Res.* 96(E3), 18861–18882.

Kodama, S., Ohtake, M., Yokota, Y., et al. (2010). Characterization of Multiband Imager aboard SELENE: Preflight and in-flight radiometric calibration. *Space Sci. Rev.* 154(1–4), 79–102.

Korotev, R.L. (1981). Compositional trends in Apollo 16 soils. *Proc. Lunar Planet. Sci. Conf. 12th*, 577–605.

Korotev, R.L., Kremser, D.T. (1992). Compositional variations in Apollo 17 soils and their relationship to the geology of the Taurus–Littrow Site. *Proc. Lunar Planet Sci.* 22, 275–301.

Korotev, R.L., Jolliff, B.L., Zeigler, R.A., et al. (2003). Feldspathic lunar meteorites and their implications for compositional remote sensing of the lunar surface and the composition of the lunar crust. *Meteorit. Planet. Sci.* 67(24), 4895–4923.

Lawrence, D.J., Feldman, W.C., Elphic R.C., et al. (2002). Iron abundances on the lunar surface as measured by the Lunar Prospector gamma-ray and neutron spectrometers. *J. Geophys. Res.* 107(E12), 5130.

Li, L. (2006). Partial least squares modeling to quantify lunar soil composition with hyperspectral reflectance measurements. *J. Geophys. Res.* 111, E04002.

Lucey, P.G., Taylor, G.J., Malaret, E. (1995). Abundance and distribution of iron on the moon. *Science* 268(5214), 1150–1153.

Lucey, P.G., Blewett, D.T., Johnson, J.L., et al. (1996). Lunar titanium content from UV -VIS measurements (abstract). *Lunar Planet. Sci.* XXVII, 781–782.

Lucey, P.G., Blewett, D.T., Hawke, B.R. (1998). Mapping the FeO and TiO$_2$ content of the lunar surface with multispectral imagery. *J. Geophys. Res.* 103(2), 3679–3699.

Lucey, P.G., Blewett, D.T., Jolliff, B.L. (2000). Lunar iron and titanium abundance algorithms based on final processing Clementine UVVIS images. *J. Geophys. Res.* 105(E8), 20297–20305.

Melendrez, D.E., Johnson, J.R., Larson, S.M., et al. (1994). Remote sensing of potential lunar resources, 2, High spatial resolution mapping of spectral reflectance ratios and implications for near side mare TiO2 content. *J. Geophys. Res.* 99(E3), 5601–5619.

Pieters, C.M. (1993). Compositional diversity and stratigraphy of the Lunar crust derived from reflectance spectroscopy. In *Remote Geochemical Analysis: Elemental and Mineralogical Composition* (eds. C. Pieters and P. Englert). Cambridge University Press, Cambridge, U.K., pp. 309–339.

Pieters, C.M., Pratt, S. (2000). Earth-based near-infrared collection of spectra for the Moon: A new PDS data set. *Lunar Planet. Sci.* XXXI, Abstract, 2059.

Pieters, C.M., Stankevich, D.G., Shkuratov, Y.G., et al. (2002). Statistical analysis of the links among lunar mare soil mineralogy, chemistry, and reflectance spectra. *Icarus* 155(2), 285–298.

Prettyman, T.H., Hagerty, J.J., Elphic, R.C., et al. (2006). Elemental composition of the lunar surface. Analysis of gamma ray spectroscopy data from Lunar Prospector. *J. Geophys. Res.* 111, E12007.

Rhodes, J.M., Rodgers, K.V., Shih, C., et al. (1974). The relationship between geology and soil chemistry at the Apollo 17 landing site. *Proc. Lunar Sci. Conf.* 5, 1097–1117.

Rhodes, J.M., Blanchard, D.P. (1981). Apollo 11 breccia and soils: Aluminous mare basalts or multi-component mixtures. *Proc. Lunar Planet. Sci. Conf.* 12, 607–618.

Robinson, M.S., Hapke, B.W., Garvin, J.B., et al. (2007). High resolution mapping of TiO$_2$ abundances on the moon using the Hubble space telescope. *Geophys. Res. Lett.* 34, L13203.

Robinson, M.S., Brylow, S. M., Tschimmel, M., et al., (2010). Lunar Reconnaissance Orbit Camera (LROC) instrument overview, 2010. *Space Sci. Rev.* 150, 81–124.

Rose, H.J. Jr., Cuttitta, F., Annell, C.S., et al. (1972) Compositional data for twenty-one Fra Mauro lunar materials. *Proc. Lunar Sci. Conf.* 3, 1215–1229.

Rose, H.J. Jr., Cuttitta, F., Berman, S., et al. (1974). Chemical compositional of rocks and soils at Taurus-Littrow. *Proc. Lunar Sci. Conf.* 5, 1119–1133.

Shkuratov, Y., Pieters, C., Omelchenko, V., et al. (2003). Estimates of the lunar surface composition with Clementine images and LSCC data. *Lunar Planet. Sci. Conf.* 34, abstract, 1258.

Shkuratov, Y., Kaydash, V.G., Stankevich, D.G., et al. (2005). Derivation of elemental abundance maps at intermediate resolution from optical interpolation of lunar prospector gamma ray spectrometer data. *Planet. Space Sci.* 53(12), 1287–1300.

Taylor, G.J., Warren, P., Ryder, G., et al. (1991). Lunar rocks. In *Lunar Sourcebook* (eds. G. Heiken, D. Vaniman and B. French). Cambridge University Press, New York, pp. 183–284.

Taylor, S.R. (1987). The unique lunar composition and its bearing on the origin of the Moon. *Geochim. Cosmochim. Acta* 51(5), 1297–1309.

Thomson, B.J., Grosfils, E.B., Bussey, D.B.J., et al. (2009). A new technique for estimating the thickness of mare basalts in Imbrium Basin. *Geophys. Res. Lett.* 36(12), L12201.

Tompkins, S., Pieters, C.M. (1999). Mineralogy of the lunar crust: Results from Clementine. *Meteorit. Planet. Sci.* 34(1), 25–41.

Thuillier, G., Floyd, L., Woods, T.N., et al. (2004). Solar irradiance reference spectra. *Geophys. Monogr.* 41(1), 171–194.

Whitaker, E.A. (1972). Lunar color boundaries and their relationship to topographic features: A preliminary survey. *Moon* 4(3–4), 523–530.

Wood, J.A., Dickey J.S., Marvin, U.B. et al. (1970). Lunar anorthosites and a geophysical model of the Moon. *Proc. Apollo 11 Lunar Sci. Conf.* 1, 965.

Wu, Y.Z., Xue, B., Zhao, B.C., et al. (2012). Global estimates of lunar iron and titanium contents from the Chang'E-1 IIM data. *J. Geophys. Res.*, 117, E02001.

Zhang, X.Y., Li, C.L., Lü, C. (2009). Quantification of the chemical composition of lunar soil in terms of its reflectance spectra by PCA and SVM. *Chin. J. Geochem.* 28(2), 204–211.

Zhao, B.C., Yang, J.F., Xue, B., et al. (2009). Optical design and on-orbit performance evaluation of the imaging spectrometer for Chang'E-1 lunar satellite. *Acta Photon. Sin.* 38(3), 479–483

Zhao, B.C., Yang, J.F., Xu, B., et al. (2010). Calibration of Chang'E-1 satellite interference imaging spectrometer. *Acta Photon. Sinica* 39(5), 769–775.

7

Lunar Clinopyroxene Abundance Retrieved from M³ Data Based on Topographic Correction

Pengju Guo, Shengbo Chen, Jingran Wang, Yi Lian, Ming Ma, and Yanqiu Li

CONTENTS

7.1 Introduction 158
    7.1.1 Lunar Minerals 158
    7.1.2 Lunar Mineral Inversion Methods 158
    7.1.3 Lunar Topography 158
    7.1.4 Topographic Correction Models 159
    7.1.5 About This Chapter 159
7.2 Sandmeier Model 159
7.3 Parameters of Lunar Sandmeier Model 162
    7.3.1 Slope e and Aspect φ 162
    7.3.2 Incidence Angle i_s and Emergence Angle i_v 163
    7.3.3 Terrain-View Factor V_t 164
    7.3.4 Direct Component of Irradiance on a
              Horizontal Surface E_s^h 165
    7.3.5 Total Irradiance on a Horizontal Surface E^h 165
    7.3.6 Binary Coefficient to Control Cast Shadow Θ 167
7.4 Elevation of the Topographic Correction Results 167
7.5 Albedo Solving Based on Hapke's Model 168
7.6 Spectral Mixing Analysis 168
7.7 Results 169
7.8 Conclusion and Discussion 170
Symbols 171
References 172

7.1 Introduction

7.1.1 Lunar Minerals

Lunar minerals have directly led to the early evolution of lunar crust and interacting geological processes. But the composition of lunar surface is relatively simple, primarily dominated by clinopyroxene, orthopyroxene, plagioclase, olivine, ilmenite, and agglutinitic glass (Papike et al. 1991; Taylor et al. 2001). The recent lunar missions, for example, SELenological and ENgineering Explorer (SELENE) Kaguya, Chang'E-1, and Chandrayaan-1, provided new opportunities to explore and understand these issues (Jin et al. 2013; Wei et al. 2013).

7.1.2 Lunar Mineral Inversion Methods

The methods for lunar mineral inversion include lookup table (Lucey 1998, 2003, 2004), spectral mixing analysis (SMA) (Li 2009), principal component analysis (PCA) (Pieters 1982), and multiple regression analysis (Pieters 2002). The first two methods are established according to the Hapke model and applied to lunar global Clementine ultraviolet-visible (UV-VIS) multispectral data; the other two are statistical methods, which were applied to Lunar Soil Characterization Consortium (LSCC) data. Hapke proposed a bidirectional reflectance model based on Chandrassekhar's radiative transfer theory, known as the original Hapke model (Hapke 1981; Chandrassekhar 1960). The effect of general macroscopic roughness was then introduced in the original Hapke model (Hapke 1984) and coherent backscattering was also incorporated. Hapke improved the original model and established the two stream approximation model (Hapke 2002). All the Hapke models have not taken the topographic effect of lunar surface into account.

7.1.3 Lunar Topography

The Moon is covered with widely distributed small to massive craters. Its topographic relief inevitably affects the quality of spectrum data. In rugged terrain, measurements from optical satellite imagery are greatly affected by a series of topographic effects. Change in slope and aspect, as a function of the relative geometry between Sun and the Moon, modifies the local illumination. On the other hand, in mountainous area, the multiple reflections between a Moon target and the surrounding terrain would complicate the radiative transfer model. Therefore, it is necessary to correct the topographic effect of lunar surface.

7.1.4 Topographic Correction Models

Three kinds of topographic correction models are used on the Earth:

1. The model based on sun-terrain-sensor (STS) geometric relationships, including cosine model (Teillet et al. 1982), C model (C is a semi-empirical moderator) (Teillet et al. 1982), and Minnaert model (Smith et al. 1980).
2. The model based on sun-canopy-sensor (SCS) geometric relationship (Gu and Gillespie 1998; Scott et al. 2005).
3. The model based on irradiance, including Sandmeier model (Sandmeier and Itten 1997) and Proy model (Proy et al.).

Comparing these models, the SCS model is developed for topographic effect correction on forest images (Gu and Gillespie 1998), which is not suitable for the Moon. The cosine correction model only shows the direct part of the irradiance. On weakly illuminated areas, the model has a disproportionate brightening effect (Meyer et al. 1993). C correction model improves the overcorrection effect but the physical analogies are not definite (Meyer et al. 1993). The value of k in Minnaert correction model is wavelength dependent and presents one value for each spectral band (Ekstrand 1996). The computation of Proy model has the disadvantage of being time-consuming (Proy et al. 1989). So the Sandmeier model is suitable for the lunar topographic correction.

7.1.5 About This Chapter

In this chapter, the Sandmeier model is selected for the topographic correction of lunar surface. Without considering backscatter, the Hapke model is used to calculate the lunar surface albedo from the topographic corrected reflectance for Moon Mineralogy Mapper (M3) data obtained by Indian Chandrayaan-1 satellite. SMA is then applied to the spectra unmixing of lunar surface albedo to retrieve the clinopyroxene abundance in Sinus Iridum edge. The reflectance and retrieved clinopyroxene abundance are compared with nontopographic correction results.

7.2 Sandmeier Model

On Earth, irradiance entering one target is divided into three different parts:

1. The solar direct irradiance.
2. Sky-diffused irradiance.
3. Adjacent terrain-reflected irradiance.

These three components of irradiance will be directionally reflected to sensor by land surface. According to the Sandmeier model, total irradiance on the Earth can be expressed as:

$$E = \Theta \cdot E_s^h \cdot \frac{\cos i_s}{\cos \theta_s} + E_d^h \cdot \left\{ k \cdot \frac{\cos i_s}{\cos \theta_s} + (1-k) \cdot V_d \right\} + E^h \cdot V_t \cdot \rho_{adj} \qquad (7.1)$$

where E is total irradiance on an inclined surface, Θ is binary coefficient to control cast shadow, E_s^h is direct component of irradiance on a horizontal surface, i_s is solar incidence angle, θ_s is solar zenith angle, E_s^h is diffuse component of irradiance on a horizontal surface, k is anisotropy index, V_d is sky-view factor, E^h is total irradiance on a horizontal surface, V_t is terrain-view factor, ρ_{adj} is average reflectance of adjacent objects.

The first term on the right of Equation 7.1 is the cosine law for solar direct irradiance, the second term is the sky diffuse irradiance, and the third term refers to terrain irradiance.

Because there is no atmosphere on the Moon, the sky diffuse irradiance can be ignored, the Sandmeier model applied on the Moon is described in Equation 7.2:

$$E = \Theta \cdot E_s^h \cdot \frac{\cos i_s}{\cos \theta_s} + E^h \cdot V_t \cdot \rho_{adj} \qquad (7.2)$$

The ultimate objective of topographic correction is to get the flat surface directional-directional reflectance. Directional reflectance is divided into two different parts: one is the directional-directional reflectance R_T and the other is hemispheric-directional reflectance R_{DT}. The surface reflectance for solar direct radiance is a directional-directional reflectance, which is illuminated in Figure 7.1.

But surface reflectance for terrain-reflected radiance is a hemispheric-directional reflectance, as shown in Figure 7.2.

Changing the irradiance E to the radiance at satellite level L, Equation 7.2 can be transformed to Equation 7.3:

$$L = \Theta \frac{E_s^h \cos i_s}{\pi \cos \theta_s} R_T + \frac{E^h \cdot V_t \cdot \rho_{adj}}{\pi} R_{DT} \qquad (7.3)$$

The radiance on the surface comes from a hemispheric space and is reflected to the direction of the sensor. Assuming that reflections from various directions are homogeneous, Nicodemus proposed an integral of the directional-directional reflectance over all incidence angles (Nicodemus et al. 1977), as R_{DT} in the equation,

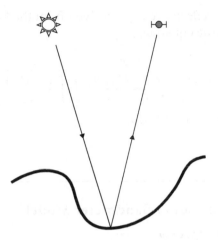

FIGURE 7.1
Downward directional-directional reflectance R_T received in a mountainous region.

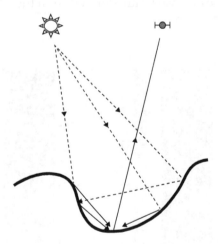

FIGURE 7.2
Downward hemispheric-directional reflectance R_D received in a mountainous region.

$$R_{DT} = \frac{1}{\pi} \int\limits_{2\pi} \int\limits_{\pi/2} R_T \cos(i_s) \sin(i_s) di_s \, d\varphi_s \qquad (7.4)$$

Dymond puts forward the method to calculate the reflectance of flat surface R_T (Dymond and Shepherd 1999), which can be expressed as Equation 7.5

$$R_T = \frac{\cos(\theta_s) + \cos(\theta_v)}{\cos(i_s) + \cos(i_v)} \cdot R_H \qquad (7.5)$$

Lunar flat surface reflectance R_H is derived from the Equations 7.3, 7.4, and 7.5, as illustrated in the equation,

$$R_H = \frac{L \cdot \pi}{\frac{\Theta E_s^h \cos i_s}{\cos \theta_s} \cdot \frac{\cos(\theta_s) + \cos(\theta_v)}{\cos(i_s) + \cos(i_v)} + \frac{E^h V_t \rho_{adj}}{\pi} \int_{2\pi} \int_{\pi/2} \frac{\cos(\theta_s) + \cos(\theta_v)}{\cos(i_s) + \cos(i_v)} \cdot \cos(i_s) \sin(i_s) \, di_s \, d\varphi_s}$$

(7.6)

7.3   Parameters of Lunar Sandmeier Model

7.3.1   Slope e and Aspect φ

Slope and aspect are two of the most basic terrain factors. Slope reflects the degree of topographic relief, and aspect reflects the topographic distribution. Both of the slope and aspect are derived from the OBS file of M³ data, as shown in Figure 7.3a and b.

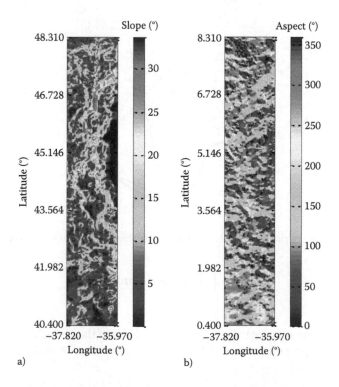

FIGURE 7.3
Slope a) and aspect b).

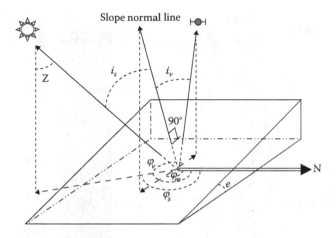

FIGURE 7.4
Three-dimensional vector space which takes slope point as origin.

7.3.2 Incidence Angle i_s and Emergence Angle i_v

When the lunar surface is flat, the zenith direction coincides with the slope normal line, solar zenith angle is the incidence angle, and M^3 zenith angle is the emergence angle. And when there is topographic relief, the zenith direction no longer coincides with the slope normal line, incidence angle is the angle between the normal to the ground surface and the Sun's rays, and emergence angle is the angle between the normal to the ground surface and the reflected rays, as illustrated in Figure 7.4.

In Figure 7.4, Z is solar zenith angle, i_s is the incidence angle, i_v is the emergence angle, e is slope, φ_m is aspect, φ_s is solar azimuth angle, φ_v is M^3 azimuth angle, and N represents the orientation of the North.

i_s is defined by Wang et al. (2000) as

$$i_s = \arccos\left(\begin{bmatrix} \sin Z \cos \varphi_s \\ \sin Z \sin \varphi_s \\ \cos Z \end{bmatrix} \cdot \begin{bmatrix} \sin e \cos \varphi_m \\ \sin e \sin \varphi_m \\ \cos Z \end{bmatrix} \right) \tag{7.7}$$

that is

$$i_s = \arccos[\cos e \cos Z + \sin e \sin Z \cos(\varphi_m - \varphi_s)] \tag{7.8}$$

In a similar way, i_v is defined as

$$i_s = \arccos[\cos e \cos \theta_v + \sin e \sin \theta_v \cos(\varphi_m - \varphi_v)] \tag{7.9}$$

where θ_v is M^3 zenith angle.

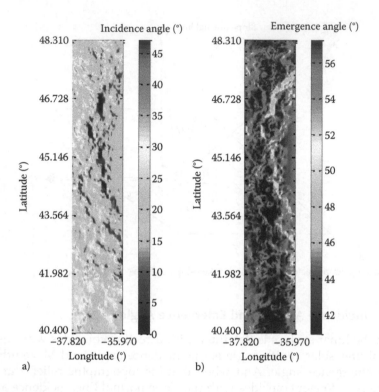

FIGURE 7.5
Incidence angle a) and emergence angle b).

Figure 7.5a and b show the incidence angle i_s and emergence angle i_v derived from Equations 7.8 and 7.9.

7.3.3　Terrain-View Factor V_t

Terrain-view factor V_t can be calculated by a simplified trigonometric approach described by Kondratyev 1969:

$$V_t = \frac{1 - \cos(e)}{2} \tag{7.10}$$

The slope e is the exclusive parameter to estimate the amount of sky and terrain seen from a point, and V_t differs from zero to one. For a vertical plane, the approach reveals the V_t of 0.5, while for a horizontal plane it turns to zero. Figure 7.6 shows the terrain-view factor calculated according to Equation 7.10.

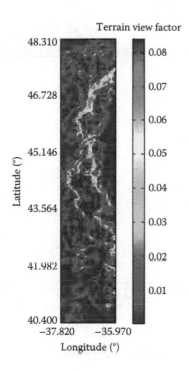

FIGURE 7.6
Terrain-view factor.

7.3.4 Direct Component of Irradiance on a Horizontal Surface E_s^h

The Sun spectrum F refers to the solar radiation energy vertical to the sunlight at the distance of 1 au from the Sun. The direct component of irradiance on a horizontal surface E_s^h at any distance R can be expressed as:

$$E_s^h = F \times \left(\frac{\bar{R}}{R} \right)^2 \tag{7.11}$$

where F is the solar spectrum, $\bar{R}$ is the average distance from Earth to the Sun, R is the distance of one position to the Sun, the unit is au.

According to Equation 7.11, E_s^h is calculated, as Figure 7.7 shows.

7.3.5 Total Irradiance on a Horizontal Surface E^h

Because there is no atmosphere on the Moon, the Sun's radiation is not affected by atmospheric absorption and scattering, and the total irradiance on a horizontal surface E^h is also the direct component of irradiance on a horizontal surface E_s^h.

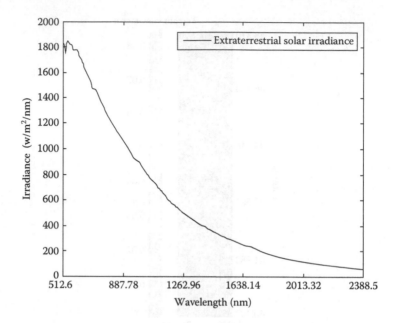

FIGURE 7.7
Direct component of irradiance on a horizontal surface.

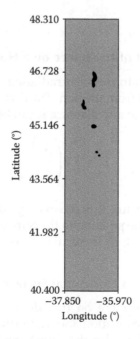

FIGURE 7.8
Binary coefficient to control cast shadow.

7.3.6 Binary Coefficient to Control Cast Shadow Θ

When solar rays are intercepted by the surrounding terrain or slope surface of the target, the direct solar radiation cannot reach the lunar surface. And this characteristic can be attributed to the binary coefficient Θ. When the target can be illuminated by the solar rays, Θ is one. Contrarily, Θ is equal to zero, as shown in Figure 7.8.

In Figure 7.8, area colored by grey can be illuminated by solar rays, while area colored by black cannot.

7.4 Elevation of the Topographic Correction Results

In order to evaluate the lunar topographic correction results, we randomly selected some pixels facing towards and away from the Sun, as point a and b in Figure 7.9a. Comparing the reflectance before and after correction, shown

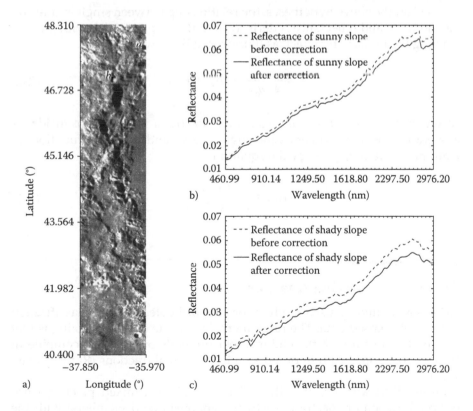

FIGURE 7.9
a) Location of selected pixels of sunny slope a and shady slope b; b) Reflectance of sunny slope before and after correction; c) Reflectance of the slope in shade before and after correction.

in Figure 7.9b and c, it is obvious that after correction, the high spectral reflectance facing the Sun is decreased, while low spectral reflectance away from the Sun is compensated. That is to say, after correction, the reflectance of sunny and shady slope is balanced.

7.5 Albedo Solving Based on Hapke's Model

The Hapke model is the foundation of our study. In this model, the single particle function, $P(g)$, describes the scattering properties of a particle as a function of phase angle. If the particles are assumed to be much larger than the wavelength of light and are randomly oriented on the surface, the surface scatters light isotropically (Mustard and Pieters 1987). And Hapke found that for an isotropic surface, $P(g) = 1$ (Hapke 1981). $B(g) = 0$ at intermediate phase angles (between 15° and 40°), commonly used by imaging spectrometers (Mustard and Pieters 1987).

Based on the above hypotheses, the relationship between single-scattering albedo (SSA) and reflectance is derived from the Hapke model, without considering the backscatter effect, as illuminated by the equation

$$r_c = \frac{w}{4} \frac{1}{\mu_0 + \mu} H(\mu_0) H(\mu) \tag{7.12}$$

where w is SSA spectra, r_c is reflectance spectra, μ is cosines of incidence angle, and μ_0 is cosines of emission angle. H is Chandrasekhar's function for isotropic scattering, given by the equation

$$H(x) = \frac{1 + 2x}{1 + 2x\sqrt{1 - w}} \tag{7.13}$$

7.6 Spectral Mixing Analysis

SMA is a commonly used method for subpixel detection and classification of remotely sensed data. The fundamental principle of linear mixing is that the spectral features of the end-member minerals overlap and combine in the composite spectrum in proportion to their areal fractions (Ramsey et al. 1998).

Linear SMA assumes that the reflectance of each mixed pixel is a linear combination of spectra of distinct component (end members) with the weights representing the abundances of end-members resident in a mixed pixel (Li and Lucey 2009).

Based on the above theory, a linear equation about mineral albedo and abundance of five kinds of minerals (clinopyroxene, orthopyroxene, plagioclase, olivine, and ilmenite) is established:

$$w_i = \sum_{j=1}^{5} (a_{ij}x_j) \tag{7.14}$$

where i represents band i, j is end member mineral j, w is albedo of mixed mineral, a is albedo of end member mineral, x is abundance of each end member mineral.

Then using the least square method, the clinopyroxene abundance of Sinus Iridum edge is calculated.

7.7 Results

In our study, Band 2, 9, 12, 22, 34, 47, 56, and 61 of M^3 data are selected for the topographic correction. Then according to Equation 7.6, and the homologous parameters, the corrected reflectance can be obtained. The reflectance before and after correction of band 61 is shown in Figure 7.10a and b as an example.

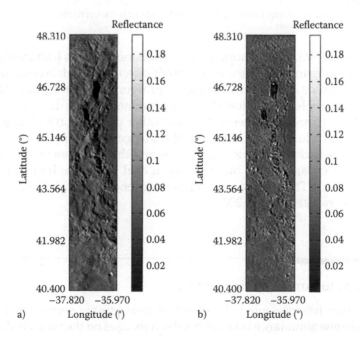

FIGURE 7.10
Reflectance before a) and after b) topographic correction of Band 61.

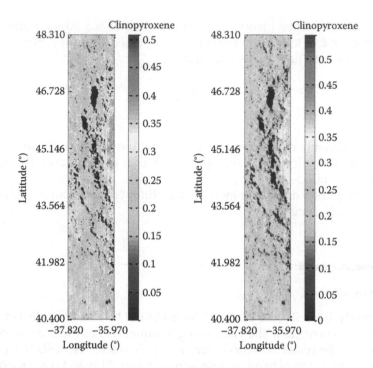

FIGURE 7.11
Abundance of clinopyroxene before a) and after b) topographic correction.

Based on the SMA, the clinopyroxene abundance of Sinus Iridum edge is calculated. In order to analyze the topographic effect, the result is compared with the clinopyroxene abundance before spectrum correction, in Figure 7.11a and b.

After topographic correction, the high abundance area is decreased, and the low abundance area becomes higher, the other part has little change. Inside of the Sinus Iridum, there is a high abundance of clinopyroxene, and it is expected due to the proportional decrease in lithic components and increase in agglutinitic glass as soils mature (Taylor et al. 2001). The highlands have a low abundance. The abundance of clinopyroxene and plagioclase are highly inversely correlated (Lucey 2004).

7.8 Conclusion and Discussion

This chapter has presented the effect of topographic correction for lunar clinopyroxene abundance inversion, which focuses on the Sinus Iridum edge rather than the global scale. Sandmeier model was firstly used for the topographic correction and then the SMA was applied to retrieve clinopyroxene abundance of Sinus Iridium edge. In order to analyze the topographic effect, the result was compared with the abundance before spectrum correction. Based on the observed results, SMA is found to be suitable for the retrieving abundance of clinopyroxene. The topography will affect the clinopyroxene abundance retrieval, but not the distribution. After the topographic correction, the effect of terrain on the observed spectrum will be decreased. Therefore, before retrieving clinopyroxene abundance, topographic effect should be taken into account.

The correction of spectral mixing effect using the Hapke model is still not thorough. In this chapter, the multiple-scattering isotropic Hapke model with a relatively lower model precision is used. The multiple-scattering two stream approximation Hapke model has higher simulation precision for the energy from multiple scattering, which will be taken into account in our future work. On the other hand, the influence of space weathering and the nonlambert body are not considered, which should be further investigated.

Symbols

The notation used in this chapter is listed for convenience.

E	Total irradiance on an inclined surface
Θ	Binary coefficient to control cast shadow
E_s^h	Direct component of irradiance on a horizontal surface
i_s	Solar incidence angle
θ_s/Z	Solar zenith angle
E_d^h	Diffuse component of irradiance on a horizontal surface
k	Anisotropy index
V_d	Sky-view factor
E^h	Total irradiance on a horizontal surface
V_t	Terrain-view factor
ρ_{adj}	Average reflectance of adjacent objects
R_T	Directional-directional reflectance
R_{DT}	Hemispheric-directional reflectance
L	Entrance pupil radiance
R_H	Lunar flat surface reflectance
e	Slope
φ/φ_m	Aspect
i_v	Emergence angle
φ_s	Solar azimuth angle
φ_v	M^3 azimuth angle
θ_v	M^3 zenith angle

N	The orientation of the North
F	Sun spectrum
$\bar{R}$	The average distance from Earth to the Sun
R	The distance of one position to the Sun
r	Bidirectional reflectance observed by the satellite
w	Average single scattering albedo
μ_0	Cosine of incidence angle
μ	Cosine of emergence angle
H(x)	Multiple-scattering function

References

Chandrasekhar S (1960). *Radioactive Transfer*. New York: Dover.
Dymond JR, Shepherd JD (1999). Correction of the topographic effect in remote sensing. *IEEE Trans. Geosci. Remote Sens.* 37(5), 2618–2620.
Ekstrand S (1996). Landsat TM-based forest damage assessment: Correction for topographic effects. *Photogr. Eng. Remote Sens.* 62(2), 151–162.
Gu D, Gillespie A (1998). Topographic normalization of Landsat TM images of forest based on subpixel sun canopy sensor geometry. *Remote Sens. Env.* 64(2), 166–175.
Hapke B (1981). Bidirectional reflectance spectroscopy: 1. Theory. *J. Geophys. Res.* 86(B4), 3039–3054.
Hapke B (1984). Bidirectional reflectance spectroscopy: 3. Correction for macroscopic roughness. *Icarus*, 59(1), 41–59.
Hapke B (1986). Bidirectional reflectance spectroscopy: 4. The extinction coefficient and the opposition effect. *Icarus* 67(2), 264–280.
Hapke B (2002). Bidirectional reflectance spectroscopy: 5. The coherent backscatter opposition effect and anisotropic scattering. *Icarus* 157(2), 523–534.
Jin SG, Arivazhagan S, Araki H (2013). New results and questions of lunar exploration from SELENE, Chang'E-1, Chandrayaan-1 and LRO/LCROSS. *Adv. Space Res.* 52(2), 285–305.
Kondratyev KY (1969). *Radiation in the Atmosphere*. London, UK: Academic.
Li L, Lucey PG (2009). Use of multiple end member spectral mixture analysis and radiative transfer model to derive lunar mineral abundance maps. *40th Lunar and Planetary Science Conference*. Woodlands.
Lucey PG (1998). Quantitative mineralogical and elemental abundance from spectroscopy of the moon: Status, prospects and limits and a plea. *Workshop on New Views of the Moon: Integrated Remotely Sensed, Geophysical and Sample Datasets*, 53–54.
Lucey PG (2004). Mineral maps of the moon. *Geophys. Res. Lett.* 31(8), cite ID L08701.
Lucey PG, Steutel D (2003). Global mineral maps of the moon. *Lunar and Planetary Science XXXIV*. Houston, Tex: Lunar and Planetary Institute.
Meyer P, Itten KI, Kellenberger T et al. (1993). Radiometric corrections of topographically induced effects on Landsat TM data in an alpine environment. *ISPRS J. Photogr. Remote Sens.* 48(4), 17–28.

Mustard JF, Pieters CM (1987). Quantitative abundance estimates from bidirectional reflectance measurements. *J. Geophys. Res.* 92(B4), E617–E626.

Nicodemus F, Richmond J, Hsia J et al. (1977). Technical report, Geometrical accounts and nomenclature for reflectance. NBS, US Department of Commerce, Washington, D C.

Papike J, Taylor LA, Simon S (1991). Lunar minerals. In: Heiken, G., Vaniman, D., French, B. (Eds.), *The Lunar Sourcebook.* Cambridge University Press, New York, pp.131–181.

Pieters CM (1982). Conspicuous crater central peak: Lunar mountain of unique composition. *Science* 215(4528), 59–61.

Pieters CM, Stankevich DG, Shkuratov YG et al. (2002). Statistical analysis of the links between lunar mare soil mineralogy, chemistry and reflectance spectra. *Icarus,* 155(2), 285–298.

Proy C, Tanre D, Deschamps PY (1989). Evaluation of topographic effects in remotely sensed data. *Remote Sens. Env.* 30, 21–32.

Sandmeier S, Itten KI (1997). A physically-based model to correct atmospheric and illumination effects in optical satellite data of rugged terrain. *IEEE Trans. Geosci. Remote Sens.* 35(3), 708–717.

Soenen SA, Peedle DR, Coburn CA (2005). SCS+C: A modified sun-canopy-sensor topographic correction in forested terrain. *IEEE Trans. Geosci. Remote Sens.* 43(9), 2148–2159.

Smith J, Lin T, Ranson K (1980). The Lambertian assumption and Landsat data. *Photogr Eng Remote Sens.* 16(9), 1183–1189.

Taylor LA, Pieters CM, Keller LP et al. (2001). Lunar mare soils: Space weathering and the major effects of surface-correlated nanophase Fe. *J. Geophys. Res.* 106(E10), 27985–27999.

Teillet PM, Guindon B, Goodeonugh DG (1982). On the slope-aspect correction of multispectral scanner data. *Can. J. Remote Sens.* 8(2), 84–106.

Wang J, White K, Robinson GJ (2000). Estimating surface net solar radiation by use of Landsat-5 TM and digital elevation models. *Int. J. Remote Sens.* 21(1), 31–43.

Wei E, Yan W, Jin SG, Liu J, Cai J (2013). Improvement of Earth orientation parameters estimate with Chang'E-1 VLBI observations. *J. Geodyn.* 72, 46–52.

8

Martian Minerals and Rock Components from MRO CRISM Hyperspectral Images

Yansong Xue and Shuanggen Jin

CONTENTS

8.1 Observation History 175
    8.1.1 Early and Telescopic Observations.. 175
    8.1.2 Spacecraft Missions 176
8.2 Methods of Mineral Analysis 180
    8.2.1 Mineral Information from Remote Sensing 182
        8.2.1.1 Reflection Spectroscopy Remote Sensing.................... 182
        8.2.1.2 Other Remote Sensing Methods.................................... 185
    8.2.2 Mineral Information from Martian Meteorite 185
    8.2.3 Mineral Information from In Situ Analysis............................... 186
8.3 Mineral Components at Gale Region from MRO CRISM Images 188
    8.3.1 Geologic Setting in Study Area.. 189
    8.3.2 Observation and Data Processing 191
        8.3.2.1 Instrument Description.. 191
        8.3.2.2 Data Preprocessing 191
        8.3.2.3 Spectral Analysis Method... 192
    8.3.3 Results and Discussion 193
        8.3.3.1 Mineral Species Identification....................................... 193
        8.3.3.2 Mineral Classes Distribution... 197
8.4 Summary of Martian Mineral.. 201
Acknowledgments 202
References.. 202

8.1 Observation History

8.1.1 Early and Telescopic Observations

As a neighbor of Earth, Mars has been the focus of scientific interest throughout recorded history. Before the advent of the telescope in 1609, early astronomic observers deduced two things about this planet. First, the sidereal period of Mars has been determined to be 687 Earth days (1.88 Earth years); second, the retrograde motion of Mars has been observed.

Since Galileo's small telescope revealed Mars' reddish-orange color and gibbous phase, larger telescopes acquired more information about the planet. In 1659, Christian Huyens discovered the diversity of Martian albedo. The identification of surface albedo allowed astronomers to deduce that the Martian rotation period was approximately 24 hours. The brighter polar caps were not noticed until Giovanni Cassini discovered them in 1666. An observation of Sir William Herschel from 1777 to 1783 indicated that Mars experiences four seasons like Earth and determined a more precise Martian rotation period to be 24 hours 39 minutes 21.67 seconds. Herschel also deduced that Mars possesses a thin atmosphere that is primarily composed of white clouds of ice particles. Martian yellow clouds were found in 1809 by Honore Flaugergues. The first Martian map with a latitude–longitude system was published in 1840 by Jonhan von Madler and Wilhelm Beer. Mars' two moons, Phobos and Deimos, were discovered by Asaph Hall in 1877, when Earth and Mars were very close.

Advances in telescopic size and technology have greatly enhanced the quality and type of astronomical observations, and Martian mineral studies have been one of the areas to reap the advances of telescope technology. Infrared observations of Mars from Earth-based telescopes and the Hubble Space Telescope (HST) confirmed the mineralogical variety across the surface, including the existence of hydrated minerals. Adaptive optics provided dramatic improvements in resolution. Radar investigations from ground-based radio telescopes have provided information on surface roughness. These Martian features were only seen previously by orbiting spacecraft.

8.1.2 Spacecraft Missions

Mars has been a major destination of spacecraft since the early days of space exploration. Mars exploration is not easy. Approximately two-thirds of all spacecraft missions to date have failed. A list of successful missions that have been launched through 2012 is given in Table 8.1.

The first Martian mineral detection began with Phobos 2 launched by the Soviet Union on July 12, 1988. Phobos 2 entered into Mars orbit on January 29, 1989, and obtained information about topography and mineralogy of the Martian surface using near-infrared mapping spectrometer (ISM). Mars Global Surveyor (MGS) (Albee et al. 2001) was launched on November 7, 1996, and reached the Mars orbit on September 11, 1997. The Mars orbiter camera (MOC) consists of three cameras each with an independent angle of view that can get images of different resolutions. The Mars orbiter laser altimeter (MOLA) investigated topography and roughness using laser pulses. The magnetometer (MAG) measured the first detailed Martian magnetic field and revealed remnant magnetization in Martian surface rocks. Mineralogical investigation has been made by thermal emission spectrometer (TES). The TES instrument uses the natural

TABLE 8.1

Mars Exploration Missions

Mission	Country	Launch Date	Type of Mission	Result
Mariner 4	US	November 28, 1964	Flyby	Flew by July 14, 1965
Mariner 6	US	February 24, 1969	Flyby	Flew by July 31, 1969
Mariner 7	US	March 27, 1969	Flyby	Flew by August 5, 1969
Mars 3	USSR	May 28, 1971	Orbiter/ lander	Arrived on December 3, 1971; got some data
Mariner 9	US	May 30, 1971	Orbiter	Lasted until November 13, 1971–October 27, 1972
Mars 5	USSR	July 25, 1973	Orbiter	Arrived on February 9, 1974; lasted a few days
Mars 6	USSR	August 5, 1973	Orbiter/ lander	Arrived on March 12, 1974; got little data
Mars 7	USSR	August 9, 1973	Orbiter/ lander	Arrived on March 9, 1974; got little data
Viking 1	US	August 20, 1975	Orbiter/ lander	Orbiter lasted from June 19, 1976 to August 7, 1980; Lander operated from July 20, 1976 to November 13, 1982
Viking 2	US	September 9, 1975	Orbiter/ lander	Orbiter lasted from August 7, 1976 to July 25, 1978; Lander operated from September 3, 1976 to August 7, 1980
Phobos 2	USSR	July 12, 1988	Orbiter/ lander	Lasted till January 29, 1989
MGS	US	November 7, 1996	Orbiter	Lasted from September 12, 1997 to November 2, 2006
MPF	US	December 4, 1996	Lander/ rover	Operated from July 4, 1997 to September 27, 1997
Mars Odyssey	US	April 7, 2001	Orbiter	Arrived on October 24, 2001; still operating
Mars Express	ESA	June 2, 2003	Orbiter/ lander	Arrived on December 25, 2003; orbiter still operating; Lander failed
Spirit, MER-A	US	June 10, 2003	Rover	Arrived on January 3, 2004; still operating
Opportunity, MER-B	US	July 7, 2003	Rover	Arrived on January 25, 2004; still operating
MRO	US	August 12, 2005	Orbiter	Arrived on March 10, 2006; still operating
Phoenix	US	August 4, 2007	Lander	Arrived on May 25, 2008– November 2, 2008
Curiosity	US	November 26, 2011	Rover	Arrived on August 6, 2012; still operating

harmonic vibrations of chemical bonds to determine surface composition and thermal properties.

The first successful rover mission, Mars Pathfinder (MPF) (Golombek et al. 1999) set down in Ares Vallis, a channel on Mars, on July 4, 1997. The mission consisted of a lander and rover called Sojourner. The lander with magnets attached has estimated the magnetic properties of dust. The rover carried a navigation camera and an alpha particle x-ray spectrometer (APXS) to determine soil compositions.

Mars Odyssey entered Mars orbit on October 24, 2001. It carried a gamma ray spectrometer (GRS) to measure gamma rays and neutrons from the surface, which is a replacement instrument carried by failed spacecrafts called Mars Observer. GRS provided information about abundances of a variety of elements within the upper surface, such as H_2O and CO_2. A thermal emission imaging system (THEMIS) onboard Odyssey was used to detect the mineralogical variation that occurs across the Martian surface. The THEMIS observes the planet in both visible and infrared wavelength, and it can work during both day and night.

Mars Express (Chicarro 2002), funded by ESA, consisted of an orbiter and the Beagle 2 surface lander. Although the Beagle 2 lander failed, the Mars Express orbiter has been very successful. All of its instruments have got corresponding data, including visible and infrared mineralogical mapping spectrometer (OMEGA), ultraviolet and infrared atmospheric spectrometer (SPICAM), sub-surface sounding radar altimeter (MARSIS), planetary Fourier spectrometer, high-resolution stereo camera (HRSC), Mars Express lander communications (MELACOM), and Mars Radio Science Experiment (MaRS). Among these instruments, OMEGA has provided evidence of water ice in the Martian polar caps and mineral distribution across its surface.

The two Mars Exploration Rover (MER) missions, named Spirit and Opportunity, were sent to Mars in 2003. On January 3, 2004, Spirit landed in Gusev crater, which was interpreted to be the site of a paleolake (Squyres et al. 2004a). Opportunity followed closely, landing in Meridiani Planum on January 25, 2004, a region comprising an abundance of the mineral hematite that often forms in an aqueous environment (Squyres et al. 2004b). The rovers each carry various scientific instruments, including a panoramic camera (Pancam), a miniature TES (Mini-TES) a rock abrasion tool (RAT), a microscopic imager (MI) to obtain magnified views of rocks and soils, an APXS, and a Mössbauer spectrometer. These instruments have provided the best mineralogical analysis of surface materials to date.

The Mars Reconnaissance Orbiter (MRO), which was launched on August 12, 2005, attained the Martian orbit on March 10, 2006. MRO contains a host of scientific instruments such as a high-resolution imaging science experiment camera (HiRISE), context camera (CTX), Mars color imager, compact reconnaissance imaging spectrometer for Mars (CRISM), Mars climate sounder (MCS), and shallow subsurface radar (SHARAD), which are used to analyze the landforms, stratigraphy, minerals, and ice of Mars (Figure 8.1). MRO joined other active spacecrafts in orbit including MGS, Mars Express, and Mars Odyssey and is planned to return over three times as much data as they put together. This progress profits from the more advanced MRO's telecommunications system that can transfer more data back to Earth than all previous interplanetary missions combined. It paves the way for future spacecraft and can serve as a highly capable relay satellite for future missions.

Phoenix is a lander mission under the Mars Scout program. The Phoenix lander was launched on August 8, 2007, and descended on Mars on May 25, 2008. The landing site was Green Valley of Vastitas Borealis. Phoenix was the first successful landing in the Martian polar region. It carries improved panoramic cameras and a volatiles-analysis instrument. The payload includes the robotic arm camera (RAC), surface stereo imager (SSI), thermal and evolved gas analyzer (TEGA), Mars descent imager (MARDI), microscopy, electrochemistry, conductivity analyzer (MECA), thermal and electrical conductivity probe (TECP), wet chemistry lab (WCL), and various microscopes. The mission includes a study of the geologic history of water, which is the key to unlocking the evolution of past climate change and to evaluate habitability in the ice soil boundary, past or present.

Curiosity is a robotic rover exploring mission as part of Mars Science Laboratory mission (MSL). Curiosity was launched on November 26, 2011, and successfully landed on Aeolis Palus in Gale Crater on August 6, 2012. The rover carries various sophisticated scientific instruments for sampling, detection, and analysis, including mast camera (MastCam), chemistry and camera

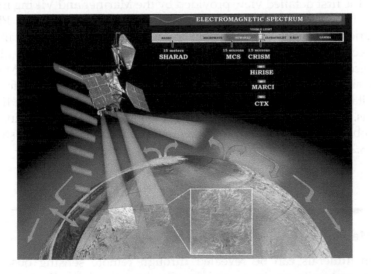

FIGURE 8.1
Diagram of instrumentation aboard MRO. (NASA/JPL PIA07241, 2005.)

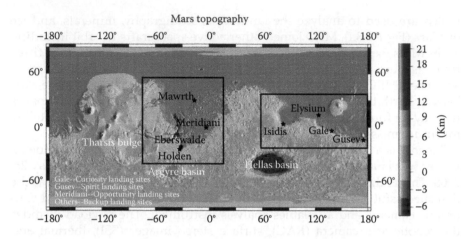

FIGURE 8.2
Martian topography and craters marked by stars.

complex (ChemCam), navigation cameras (navcams), rover environmental monitoring station (REMS), hazard avoidance cameras (hazcams), Mars hand lens imager (MAHLI), APXS, chemistry and mineralogy (CheMin), sample analysis at Mars (SAM), dust removal tool (DRT), radiation assessment detector (RAD), dynamic albedo of neutrons (DAN) and Mars descent imager (MARDI). The goals include investigation of the Martian geology and climate; estimation of whether the Gale Crater can offer conditions favorable for microbial life, such as the role of water; and planetary habitability in preparation for future human exploration.

Since the first detailed view provided by the Mariner and Viking missions (Soffen, 1977; Kieffer et al., 1992), current missions (MER, MRO, Phoenix, and Curiosity) greatly expanded our knowledge about the planet in different aspects. The global-scale topography and mineral analysis is allowing us to reconstruct the more conceivable evolutionary history of Mars surface; in situ detection and measurement provide more precise information for habitability of Mars, which was impossible to conceive of with the early missions. Our knowledge of Mars has increased dramatically due to the recent observations. Martian topography and characteristics are shown in Figure 8.2, including landing sites and craters marked by stars.

8.2   Methods of Mineral Analysis

Martian minerals can be revealed through remote sensing observation, Martian meteorite analysis, and in situ investigations. Remote sensing data can be obtained from Earth-based telescope, HST, and Mars orbiting spacecraft. All of these techniques utilize reflection spectroscopy to identify mineral information. Due to vibrational motions within the crystal lattices of minerals comprising the planet's surface, solar radiation reflected by planet carries the information on mineral constitution of the planet (Christensen et al. 2001). Removing the solar spectrum and atmosphere absorption, the absorption spectrum of the planet can be retrieved. The wavelength of these absorptions can be used to determine which mineral is absorbed at a specific wavelength.

Many of the diagnostic mineral absorptions occur in the infrared (IR) region (Hanel et al. 2003), where the wavelengths vary between 1 μm and 1 mm. Infrared consists of near infrared (NIR, 0.7 to 5 μm), middle infrared (MIR, 5 to 30 μm), and far infrared (FIR, 30 to 350 μm). The spectral region of Earth-based IR observations are limited to the NIR since H_2O in the Earth's atmosphere can absorb longer wavelengths. The near infrared camera and multi-object spectrometer (NICMOS) instrument aboard on HST also is limited to the NIR, in the range from 0.8 to 2.5 μm. Mars Orbiter can operate in a wider range of wavelengths and get higher-resolution data due to thin Martian atmosphere and proximity to Mars. Most of the spectrometers onboard Mars Orbiter observe only specific band range; these bands were selected to determine the minerals that are of particular interest, such as hydrous silicates, salts, and carbonates. CRISM is the first hyperspectral remote sensor, which was observed at a range of 0.362–3.92 μm (Table 8.2 lists the wavelength ranges covered by spectrometers on several Mars missions).

In addition to reflection spectroscopy, there are other techniques used to obtain the compositional information. The Mars Orbiter GRS system consists of a gamma ray spectrometer (GRS), neutron spectrometer (NS), and a high-energy neutron detector (HEND). GRS detects radioactive elements such as potassium (K), uranium (U), and thorium (Th), which can emit gamma rays through decay, and nonradioactive elements such as chlorine (Cl), iron (Fe), and carbon (C), which also produce gamma rays when cosmic rays interact with them. The NS and HEND detect the diverse energy-level neutrons emitted through cosmic ray interaction with the minerals of the Martian surface. The comparison of diverse energy-levels provides information on the mineral composition within 1 m of the surface.

The rovers on MPF and MER each carried an APXS and Mössbauer spectrometer to determine mineral compositions of the Martian surface (Rieder et al. 1997a, 2003). The APXS bombards a sample with alpha particles and x-rays. The alpha particles backscattered by sample and x-rays emitted by ionization of sample are diagnostic of the composition of the sample. MER Mössbauer spectrometer bombards samples with gamma rays. The gamma rays reemitted from the sample are used to measure and investigate minerals (Klingelhöfer et al. 2003).

TABLE 8.2

Wavelengths of Observations by Selected Spacecraft Missions

Mission/Instrument	Wavelengths
Mariner 6/7	
UV spectrometer	110–430 nm
IR spectrometer	1.9–14.3 nm
Mariner 9	
UV spectrometer	110–352 nm
IRIS (infrared interferometer spectrometer)	6–50 µm
Viking 1/2 Oribiter	
IRTM (infrared thermal mapper)	15 µm
Phobos 2	
ISM (infrared spectrometer)	0.7–3.2 µm
KRFM	0.3–0.6 µm
Thermoscan	
Visible	0.5–0.95 µm
Infrared	8.5–12 µm
MGS	
TES	
Spectrometer	6.25–50 µm
Bolometer	4.5–100 µm
Albedo	0.3–2.7µm
Mars Odyssey	
THEMIS	6.62, 7.78, 8.56, 9.30, 10.11, 11.03, 11.78, 12.58, 14.96 µm
	423, 553, 652, 751, 870 nm
Mars Express	
OMEGA	1.0–5.2 µm, 05–1.1 µm, 1.2–5 µm
SPICAM	118–320 nm 1.0–1.7 µm
MRO	
CRISM	0.362–3.92 µm

8.2.1 Mineral Information from Remote Sensing

8.2.1.1 Reflection Spectroscopy Remote Sensing

Long-term continuous Earth-based telescopic observations have provided primitive insights into Mars although it has limited wavelength range and relatively low resolution (Singer 1985; Erard 2000). Figure 8.3 shows reflectance spectra of the bright and dark regions of Mars using data from the US Geological Survey Spectral Laboratory (peclab.cr.usgs.gov).

The reflectance increases between 0.3 and 0.75 µm in the bright region spectrum is attributed to absorption near 0.86 µm by ferric iron (Fe^{3+}). Previous researches consider it to be amorphous forms of iron oxides such

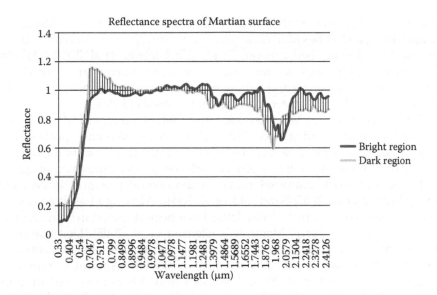

FIGURE 8.3
Reflectance spectra of the bright versus dark regions of Mars.

as palagonite since the spectrum is discrepant with Fe^{3+} occurring in crystalline form (Singer 1985).

The peak of reflectance at 0.75 μm in the dark region spectrum denotes the presence of ferrous iron (Fe^{2+}), indicating differences in oxidation states between the two regions. Absorptions near 1 μm are indicative of Mafic minerals such as pyroxenes. The presence of clay is suggested by the absorptions at 1.4, 1.9, and 3.0 μm that are diagnostic of OH and H_2O.

HST observations can provide more spatial resolution than ground-based observation (Bell et al. 1997). The ratio of color obtained from HST revealed Martian regional variation in composition. These results include the higher or fresher pyroxene concentrations in Syrits Major than Acidalia and Utopia Planitae and hemispheric differences in the distribution of hydrated mineral (Bell et al. 1997; Noe Dobrea et al. 2003).

Orbiting spacecraft can provide the highest resolution pictures of Mars. The thermal spectrometer (TES) onboard on MGS performed the first detailed mineralogic survey of the entire Martian surface (Christensen et al. 2001). TES is mainly designed to detect volcanic materials, particularly basalt (Christensen et al. 2000). The result is consistent with Earth-based observation and HST observations. Information about bedrock is obtained sparsely in bright regions since it is covered by dust (Bandfield 2002). Plagioclase and clinopyroxene-rich basalt dominate the dark region of the southern hemisphere. Dark regions of the northern hemisphere have been diagnostic of unaltered basaltic andesite, andesite, or weathered basalt (Mustard and Cooper 2005; Wyatt and McSween 2002; McSween et al. 2003). TES discovered outcrop of crystalline gray hematite (a-Fe_2O_3) in Terra Meridiani, Aram Chaos, and Valles Mariners, which were interpreted as a chemical precipitate from Fe-enriched aqueous fluids (Christensen et al. 2001). TES also determined olivine and orthopyroxenite, which occurred in regional concentrations, but carbonate and sulfate were not identified (Hoefen et al. 2003; Hamilton et al. 2003; Christensen et al. 2001; Bandfield 2002).

THEMIS onboard on Mars Odyssey and OMEGA onboard Mars Express extend the understanding from TES with higher spatial and spectral resolutions. More mineral diversity has been found. These results include the spatial distribution sparsity of olivine-rich basalt and exposition of quartz and plagioclase-rich granitoid rocks in the central peaks of impact craters (Mustard et al. 2005; Bandfield et al. 2004). Although basalt comprises much of the Martian crust, dacite lavas have been detected in some regions like Nili Patera in Syrtis Major (Christensen et al. 2005). Domination of high-calcium pyroxene in low-albedo volcanic regions and exposition of low-calcium pyroxene in bright outcrops within the ancient terrain were discovered (Mustard et al. 2005). Hydrated minerals are exposed in many of the ancient terrain units (Bibring et al. 2005; Poulet et al. 2005). Hydrated sulfates have been detected in the layered deposits within north polar region (Langevin et al. 2005; Gendrin et al. 2005). In the light of these findings, a new mineralogy-based Martian geological time scale has been suggested (Bibring et al. 2006). The Phyllosian period spans early to middle Noachian, when the planet experienced intense aqueous alteration, producing phyllosilicates discovered in the ancient terrains. The Phyllosian period was followed by the Theiikian period, which extended from the late Noachian through the early Hesperian, when characterized by acidic aqueous alteration processes that formed the sulfur deposits found in localized regions of Mars. The Siderikian period extends from the end of the Hesperian to the present day, when the Martian climate is characterized by lack of liquid water and alteration dominated by weathering and oxidation from atmospheric peroxides, which produced iron oxides that give the planet its familiar red color.

The CRISM is a visible-infrared hyperspectrometer aboard the MRO for detecting the mineral evidence for water at past or present. CRISM has provided most precise mineralogical distribution and analysis on large scale. Carbonate-bearing rocks were firstly identified in the Nili fossae region, which suggests that waters were neutral to alkaline when it formed and indicates that acidic weathering did not dominate all aqueous environments (Ehlmann et al. 2008). In 2009, an article in *Science* reported the ice exposition in new craters, which have been excavated by impact. These new craters and outcrop were discovered by CTX camera, but the CRISM finally has confirmed that the outcrop is composed of ice (Byrne et al. 2009). In addition, CRISM expanded greatly classes of hydrous minerals. Different types of clay mineral (also called phyllosilicates) were found in many places, including Fe/Mg smectite, chlorite, prehnite, kaolinite, serpentine, illite/muscovite, analcime, and others (Ehlmann et al. 2009).

8.2.1.2 Other Remote Sensing Methods

In addition to reflection spectroscopy, there are other methods that can be used to obtain mineral information. The GRS system aboard on Mars Odyssey consists of a GRS, a NS, and HEND. They can detect the H within the upper meter of the Martian surface based on thermal, epithermal, and fast neutron analysis (Feldman et al. 2004).

Figure 8.4 shows the distribution of H_2O, as indicated by H detected by the GRS system in low- and mid-latitude region of Mars (Boynton et al. 2004). The highest equatorial concentrations of hydrogen are found around and to the east of Apollineris (left and right center of map) and centered around Arabia Terra (center of map). This hydrogen may be in the form of hydrated minerals or buried ice deposits, but the former is more likely. The white sections at the top and bottom of the map represent regions of the planet with high hydrogen concentration due to large amounts of buried water ice; the locations of the five successful lander missions are marked: Viking 1 (VL1), Viking 2 (VL2), Pathfinder (PF), Spirit at Gusev (G), and Opportunity at Meridiani (M).

8.2.2 Mineral Information from Martian Meteorite

Information about Mars' evolution can be obtained from an analysis of Martian rocks, but none of the spacecraft missions to Mars has yet returned soil or rock samples. However, Martian meteorites provided scientists with Martian surface samples that can be analyzed in terrestrial laboratories.

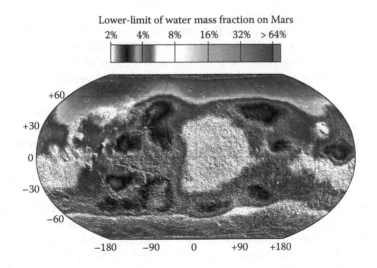

FIGURE 8.4
H abundance detected by GRS system. (NASA/JPL/Los Alamos National Laboratory PIA04907, 2003.)

The Martian meteorites can be subdivided into an orthopyroxenite (ALH84001), basaltic shergottites, olivine–phyric shergottites, lherzolitic shergottites (plagioclase-bearing peridotites), olivine-rich nakhlites, and dunites (chassignites) (McSween et al. 2003).

Geochemical analysis of Martian meteorites provides important insights into the early history of the Martian interior. All of the Martian meteorites have higher concentrations of pyroxene than plagioclase, which is opposite of the TES results for the dark regions. This result indicates that the SNCs originated from the dust-covered young volcanic regions (McSween 2002).

The compositional differences among the shergottites have been interpreted as contamination of the magma through inclusion of early crustal material (Borg et al. 1997). Alteration products resulting from fluid–rock interactions and carbonates formed by deposition of brines, evaporation of brines, or hydrothermal fluids reactions have been detected in the Martian meteorites. These results indicate that the Martian meteorites primarily come from younger volcanic regions that have interacted with groundwater and/or surface water throughout their histories.

8.2.3 Mineral Information from In Situ Analysis

In situ investigation comprises rover and lander, the rovers such as Spirit, Opportunity, and recently Curiosity, which all can move on the surface of Mars and directly sample the surface to get wider information than the landers like Viking and Pathfinder, which are unmovable. Whether lander or rover, the measurements are done in situ, so they can obtain more accurate measurementsa than Orbiter.

Six missions returned surface information from the Mars: Viking 1 lander (VL1) in Chryse Planitia, Viking 2 lander (VL2) in Utopia Planitia, MPF in the outwash region of Ares Valles, MER Spirit rover in Gusev Crater, MER Opportunity rover in Meridiani Planum, and MSL Curiosity rover in Gale crater.

VL1 and VL2 analyzed the soil with x-ray fluorescence spectrometers (Clark et al. 1977, 1982). The results suggested that the soil contained higher concentrations of Fe, S, Cl but lower Al than Earth, suggesting creation from mafic to ultramafic rocks. Soil composition investigated by each lander was almost identical due to deposition of airborne dust resulting from global dust storms (Rieder et al. 1997b; Yen et al. 2005).

MPF was the first mission to analyze Martian rocks. An APXS loaded on MPF can identify the elements out of range of the VL1 and VL2 detection limits. The surfaces of rock have been weathered and covered with dust (McSween et al. 1999; Morris et al. 2000). The rocks investigated by MPF appear to be of andesitic composition formed by fractionation of tholeiitic basaltic magmas during early melting of the Martian mantle (McSween et al. 1999). Rocks and adjacent soil have different composition, which suggested that the soils resulted from hydrous alteration of basalt mixed with material derived from the andesitic rocks at the landing site (Bell et al. 2000).

Gusev Crater was selected as the Spirit landing site because geomorphic analysis suggested that it was the site of a paleolake (Golombek et al. 2003). Rocks investigated by Spirit in Gusev Crater indicate the composition of volcanic rocks range from basalts to andesites and they were not derived from sedimentary origin but from olivine-rich basaltic lavaflows (Christensen et al. 2004; Gellert et al. 2004 and 2006; McSween et al. 2004 and 2006; Morris et al. 2004 and 2006). The rocks in Gusev have almost identical composition to the olivine-phyric shergottites, but different from the basaltic shergottites studied at the MPF landing site. These results indicate that the Gusev plains originated from the magmas, which generated in great depth the Martian mantle without experiencing fractionation (McSween et al. 2006). After roving volcanic plains, the Spirit rover climbed into the Columbia Hills. This elevated region is composed of basalts, ultramafic sedimentary rocks mixed with sulfates and clastic rocks, which undergo varying degrees of hydrous alteration (Squyres et al. 2006). The trenches dug by Spirit's wheels reveal a bright material, which contained high concentrations of salts suggestive of water activity within this region in the past.

Opportunity landed in Meridiani Planum where the largest outcrop of gray crystalline hematite has been detected by TES (Golombek et al. 2003). The cross-stratification, ripple patterns, and hematite spherules provided evidence that water had existed in this area (Squyres et al. 2004c). The hematite-rich spherules embedded in the sedimentary rocks are the unique compositional signature of this region. Hematite concretions like this maybe a result of the precipitation of ferrous fluids mixed with oxidizing undergroundwater (Chan et al. 2004; Morris et al. 2005). Vugs in the rock are probably molds of salt crystals formed in the rock and subsequently fell out or dissolved (Herkenhoff et al. 2004).

Compositional analysis revealed that most of the fine-grained materials in Meridiani Planum were derived from basalt, which contained high concentrations of various salts (Cl, Br, S) and sulfur minerals including jarosite $(NaFe_3(SO_4)_2(OH)_6)$ (Rieder et al. 2004; Klingelhöfer et al. 2004). These results suggested that Meridiani Planum have experienced acid–sulfate weathering in aqueous environment (Golden et al. 2005).

Phoenix is the first spacecraft to be operated in the Martian polar region. The WCL on the Phoenix Mars Lander performed many chemical analyses of Martian soil around the Phoenix landing site and identified various mineral classes, including chloride, bicarbonate, magnesium, sodium, potassium, calcium, carbonate, and perchorate. The presence of calcium carbonate in the soil indicates the possibility of a buffered system condition on Mars' surface, which plays an important role in habitability of the planet (Boynton 2009). In addition, the chemical analysis of Martian soil by Phoenix has revealed that perchlorate is the dominant form of chlorine in the Martian soil, which is relevant to water sequestration, soil, and atmospheric control, and resource utilization for human exploration (Hecht et al. 2009).

Curiosity was sent to explore the Gale crater, where a broad diversity of minerals was detected from orbit. An APXS analyzed the rock called "Jake_M" and found it is compositionally similar to terrestrial mugeraites, which suggests that alkaline magmas may be more abundant on Mars than Earth (Stopler et al. 2013). The ChemCam instrument identified two principal Martian soil types along the rover traverse: a fine-grained mafic type and a locally derived, coarse-grained felsic type (Meslin et al. 2013). Analysis of the soil with CheMin and XRD instruments identified plagioclase, forsteritic olivine, augite, and pigeonite, with minor K-feldspar, magnetite, quartz, anhydrite, hematite, and ilmenite (Bish et al. 2013). In addition, Curiosity has analyzed the Aeolian deposit in the region called Rocknest with various scientific instruments and has got similar features analyzed by Spirit and Opportunity in Gusev and Meridiani Planum (Leshin et al. 2013). The similarity among different Mars sites implies global-scale mixing of basaltic material or they have the similar basaltic source at all three locations (Blake et al. 2013).

8.3 Mineral Components at Gale Region from MRO CRISM Images

The aqueous alteration on Mars is the focus of Mars research, which is directly related to the existence of life on the Mars at past or present. Martian mineral detection and mapping can provide important information and constraints in Martian aqueous history, which can be used to assess the potential habitability of Mars. Degrees of addressing the key question for Martian aqueous alteration are dictated by the depth and extent of grasping the Martian hydrous mineral. It is important to know detailed minerals and chemical induction of the existence of water on the Martian surface at past or present in a large scale. In situ observations of the Martian rovers, such as Spirit, Opportunity, and Curiosity have provided the mineralogical analysis of Martian surface, but restricting in limited areas.

The TES onboard MGS has provided the first detailed mineralogical survey (Christensen et al. 2000). TES detected the outcrops of crystalline gray hematite in Meridiani Terra, Aram Chaos, and Valles Marineris, which indicates the existence of chemical precipitation of iron-enriched aqueous fluids. This argument was confirmed by Martian rover Opportunity, which was landed in the Merridiani Planum and almost immediately discovered evidence that water existed in this region (Squyres et al. 2004c).

The hydrated mineral detection by OMEGA (Observatoire pour la Mineralogie, L'Eau, Les Glaces et l'Activité) onboard Mars Express (Poulet et al. 2005) provided the mineralogical evidence of the water on early Mars, which supported preexisting geologic and isotopic arguments. Hydrated minerals, particularly phyllosilicates, were mostly exposed in the ancient terrain units, such as Noachian, which indicates the age correlation for specific mineral (Bibring et al. 2006).

All the spectrometers onboard this orbiter observe only specific bands within the wavelength range until NASA's MRO was launched from Cape Canaveral in 2005 with six payloads. The CRISM aboard the MRO is the first hyperspectral imaging spectrometer for Mars exploration, which has 544 channels covering the visible to near-infrared spectra from 0.4 to 4.0 µm, that can provide more information in spatial and time scale with enhanced spectral resolution. Using data from CRISM, more minerals can be discovered and identified.

The most common method for mineral identification is spectral analysis based on the specific feature of a given mineral, which has been used to get mineral information of the Martian surface. Although spectral analysis can be used to ascertain the composition of the target, more precise classification are needed to check the spectral angle (Ehlmann et al. 2009). The hyperspectral remote sensor makes calculation of spectral angle possible. Since the spectral angle method depends on the ship of spectral profile to classify the mineral rather than the diagnostic spectral absorption, using spectral angle method does not result in sufficient accuracy. In this chapter, combining spectral features with spectral angle methods, CRISM near-infrared spectral data are used to identify mineral classes and distribution at Martian Gale region.

8.3.1 Geologic Setting in Study Area

Gale Crater on Mars was chosen as the landing site of the Curiosity rover. The diameter of the Gale Crater is about 152 km, centered at 5.3°S, 222.3°W, in the northwest part of the Aeolis quadrangle region on Mars (Figure 8.1). The geologic terrain in Gale Crater spans the period from Noachian–Hesperian boundary to Amazonian (Cabrol 1999). Hence, mineral research for this region can provide information on the alteration of environments and geological evolution on Mars.

The Martian geologic history has been divided into Noachian, Hesperian, and Amazonian from ancient to present through studies of impact crater densities on the Martian surface. The Noachian epoch spans the period from the solidification of Martian crust to around 3.6 Ga (Hartmann and Neukum 2001). The detection of phyllosilicates in Noachian terrain indicates that liquid water was abundant on the Martian surface from early to middle Noachian. The corresponding geological process for phyllosilicate formation, such as leaching, pedogenesis, and hydrothermal precipitation, requires extensive time scale in aqueous and alkaline environment. Domination of phyllosilicates characterizes the early-middle Noachian as the Phyllosian period (Bibring et al. 2006).

The late Noachian was a mineralogical transitional period where the dominant mineral changed from the phyllosilicate to sulfate. Discovery of sulfate on Hesperian terrain indicates the precipitation of salts from acidic waters in the Hesperian (Gendrin et al. 2005; Bibring et al. 2006). Increasing volcanic activity during Late Noachian and Early Hesperian periods outgassed a large amount of sulfur dioxide into Martian atmosphere, which dissolved into water and created the sulfuric acid–rich environment. The mineralogical evidence for highly acidic and saline aqueous conditions discovered by Opportunity can be dated from this period (Squyres et al. 2004c). The formation of hydrated sulfates occurred in this acid environment, which characterizes the Theiikian Period.

Ferric oxides are the predominant mineral in late Hesperian and Amazonian. By the end of the early Hesperian, the climatic condition became arid and cool as the Martian atmosphere thinned and surface water diminished, which has continued until the present. The dominant geological process has been oxidation of the iron-rich rocks by atmospheric peroxides. The iron oxides have been formed in this epoch. Therefore, the mineral associated with liquid water was concentrated on Noachian and Hesperian terrains. Impact highlands on south equatorial Mars have most ancient terrains, which experienced less weathering in comparison with the Amazonian terrain.

Gale Crater has an enormous mound around its central peak, officially named Aeolis Mons with approximately 45 km by 90 km in area. The mound shows no unambiguous volcanic landforms (e.g., lava flows, vents, cones) and thus the rocks are considered to be sedimentary in origin (Malin and Edgett 2000). The presence of an Amazonian impact crater lake in Gale Crater indicates that water was present on the surface of Mars in relatively recent past (Scott and Chapman 1995). Studies of impact crater densities in Gale Crater have revealed that the formation age of the mound of deposit ranged from 3.6 to 3.8 Ga, which covered the Noachian/Hesperian boundary.

Gale Crater possesses a sequence of layered deposits that exceed several kilometers in depth. This type of layered sedimentary sequences is of particular interest in planetary exploration since they constitute one of the key discrepancies between bodies that retain an atmosphere/hydrosphere and those that lack them (Thomson et al. 2011). Specifically, thick and laterally extensive sedimentary deposits require long and suitable transportation and precipitation by wind or liquid water. The mound of layered material has a maximum relief of 5.2 km and an average height of 3.8 km, which is more than twice the average depth of the Grand Canyon on Earth of 1.6 km (Thomson et al. 2011). Therefore, the Gale was considered as being generated by aquifer/surface drainage and water infiltration in recent geological times (Newsom and Landheim 1999). Sedimentary sequence of mound in Gale Crater exhibits stratigraphic changes in lithology that are consistent with major mineralogical and climatic changes proposed by Bibring et al. (2006).

8.3.2 Observation and Data Processing

8.3.2.1 Instrument Description

The CRISM aboard the MRO is a first hyperspectral imager for Mars exploration. There are 544 channels covering the visible to near-infrared spectra from 0.4 to 4.0 µm. The CRISM instrument consists of two spectrometers ("S" and "L"). The "S" spectrometer covers the wavelength range from 0.36 to 1.05 µm and the "L" spectrometer covers the wavelength range from 1.0 to 3.9 µm. Spectral resolution is approximately 6.55 nm/channel and spatial resolution in the "full resolution targeted" or FRT mode is 15–19 m/pixel (Murchie et al. 2007).

CRISM is used to search evidence of aqueous or hydrothermal activity, and to map and characterize the composition, geology, and stratigraphy of surface features, particularly the presence of indicative minerals of aqueous alteration, for example, phyllosilicates, carbonate, sulfate, salt and oxides, which can be chemically altered or formed in the presence of water. All of these minerals have specific features in their visible-infrared spectral profile and can be identified by CRISM. In addition, CRISM is used to characterize seasonal variations in dust and ice particulates in the Martian atmosphere.

8.3.2.2 Data Preprocessing

We have studied numerous CRISM scenes over the Gale Crater region (Table 8.3). CRISM data from multiple spectra have been combined into three-dimensional images with two spatial and one spectral dimension, and

TABLE 8.3

Studied CRISM Scenes

frt000037df	frt0000a091	frt00016acd	frt00021d9c
frt000045f2	frt0000b5a3	frt00017327	frt000233ac
frt000058a3	frt0000b6f1	frt0001791f	frt0002376b
frt00007c09	frt0000bee7	frt00017d33	frt00024077
frt00007c09	frt0000bfca	frt00018285	frt000243ff
frt00007c09	frt0000c0ef	frt00018c29	frt0002456a
frt00007c09	frt0000c518	frt00018f9c	frt000248e9
frt00007c09	frt0000c620	frt00019dd9	frt00024c67
frt00007c09	frt0000cd48	frt0001b696	frt00021d9c
frt00007c09	frt00010820	frt0001bba1	frt000233ac
frt00007c09	frt0001117b	frt0001ca91	frt000233ac
frt00007c09	frt00012506	frt0001d889	frt0002376b
frt00007c09	frt0001422f	frt0001e238	frt000243ff
frt00007c09	frt000148b8	frt0001ebc6	frt0002456a
frt0000901a	frt0001646d	frt00020b9d	frt000248e9
frt000095ee	frt00016648	frt00021c92	

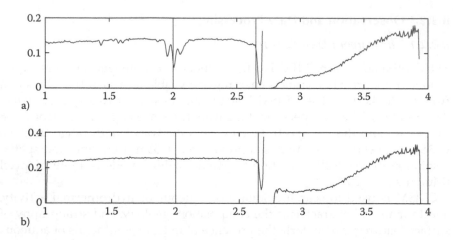

FIGURE 8.5
Spectral profile of atmospheric correction (up: original, down: corrected). The vertical line is at 2.0 μm and 2.65 μm.

the resultant data set for spectral features would allow us to identify surface mineralogy.

In order to avoid thermal emission effects, we utilize the 1.1–2.65 μm range in the "L" spectrometer data. The diagnostic spectra concentrated in the range of 1.1–2.65 μm. The most important absorption of water is 1.9 μm band depth. CRISM "L" spectrometer data are converted from calibrated radiance to apparent reflectance and to surface reflectance in a standard two-step process accounting for photometric correction and atmospheric corrections.

The CO_2 absorption effect has been removed by using a "volcano scan" atmospheric correction (Langevin et al. 2005). Figure 8.5 shows the atmospheric correction using the volcano scan method.

8.3.2.3 Spectral Analysis Method

The spectral parameters method has been successfully used in the past (Bell et al. 2000). The determination of specific mineral existence depends on the discovery of relevant spectral feature. The given spectral feature can be captured by a single parameter value. The parameters include surface parameters and atmospheric parameters. In this chapter, we focus on surface parameters, particularly highlighting the mafic mineralogy, hydrated silicates, and sulfate. Although we can capture the spectral feature of corresponding minerals using these parameters, some mistakes still cannot be avoided. Therefore, the quantitative thresholds have been developed for each parameter as the criteria. Due to other factors, for example, detection noise, aerosols, and continuum slope, the valid threshold varies image by image.

Although multiple spectral parameters can be used to determine the mineral classes, the mineral species cannot be discriminated with multiple spectral parameters alone. Therefore, in order to get a satisfactory distinction, we have utilized more parameters and specific mineral spectral feature for analysis.

A spectral angle mapper (SAM) is a physically based spectral classification that uses an n-D angle to match pixels for reference spectra. The algorithm determines the spectral similarity between two spectra by calculating the angle between the spectra and treating them as vectors in a space with dimensionality equal to the number of bands. The spectral angle can be demonstrated like this formula (Kruse et al. 1993):

$$a = \cos^{-1} \left[\sum_{i=1}^{n_b} t_i r_i \bigg/ \left(\sum_{i=1}^{n_b} t_i^2 \right)^{0.5} \left(\sum_{i=1}^{n_b} r_i^2 \right)^{0.5} \right] \qquad (8.1)$$

where t and r are spectra from spectral libraries and CRISM data cube's extraction, respectively; a is the spectral angle. SAM method compares the angle between the end-member spectrum vector and each pixel vector in n-D space. Smaller angles represent closer matches to the reference spectrum. Using this method, some minerals that are difficult to be determined by spectral feature alone can be recognized.

8.3.3 Results and Discussion

8.3.3.1 Mineral Species Identification

Based on multispectral parameters derived from reflectance in key wavelengths from CRISM observation, the main mineral classes have been determined. But it is insufficient for discriminating the mineral species with spectral parameters alone. In this section, the mineral species have been discriminated by using the spectral profile in conjunction with special spectral absorption. Table 8.4 shows the minerals species detected by CRISM.

8.3.3.1.1 Clay Mineral

Clay minerals are hydrous phyllosilicates containing variable amounts of cations, such as aluminium, iron, and magnesium. Clay minerals are common weathering and hydrothermal alteration products, which are very common in fine-grained sedimentary rocks. Clay minerals can be classified as T-O or T-O-T layer structure according to their fundamental construct of tetrahedral silicate sheets and octahedral hydroxide sheets. The T-O clay mineral includes kaolinite and serpentine group that consist of one tetrahedral sheet and one octahedral sheet, while the T-O-T clay mineral consists of an octahedral sheet sandwiched between two tetrahedral sheets like chlorites and smectite group.

TABLE 8.4

Hydrous Minerals Components in Gale Crater Detected by CRISM

Type	Group	Name	Composition	Points	%
Phyll.	Kaolins	Kaolinite	$Al_2Si_2O_5(OH)_4$	7792	0.0400
Phyll.	Serpentine	Serpentine	$(Mg, Fe)_3Si_2O_5(OH)_4$	7368	0.0379
Phyll.	Chlorite	Chamosite	$(Fe^{2+},Mg,Fe^{3+})^5Al(Si_3Al)$ $O_{10}(OH,O)_8$	28	0.00014
Phyll.	Smectite	Nontronite	$(CaO_{0.5},Na)_{0.3}Fe_2(Si,Al)_4O_{10}$ $(OH)_2\cdot nH_2O$	76	0.00039
Phyll.	Smectite	montmorillonite	$(Na,Ca)_{0.33}(Al,Mg)_2(Si_4O_{10})$ $(OH)_2\cdot nH_2O$	35	0.00018
Salt	Sulfate	Romerite/ jarosite	$Fe_3(SO_4)_4\cdot 14H_2O/$ $KFe_3(OH)_6(SO_4)_2$	3321	0.0171
Salt	Carbonate	Hydromagnesite	$Mg_5(CO_3)_4(OH)_2\cdot 4H_2O$	183887	0.9448
Salt	Carbonate	Northupite	$Na_3Mg(CO_3)_2Cl$	146130	0.7508
	Oxide	Lepidocrocite	γ-FeO(OH)	704	0.0036
	Oxide	Akaganéite	$Fe^{3+}O(OH,Cl)$	55	0.00028
	Oxide	Diaspore	AlO(OH)	31027	0.1594
	Oxide	Boehmite	γ-AlO(OH)	9	0.000046
Tectos	Zeolites	Zeolite/natrolite	$Na_2Al_2Si_3O_{10}\cdot 2H_2O$	2818	0.0145
	Phosphate	Apatite	$Ca_5(PO_4)_3(F,Cl,OH)$	2468911	12.6854
	Silicate	Pyrophyllite-talc	$Al_2Si_4O_{10}(OH)_2$ $- Mg_3Si_4O_{10}(OH)_2$	5	0.000026

There are five clay minerals detected in Gale Crater using SAM method, final identification of minerals need to comparison with laboratory spectra. The kaolinite-serpentine group: kaolinite and serpentine, the chlorite group: chamosite, and the smectite group: nontronite and montmorillonite. The dominant two clay minerals are kaolinite and serpentine, both belonging to kaolinite-serpentine group. They account for 90% of the clay minerals. In Earth, nontronite forms from the weathering of basalts, kimberlites, and other ultramafic igneous rocks, precipitation of iron and silicon-rich hydrothermal fluids and in the sea hydrothermal vents (Bischoff 1972). The presence of nontronite indicates moderate pH and reducing conditions at past in Gale Crater (Harder 1976), which are favorable for the preservation of organic material. Some evidence also suggests that microorganisms may play an important role in their formation (Köhler et al. 1994).

8.3.3.1.1.1 Kaolinite Kaolinite is a T-O clay mineral with the chemical composition $Al_2Si_2O_5(OH)_4$. This group includes kaolinite and other rare hydrated forms including dickite, nacrite, and halloysite (Giese 1988). Figure 8.6 shows the kaolinite spectral profile retrieved from CRISM data and its distribution in frt0000c518.

The two vertical full lines in Figure 8.6 are at 1.4 and 2.2 μm that are diagnostic of kaolinite. The absorption near 1.4 μm is caused by vibrations of inner

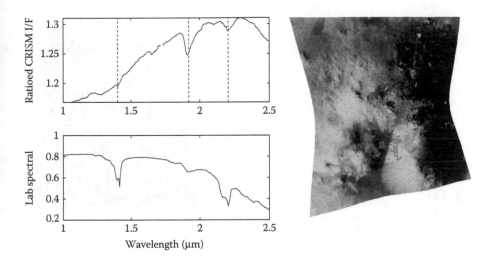

FIGURE 8.6
Kaolinite group mineral spectral profile and its distribution in frt0000ada4 (Nili Fossae).

hydroxyl groups between the tetrahedral and octahedral sheets in kaolin group minerals (Crowley and Vergo 1988; Clark et al. 1990). The absorption at 2.2 μm is due to a combination of vibrations of 2Al-OH groups in the kaolinite mineral structure (Clark et al. 1990). In addition, the absorption near 1.9μm shows the presence of crystallized water in hydrated mineral.

8.3.3.1.1.2 Smectite Smectite is a family of T-O-T phyllosilicate clays having permanent layer charge because of the isomorphous substitution in each of the octahedral sheet. The smectite group includes pyrophyllite, montmorillonite, Beidellite, nontronite, saponite, hectorite, and sauconite. They are formed by the alteration of mafic minerals, weak linkage, and substitution of cations.

The most common smectite on Earth is montmorillinite, the main constituent of bentonite derived by weathering of volcanic. In the Gale Crater region, the dominating smetite is nontronite detected from the CRISM data. Figure 8.7 shows CRISM spectra of iron smectite–bearing materials and its distribution in frt00019dd9.

Vertical lines in Figure 8.8 are placed near 1.40, 1.92, 2.30, 2.39 μm, among of which the two diagnostic spectral feature for smectite (marked as solid lines) are located at 1.92 μm and 2.30 μm. The absorption near 2.3 and 2.39 μm resulting from a combination of overtones of the Fe-OH, Mg-OH stretch, and indicates that the semetites in Gale Crater is more likely a mixture of montmorillonite and nontronite.

8.3.3.1.2 Salts: Sulfate and Carbonate
The presence of salts (sulfate and carbonate) in Gale Crater indicates that some minerals could precipitate on the surface of Mars. The most important

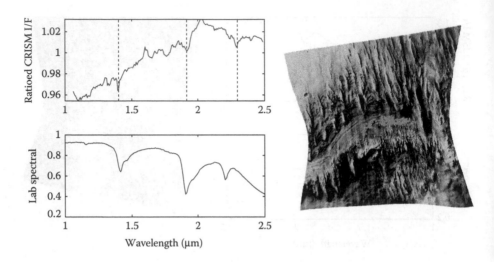

FIGURE 8.7
Smectite group mineral spectral profile and its distribution in frt00019dd9.

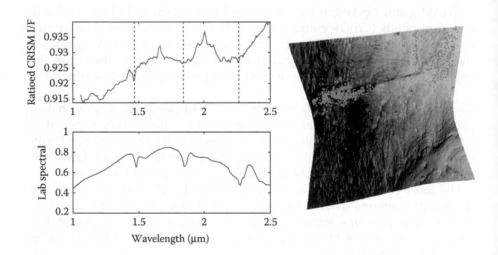

FIGURE 8.8
The spectra of jarosite and its distribution in frt0000a091.

sulfate is jarosite. It is hydrous sulfate of iron formed in ore deposits by the oxidation of iron sulfides and commonly associated with acid mine drainage and acid sulfate soil environments. There are three scenses positive for jarosite using SAM (tolerance = 0.100): frt00019dd9, hrs00004259, frt0000b5a3. Figure 8.8 shows the spectra of jarosite from CRISM scene and its distribution in frt0000a091. The vertical line is at 2.265, 2.46, 2.51, and 2.62 μm that match the jarosite spectral feature in library.

Northupite has been detected in Gale Crater using SAM (tolerance = 0.100). Northupite is an uncommon evaporate mineral. This mineral is formed in a water-rich environment. In addition, the water must evaporate for the mineral precipitation. This indicates that the water input in the environment remains below the net rate of evaporation. The presence of evaporation in Gale revealed the alteration of the environment from water-rich environment favorable to mineral dissolution to dryer climate favorable to water evaporation.

8.3.3.1.3 Oxide

Four oxides have been detected in Gale Crater using SAM (tolerance = 0.100): oxide-hydroxide, lepidocrocite, akaganéite, and diaspore. Boehmite and lepidocrocite are a kind of iron oxide-hydroxide mineral formed when iron-containing substances rust under water. Lepidocrocite is commonly found in the weathering of primary iron minerals and in iron ore deposits. It can be seen as rust scale inside old steel water pipes and water tanks. Akaganéite is an iron oxide-hydroxide/chloride mineral formed by the weathering of pyrrhotite. It has also been found in widely dispersed locations around the Earth and in rocks that were brought back from the Moon during the Apollo Project. Diaspore is an aluminium oxide hydroxide mineral. The mineral occurs as an alteration product of corundum or emery and is found in granular limestone and other crystalline rocks. Boehmite is dimorphous with diaspore. Boehmite occurs in tropical laterites and bauxites developed on alumino-silicate bedrock. It also occurs as a hydrothermal alteration product of corundum and nepheline. These oxides indicate the diversity of oxidizing environment in Martian geological history.

8.3.3.2 Mineral Classes Distribution

We have studied 50 CRISM scenes within the Gale Crater region. In this process, the spectral parameters are used to evaluate the mineral distribution (Pelkey et al. 2009). Most parameters have been designed to identify mineral classes rather than species. Six parameters are used to detect surface mineralogy in this research, which focus on features generated by mafic minerals and secondary alteration products, including phyllosilicates, sulfates, and carbonates.

Figure 8.9 shows the Frt000058a3 scene located in the northwest of Gale Crater, which has particular relevance between the landform and its spectral feature. This thick light-toned unit with its deep grooves has been eroded by the wind and may be a deposit of cemented, altered ash, or dust. It was probably deposited and eroded much later than the rest of the mound.

8.3.3.2.1 Mafic Minerals

Mafic minerals can be found in basalt, which mainly comprises Martian crust. Most mafic minerals have a dark color, which is caused by their richness in magnesium and ferric salts. Common rock-forming mafic minerals

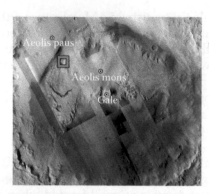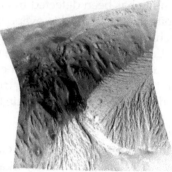

FIGURE 8.9
The scene located in the northwest of Gale Crater, and the coordinate of this scene is 4°47′52.80″ S, 137°26′38.40″E.

include olivine, pyroxene, amphibole, and biotite, which are composed of basalt, dolerite, and gabbro. On Mars, the basaltic rock comprises the flat plains of lava flows called flood basalts.

Mafic pyroclastic deposition produced by water–magma interactions or rapid ascent of deep-sourced magmas have been proposed for formation of Martian paterae, which is the oldest volcanic structure on Mars and are primarily concentrated around the Hellas impact basin (Crown and Greeley 1993; Gregg and Williams 1996). On a global scale, the pyroxenes mostly concentrate on the southern highland. The HCP dominates in low albedo volcanic regions and crater ejecta, such as Hesperian volcanic regions, while the LCP is found in moderate to bright outcrops within the ancient terrain, like Noachian crated region (Pelkey et al. 2009).

In 1976, VL1 and VL2 landed in Chryse Planitia and Utopia Planitia, which are the younger, lower-elevation northern plains. The result of soil analysis suggested that there was an accretion of mafic to ultramafic rock because of higher concentrations of iron, sulfur, and chlorine, lower aluminum than terrestrial soils (Clark et al. 1977 and 1982). HST revealed the higher or fresher pyroxene concentration in Syris Major when compared to Acidalia and Utopia Planitiae (Bell et al. 1997). In addition, the regional variations are also expressed as the distribution of hydrated minerals (Noe Dobrea et al. 2003).

TES onboard MGS confirmed the result of HST and revealed regional concentrations of olivine and pyroxenite and the plagioclase and clinopyroxene-rich basalt dominate in the dark regions of the southern hemisphere (Bandfield et al. 2000). THEMIS onboard Mars Odyssey have expanded the results from the TES with higher spectral spatial resolution and revealed that olivine-rich basalts occur in separated regions and outcrops of olivine are associated with some impact craters and basins (Christensen et al. 2005; Mustard et al. 2005).

Rocks investigation by Spirit in Gusev Crater discovered that soil on Gusev plains are derived from olivine-rich basaltic lava flows (Christensen et al. 2004). The similar composition with the olivine-phyric shergottties and distinction from the basaltic shergotties suggests that the source magmas of the Gusev plains originated from the great depth in the mantle without undergoing subsequent fractionation (McSween et al. 2006). The presence of olivine indicates that the Martian weathering has been dominated by dry process because of the rapid alteration of olivine in the presence of water. The most important parameters for detection of mafic minerals are those designed for olivine and pyroxene: OLINDEX, LCPINDEX, and HCPINDEX.

Figure 8.10 show rich pyroxenes both (LCP and HCP) in the northwestern lowland and rich olivine in the southeastern highland. Olivine and pyroxene are two important classes of rock-forming mafic minerals that have absorption bands in the visible/NIR that yield electronic crystal field transitions of iron in octahedral coordination (Burns et al. 1970). These diagnostic absorption features can be used to ascertain the minerals and their chemical composition (Adams 1974). OLINDEX was designed to detect the fayalite [Fe_2SiO_4], which is the iron-rich end-member of the olivine. It occurs in ultramafic volcanic and plutonic rocks, less common in felsic plutonic rocks. LCPINDEX and HCPINDEX were designed to detect the low-calcium (e.g., orthopyroxene) and high-calcium pyroxene (e.g., clinopyroxene).

Olivine has a solid solution series ranging from forsterite, Mg_2SiO_4 to fayalite, Fe_2SiO_4 (Deer 1963). Olivine [$(Mg,Fe)_2SiO_4$] has an absorption centered band near 1 mm, which varies in width, position, and shape with its iron content (Cloutis and Gaffey 1991). Pyroxenes [$(Ca,Fe,Mg)_2Si_2O_6$] can be recognized by the presence of two distinct absorptions centered near 1 and 2 μm, where the absorption band centers shift toward longer wavelengths with increasing calcium content. LCPs have the shorter wavelength band centers in the range from 0.9 to 1.8 μm, whereas high-calcium pyroxenes typically have longer wavelength band centers ranging from 1.05 to 2.3 μm (Burns et al. 1970).

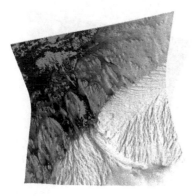 

FIGURE 8.10
Pyroxene distribution (left: LCP; right: HCP).

8.3.3.2.2 Hydrated Minerals

Hydrated minerals have H_2O or hydroxyl (OH) molecules attached to their crystal structure, also called hydrate. Although the snow line occurs beyond the Mars, close to Jupiter, hydrated minerals can form in the inner solar system. Asteroid collisions and comet impact can deliver the estimated Martian water inventory.

Infrared observation of Mars from Earth-based telescopes and HST provide evidence of presence of hydrated minerals and mineral distribution inhomogeneity on the Martian surface. The result of TES revealed that hydrated minerals, particularly phyllosilicates, are exposed in many of the older terrain units (Bibring et al. 2005; Poulet et al. 2005). Although carbonate has not been found, hydrated sulfates have been detected in layered deposits within the North Polar Region, which is the probable sedimentary origin (Langevin et al. 2005; Gendrin et al. 2005).

OMEGA's detection of hydrated minerals outcrop in Noachian indicates that the phyllosilicates formed early in Martian history by aqueous process were covered from weathering processes dominating throughout most of Martian history and have only recently been exposed. This discovery by Bibring et al. (2006) led to characterize the early to middle Noachian as the Phyllosian period when aqueous alteration is dominant on the Martian surface.

The detection of hydrated minerals relies on the use of multiple parameters BD1900 and D2300. BD1900 gauges the depth of an absorption band at 1.9 µm due to the presence of H_2O, while BD1900 alone is insufficient to make mineralogic distinctions. Therefore, the D2300 or D2400 should be included. D2300 is designed to capture spectral features observed in hydrated silicates, which gauges the decrease near 2.3 µm due to absorption of metal-OH vibrations (Clark et al. 1990). D2400 gauges the drop near 2.4 µm features and result from absorption of H_2O in a sulfate lattice. Therefore, it is more sensitive to hydrated sulfates. Figure 8.11 shows the distribution of hydrated

FIGURE 8.11
Images at 1.9 µm band from BD1900 (left) and 2.3 µm drop from D2300 (right).

minerals determined by BD1900 and D2300 (left and right) overlaid on an Frt000058a3 for context. Blue areas represent hydrated material, while the greens indicate hydrated silicates.

8.3.3.2.3 Sulfate and Carbonate

Differentiation and core formation of the Mars led to iron and sulfur compounds getting deposited in deep planetary interiors, not planetary crusts. Sulfate minerals on the Martian surface are mainly formed in late Noachian and early Hesperian periods when volcanic activity outgassed a large amount of sulfur, which interacted with liquid water to produce the sulfate, which characterizes the Theiikian Period, followed by the Phyllosian period. Spirit has detected the ultramafic sedimentary rocks cemented with sulfates in Columbia Hills. These rocks indicate diverse aqueous alteration. Compositional analysis from Opportunity found that the rocks contain high concentrations of various salts (Cl, Br, S) including sulfur minerals jarosite $(NaFe_3(SO_4)_2(OH)_6)$ (Rieder et al. 2004; Klingelhöfer et al. 2004).

8.4 Summary of Martian Mineral

Mineral information gained from remote sensing, Martian meteorite analysis, and in situ analysis from lander and rover combined with increasing computer technology have revealed that the planet has distinctive mineralogic periods, which generally correlate with the well-established geologic periods. The Fe-rich volcanic materials dominate on Mars, particularly in dust-covered bright regions. The low-albedo regions can be subdivided into basalt (southern highlands) and andesite (northern plains), which evolved from magmas resulting from fractionation or the result of weathering of basaltic materials.

Mineral component results at the Gale region from MRO CRISM images show the distribution of the important minerals classes, including mafic mineral, hydrated mineral, and sulfate and carbonate. Mafic minerals have been distributed widely. The olivine-dominated southeastern highland and the pyroxenes were distributed mainly in the northwestern region of the scene. The hydrated mineral mainly distributed on both sides along the downstream of riverbed in southeastern highland. The sulfate and carbonate dominated the northwestern region and upstream of riverbed. Furthermore, Mars has experienced geologic and climatic changes throughout its 4.5 Ga of existence. Phyllosilicates detected in ancient terrains evinced that water was abundant in the Martian early history. Sulfate deposits began to form near the end of the Noachian period when acid-sulfate weathering has dominated. The presence of olivine indicated that dry weather has dominated the Mars for the last 3 Ga, since the minerals like olivine is rapidly altered in aqueous environment.

The MRO CRISM data are used to identify mineral and produce the more precise classification, which can be used to ascertain the geological and climate alteration processed on Mars. Furthermore, more understanding of the rock and mineral distribution of Martian surface on a large scale can be useful for selecting landing sites of the MER in the future.

Acknowledgments

This research was supported by the National Basic Research Program of China (973 Program) (Grant No. 2012CB720000), Main Direction Project of Chinese Academy of Sciences (Grant No. KJCX2-EW-T03), and Shanghai Science and Technology Commission Project (Grant No. 12DZ2273300).

References

Adams, J.B. (1974). Visible and near-infrared diffuse reflectance spectra of pyroxenes as applied to remote sensing of solid objects in the solar system. *J. Geophys. Res.*, 79, 4829–4836.

Albee, A.L., Arvidson, R.E., Palluconi, F., et al. (2001). Overview of the mars global surveyor mission. *J. Geophys. Res.* 106, 23291–23316.

Bandfield, J.L. (2002). Global mineral distributions on Mars. *J. Geophys. Res.* 107, 5042.

Bandfield, J.L., Hamilton, V.E., Christensen, P.R. (2000). A global view of Martian volcanic compositions. *Science* 287, 1626–1630.

Bandfield, J.L., Hamilton, V.E., Christensen, P.R., et al. (2004). Identification of quartz-ofeldspathic materials on Mars. *J. Geophys. Res.* 109, E10009.

Bell, J.F., McSween, H.Y., Crisp, J.A., et al. (2000). Mineralogic and compositional properties of Martian soil and dust: Results from Mars Pathfinder. *J. Geophys. Res.* 105, 1721–1755.

Bell, J.F., Wolff, M.J., James, P.B., et al. (1997). Mars surface mineralogy from Hubble Space Telescope imaging during 1994–1995: Observations, calibration, and initial results. *J. Geophys. Res.* 102, 9109–9123.

Bibring, J.P., Langevin, Y., Gendrin, A., et al. (2005). Mars surface diversity as revealed by the OMEGA/Mars Express observations. *Science* 307, 1576–1581.

Bibring, J.P., Langevin, Y., Mustard, J.F., et al. (2006). Global mineralogical and aqueous Mars history derived from OMEGA/Mars Express data. *Science* 312, 400–404.

Bischoff, J.L. (1972). Clays and clay minerals at a glance. *Clay Miner.* 20, 217–223.

Bish, D.L., Blake, D.F.,Vaniman, D.F., et al. (2013). X-ray diffraction results from Mars science laboratory: Mineralogy of Rocknest at Gale Crater. *Science* 314.

Bishop, J.L., Murad, E., Dyar, M.D. (2002). The influence of octahedral and tetrahedral cation substitution on the structure of smectites and serpentines as observed through infrared spectroscopy. *Clay Miner.* 37, 617–628.

Blake, D.F., Morris, R.V., Kocurek, G., et al. (2013). Curiosity at Gale crater, Mars: Characterization and analysis of the Rocknest sand shadow. *Science* 341.

Borg, L.E., Nyquist, L.E., Taylor, L.A., et al. (1997). Constraints on Martian differentiation process from Rb–Sr and Sm–Nd isotopic analyses of the basaltic shergottite QUE94201. *Geochim. Cosmochim. Acta* 61, 4915–4931.

Boynton, W.V., Feldman, W.C., Mitrofanov, I.G., et al. (2004). The Mars Odyssey gamma-ray spectrometer instrument suite. *Space Sci. Rev.* 110, 37–83.

Boynton, W., Ming, D., Kounaves, S., Young, S., Arvidson, R., Hecht, M., Hoffman, J., Niles, P., Hamara, D., Quinn, R., et al. Evidence for calcium carbonate at the Mars Phoenix landing site. Science, American Association for the Advancement of Science, 2009, 325, 61–64.

Burns, R.G. (1970). *Mineralogic Applications of Crystal Field Theory*, Cambridge Univ. Press, Cambridge.

Byrne, S., Dundas, C.M., Kennedy, M.R., et al. (2009). Distribution of mid-latitude ground ice on Mars from new impact craters. *Science* 325, 1674–1676.

Cabrol, N.A., Grin, E.A. (1999). Distribution, classification, and ages of Martian impact crater lakes. *Icarus*, 142, 160–172.

Chan, M.A., Beitler, B., Parry, W.T., et al. (2004). A possible terrestrial analogue for haematite concretions on Mars. *Nature* 429, 731–734.

Chicarro, A. (2002). Mars Express mission and astrobiology. *Solar Syst. Res.* 36, 487–491.

Christensen, P.R., Bandfield, J.L., Clark, R.N., et al. (2000). Detection of crystalline hematite mineralization on Mars by the thermal emission spectrometer: Evidence for near-surface water. *J. Geophys. Res.* 105, 9632–9642.

Christensen, P.R., Bandfield, J.L., Hamilton, V.E., et al. (2001). Mars global surveyor thermal emission spectrometer experiment: Investigation description and surface science results. *J. Geophys. Res.* 106, 23823–23871.

Christensen, P.R., McSween, H.Y., Bandfield, J.L., et al. (2005). Evidence for magmatic evolution and diversity on Mars from infrared observations. *Nature* 436, 504–509.

Christensen, P.R., Ruff, S.W., Fergason, R.L., et al. (2004). Initial results from the Mini-TES experiment in Gusev Crater from the Spirit rover. *Science* 305, 837–842.

Clark, B.C., Baird, A.K., Rose, H.J., et al. (1977). The Viking x-ray fluorescence experiment: Analytical methods and early results. *J. Geophys. Res.* 82, 4577–4594.

Clark, B.C., Baird, A.K., Weldon, R.J., et al. (1982). Chemical composition of Martian fines. *J. Geophys. Res.* 87, 10059–10067.

Clark, R.N., Swayze G.A., Singer, R.B., et al. (1990). High-resolution reflectance spectra of Mars in the 2.3-micron region: Evidence for the mineral scapolite. *J. Geophys. Res.* 95, 14463–14480.

Cloutis, E.A., Gaffey, M.J. (1991). Pyroxene spectroscopy revisited: Spectral-compositional correlations and relationship to geothermometry. *J. Geophys. Res.* 96, 22809–22826.

Crowley, J.K., Vergo N. (1988). Near-infrared reflectance spectra of mixtures of kaolin-group minerals: Use in clay mineral studies, *Clays Clay Miner.* 36, 310–316.

Crown, D.A., Greeley, R. (1993). Volcanic geology of Hadriaca Patera and the eastern Hellas region of Mars. *J. Geophys. Res.* 98, 3431–3451.

Deer, W.A., Howie, R.A. Zussman, J. (1963). Orthosilicates, vol. 1. *A of Rock Forming Minerals*. Longmans Green, London.

Ehlmann, B.L., Mustard, J.F., Murchie, S.L., et al. (2008). Orbital indentification of carbonate-bearing rocks on Mars. *Science* 322, 1828–1832.

Ehlmann, B.L., Mustard, J.F., Swayze, G.A., et al. (2009). Identification of hydrated silicate minerals on Mars using MRO-CRISM: Geologic context near Nili Fossae and implications for aqueous alteration. *J. Geo. Res.* 114, E00D08, 1–33.

Erard, S. (2000). The 1994–1995 apparition of Mars observed from Pic-du-Midi. *Planet. Space Sci.* 48, 1271–1287.

Feldman, W.C., Prettyman, T.H., Maurice, S., et al. (2004). Global distribution of near-surface hydrogen on Mars. *J. Geophys. Res.* 109, E09006.

Frost, R.L., Kloprogge, J.T., Ding, Z. (2002). Near-infrared spectroscopic study of non-tronites and ferruginous smectite. *Spectrochim. Acta A* 58, 1657–1668.

Gellert, R., Rieder, R., Anderson, R.C., et al. (2004). Chemistry of rocks and soils in Gusev Crater from the alpha particle x-ray spectrometer. *Science* 305, 829–836.

Gellert, R., Rieder, R., Brückner, J., et al. (2006). Alpha particle x-ray spectrometer (APSX): Results from Gusev Crater and calibration report. *J. Geophys. Res.* 111, E02S05.

Gendrin, A., Mangold, N., Bibring, J.P., et al. (2005). Sulfates in Martian layered terrains: The OMEGA/Mars Express view. *Science* 307, 1587–1591.

Giese, R.F. (1988). Kaolin minerals: Structures and stabilities, In *Reviews in Mineralogy*, vol. 19, Hydrous phyllosilicates (Exclusive of Micas), edited by S. W. Bailey, pp. 29–62, Mineral. Soc. of Am., Washington, D.C.

Golden, D.C., Ming, D.W., Morris, R.V., et al. (2005). Laboratory-simulated acid-sulfate weathering of basaltic materials: Implications for formation of sulfates at Meridiani Planum and Gusev Crater, Mars. *J. Geophys. Res.* 110, E12S07.

Golombek, M.P., Anderson, R.C., Barnes, J.R., et al. (1999). Overview of the Mars Pathfinder Mission: Launch through landing, surface operations, data sets, and science results. *J. Geophys. Res.* 104, 8523–8553.

Golombek, M.P., Grant, J.A., Parker, T.J., et al. (2003). Selection of the Mars Exploration Rover landing sites. *J. Geophys. Res.* 108, 8072.

Gregg, T.K.P., Williams, S.N. (1996). Explosive mafic volcanoes on Mars and Earth: Deep magma sources and rapid rise rate. *Icarus* 122, 397–405.

Hamilton, V.E., Christensen, P.R., McSween, H.Y., et al. (2003). Searching for the source regions of martian meteorites using MGS TES: integrating Martian meteorites into the global distribution of igneous materials on Mars. *Meteorit. Planet. Sci.* 38, 871–885.

Hanel, R.A., Conrath, B.J., Jennings, D.E., et al. (2003). *Exploration of the Solar System by Infrared Remote Sensing*, 2nd edn. Cambridge, UK: Cambridge University Press.

Harder, H., (1976). Nontronite synthesis at low temperatures. *Chem. Geol.* 18, 169–180.

Hartmann, W.K., Neukum, G. (2001). Cratering chronology and the evolution of Mars. *Space Sci. Rev.* 96, 165–194.

Hecht, M., Kounaves, S., Quinn, R., West, S., Young, S., Ming, D., Catling, D., Clark, B., Boynton, W., Hoffman, J., et al. (2009). Detection of perchlorate and the soluble chemistry of Martian soil at the Phoenix lander site. *Science, American Association for the Advancement of Science,* 325, 64–67.

Herkenhoff, K.E., Squyres, S.W., Arvidson, R., et al. (2004). Evidence from Opportunity's microscopic imager for water on Meridiani Planum. *Science* 306, 1727–1730.

Hoefen, T.M., Clark, R.N., Bandfield, J.L., et al. (2003). Discovery of olivine in the Nili Fossae region of Mars. *Science* 302, 627–630.

Kieffer, H.H., Jakosky, B.M., and Snyder, C.W. (1992). The planet Mars: From antiquity to the present. In *Mars*, ed. H.H. Kieffer, B.M. Jakosky, C.W. Snyder, and M.S. Matthews. Tucson, AZ: University of Arizona Press, pp. 1–33.

King, T.V.V., Clark, R.N. (1989). Spectral characteristics of chlorites and Mg-serpentines using high resolution reflectance spectroscopy. *J. Geophys. Res.* 94, 13997–14008.

Klingelhöfer, G., Morris, R.V., Bernhardt, B. et al. (2003). Athena MIMOS II Mössbauer spectrometer investigation. *J. Geophys. Res.* 108, 8067.

Klingelhöfer, G., Morris, R.V., Bernhardt, B., et al. (2004). Jarosite and hematite at Meridiani Planum from Opportunity's Mössbauer spectrometer. *Science* 306, 1740–1745.

Köhler, B., Singer, A., and Stoffers, P. (1994). Biogenic nontronite from marine white smoker chimneys. *Clays and Clay minerals* 42: 689–701.

Kruse, F.A., Lefkoff, A.B., Boardman, J.B., et al. (1993). The spectral image processing system (SIPS)—Interactive visualization and analysis of imaging spectrometer data. *Remote Sens. Env.* 44, 145–163.

Langevin, Y., Poulet, F., Bibring, J.P., et al. (2005). Sulfates in the north polar region of Mars detected by OMEGA/Mars Express. *Science* 307, 1584–1586.

Leshin, L.A., Mahaffy, P.R., Webster, P.R., et al. (2013). Volatile, isotope, and organic analysis of Martian fines with the Mars Curiosity Rover. *Science* 341.

Malin, M.C., Edgett, K.S. (2000). Sedimentary rocks of early Mars. *Science* 290, 1927–1937.

McSween, H.Y. (2002). The rocks of Mars, from far and near. *Meteorit. Planet. Sci.* 37, 7–25.

McSween, H.Y., Arvidson, R.E., Bell, J.F., et al. (2004). Basaltic rocks analyzed by the Spirit rover in Gusev Crater. *Science* 305, 842–845.

McSween, H.Y., Grove, T.L., Wyatt, M.B. (2003). Constraints on the composition and petrogenesis of the Martian crust. *J. Geophys. Res.* 108, 5135.

McSween, H.Y., Murchie, S.L., Crisp, J.A., et al. (1999). Chemical, multispectral, and textural constraints on the composition and origin of rocks at the Mars Pathfinder landing site. *J. Geophys. Res.* 104, 8679–8715.

McSween, H.Y., Wyatt, M.B., Gellert, R., et al. (2006). Characterization and petrologic interpretation of olivine-rich basalts at Gusev Crater, Mars. *J. Geophys. Res.* 111, E02S10.

Meslin, P.Y., Gasnault, O., Forni, O., et al. (2013). Soil diversity and hydration as observed by chemcam at Gale Crater, Mars. *Science* 341.

Morris, R.V., Golden, D.C., Bell, J.F., et al. (2000). Mineralogy, composition, and alteration of Mars Pathfinder rocks and soils: Evidence from multispectral, elemental, and magnetic data on terrestrial analogue, SNC meteorite, and Pathfinder samples. *J. Geophys. Res.* 105, 1757–1817.

Morris, R.V., Klingelhöfer, G., Bernhardt, B. et al. (2004). Mineralogy at Gusev crater from the Mössbauer spectrometer on the Spirit rover. *Science* 305, 833–836.

Morris, R.V., Klingelhöfer, G., Schröder, C., et al. (2006). Mössbauer mineralogy of rock, soil, and dust at Gusev Crater, Mars: Spirit's journey through weakly altered olivine basalt on the plains and pervasively altered basalt in the Columbia Hills. *J. Geophys. Res.* 111, E02S13.

Morris, R.V., Ming, D.W., Graff, T.G. et al. (2005). Hematite spherules in basaltic tephra altered under aqueous, acid-sulfate conditions on Mauna Kea volcano, Hawaii: Possible clues for the occurrence of hematite-rich spherules in the Burns formation at Meridiani Planum, Mars. *Earth Planet. Sci. Lett.* 240, 168–178.

Murchie, S.L., Arvidso, R., Bedini, P., et al. (2007). Compact Reconnaissance Imaging Spectrometer for Mars (CRISM) on Mars Reconnaissance Orbiter (MRO). *J. Geophys. Res.* 112, E05S03.

Murchie, S.L., Murchie, S.L., Seelos, F.P., Hash, C.D., Humm, D.C., Malaret, E., McGovern, J.A. and Choo, T.H., Seelos, K.D., Buczkowski, D.L., Morgan, M.F., et al. (2009). Compact Reconnaissance Imaging Spectrometer investigation and data set from the Mars Reconnaissance Orbiter's primary science phase, *J. Geophys. Res.*, 114, E00D07.

Mustard, J.F., Cooper, C.D. (2005). Joint analysis of ISM and TES spectra: the utility of multiple wavelength regimes for Martian surface studies. *J. Geophys. Res.* 110, E05012.

Mustard, J.F., Poulet, F., Gendrin, A., et al. (2005). Olivine and pyroxene diversity in the crust of Mars. *Science* 307, 1594–1597.

NASA/JPL PIA07241, (2005). http://photojournal.jpl.nasa.gov/catalog/PIA07241

NASA/JPL/Los Alamos National Laboratory PIA04907, (2003). http://photojournal.jpl.nasa.gov/catalog/PIA04907

Newsom, H.E., Landheim, R. (1999). Hydrogeologic evolution of Gale Crater and its relevance to the exobiological exploration of Mars. *Icarus* 139, 235–245.

Noe Dobrea, E.Z., Bell, J.F., Wolff, M.J., et al. (2003). H_2O- and OH- bearing minerals in the Martian regolith: Analysis of 1997 observations from HST/NICMOS. *Icarus* 166, 1–20.

Pelkey, S.M., Mustard, J.F., Murchie, S., et al. (2009). CRISM multispectral summary products: Parameterizing mineral diversity on Mars from reflectance. *J. Geophys. Res.* 112, E08S14.

Poulet, F., Bibring, J.P., Mustard, J.F., et al. (2005). Phyllosilicates on Mars and implications for early Martian climate. *Nature* 438, 623–627.

Rieder, R., Economou, T., Wänke, H., et al. (1997b). The chemical composition of Martian soil and rocks returned by the mobile alpha proton x-ray spectrometer: Preliminary results from the X-ray mode. *Science* 278, 1771–1774.

Rieder, R., Gellert, R., Anderson, R.C., et al. (2004). Chemistry of rocks and soils at Meridiani Planum from the alpha particle x-ray spectrometer. *Science* 306, 1746–1749.

Rieder, R., Gellert, R., Brückner, J., et al. (2003). The new Athena alpha particle x-ray spectrometer for the Mars Exploration rovers. *J. Geophys. Res.* 108, 8066.

Rieder, R., Wänke, H., Economou, T., et al. (1997a). Determination of the chemical composition of Martian soil and rocks: The alpha proton x-ray spectrometer. *J. Geophys. Res.* 102, 4027–4044.

Romanek, C.S., Grady, M.M., Wright, I.P., et al. (1994). Record of fluid–rock interactions on Mars from the meteorite ALH84001. *Nature* 37, 655–657.

Scott, D.H., Chapman, M.G. (1995). Geologic and topographic maps of the Elysium Paleolake basin, Mars. U. S. *Geol. Survey Geol. Series Map I-2397*, scale 1: 5,000,000.

Singer, R.B. (1985). Spectroscopic observation of Mars. *Adv. Space Res.* 5, 59–68.

Soffen, G.A. (1977). The Viking project. *J. Geophys. Res.* 82, 3959–3970.

Squyres, S.W., Arvidson, R.E., Bell, J.F., et al. (2004a). The Spirit rover's Athena science investigation at Gusev Crater, Mars. *Science* 305, 794–799.

Squyres, S.W., Arvidson, R.E., Bell, J.F., et al. (2004b). The Opportunity rover's Athena science investigation at Meridiani Planum, Mars. *Science* 306, 1698–1703.

Squyres, S.W., Arvidson, R.E., Blaney, D.L., et al. (2006). Rocks of the Columbia Hills. *J. Geophys. Res.* 111, E02S11.

Squyres, S.W., Grotzinger, J.P., Arvidson, R.E., et al. (2004c). In situ evidence for an ancient aqueous environment at Meridiani Planum, Mars. *Science* 306, 1709–1714.

Stopler, E.M., Baker, M.B., Newcombe, M.E., et al. (2013). The petrochemistry of Jake_M: A Martian Mugearite. *Science* 314.

Thomson, B.J., Bridges, N.T., Millike, R., et al. (2011). Constraints on the origin and evolution of the layered mound in Gale Crater, Mars using Mars reconnaissance orbiter data. *Icarus* 214, 413–432.

Warren, P.H. (1998). Petrologic evidence for low-temperature, possibly flood-evaporitic origin of carbonates in the ALH 84001 meteorite. *J. Geophys. Res.* 103, 16759–16773.

Wyatt, M.B., McSween, H.Y. (2002). Spectral evidence for weathered basalt as an alternative to andesite in the northern lowlands of Mars. *Nature* 417, 263–266.

Yen, A.S., Gellert, R., Schröder, C., et al. (2005). An integrated view of the chemistry and mineralogy of Martian soils. *Nature* 436, 49–54.

9

Anomalous Brightness Temperature in Lunar Poles Based on the SVD Method from Chang'E-2 MRM Data

Yi Lian, Sheng-bo Chen, Zhi-guo Meng, Ying Zhang, Ying Zhao, and Peng-ju Guo

CONTENTS

9.1 Introduction 209
    9.1.1 Instrument and TB Data Sets..210
9.2 Methodology and Model 211
    9.2.1 Hour Angle Correction 211
    9.2.2 Solar Solar Altitude Angle 213
    9.2.3 Terrain Screen Angle..213
    9.2.4 Microwave Radiative Transfer Simulation...............................214
    9.2.5 SVD Model..216
9.3 Results and Analysis..216
    9.3.1 The Permanently Shadowed Regions 216
    9.3.2 Brightness Temperature Map of Different Hour Angle 217
    9.3.3 Simulation of Detecting Depth 217
    9.3.4 Regional Temperature Anomalies of the Moon by SVD
         Model..218
9.4 Conclusion and Discussion 220
References..220

9.1 Introduction

Interest in the Moon has been steadily increasing since the beginning of the twenty-first century with the launch of Lunar Prospector, Clementine, LRO, SELENE detector, Chang'E-1/Chang'E-2, and Chandrayaan-1 (Jin, Arivazhagan, Araki 2013; Wei et al. 2013), particularly on the rear of the Moon and its polar regions. The problem of water and ice is one of the hot topics in current lunar science research (Neal 2009).

After the Apollo missions, a number of regional and global iron and titanium abundance measurements have been made using a variety of remote sensing techniques including laser altimeter, lunar double ground radar, aynthetic aperture radar of foundation, neutron probe, and microwave radiometer. The idea that there is water or ice on the Moon was first proposed by Watson et al. in 1961. They considered that there may be permanent shadow areas where the sunshine cannot reach at the bottom of an impact crater at the lunar polar region according to the terrain data, and water or ice may be present there. They also speculate on the existing way and state of water and ice in lunar soil. Based on lunar double ground radar, Nozette et al. (1996) found that equidirectional polarization increased exactly on the lunar permanent shadow regions, while it never happened in the area where sunshine can reach, which indicates that the increase of equidirectional polarization occurs only in the permanent shadow areas and suggests possibly the existence of lunar soil water ice.

Butler, Muhleman, and Slade (1993), and Vasavada, Paige, and Wood (1999) further explored the existing form and distribution scope of lunar polar water ice. Based on the synthetic aperture radar of foundation, Stacy mapped the lunar polar using double polarization method to search for water ice in lunar polar permanent shadow areas (Vasavada, Paige, and Wood 1999). Feldman detected hydrogen base on neutron detectors and thought that water ice content is about 0.3–1% buried at about 40 cm below the surface of the Moon at the lunar poles in permanent shadow regions (Feldman et al. 1998 and 2000). And with the successful launch of Chang'E, detecting water ice using a microwave radiometer was possible.

Passive microwave mainly gets lunar brightness temperature with certain penetrability to conduct the inversion of dielectric permittivity and soil thickness (Dobson et al. 1985; Wang et al. 2010). As the existence of water ice will affect lunar brightness temperature and dielectric permittivity, a microwave radiometer can do the study on lunar polar water ice. Based on simplified radiative transfer model and experiments, Meng has proved the feasibility of finding water ice using microwave radiometer data (Meng 2008). Zhang et al. analyzed lunar polar temperature anomalies combined with DEM data and thought that the anomalies arise mainly because of the water ice (Zhang et al. 2009).

This chapter analyzes lunar light conditions and looks for permanent shadow regions where water ice may exist by using DEM data. The different time and frequency brightness temperature data from microwave radiometer were obtained by angle correction and kriging interpolation. The different frequency detecting depths of microwave radiometer were simulated by radiation transfer model and detecting depths of 19 and 37 GHz were found near the depth of buried water ice. Then abnormal brightness temperature changes of 37 and 19 GHz channels were analyzed based on SVD model combined with the data after processing to look for the possible location where water ice may exist.

9.1.1 Instrument and TB Data Sets

CE-2, China's scientific mission to explore planetary bodies beyond the Earth, was launched on October 1, 2010. One set of microwave radiometers with four frequency channels, 3, 7.8, 19.35, and 37 GHz the same as CE-1 was onboard to measure microwave emission from the Moon Table 9.1. (Zheng et al. 2008). The multichannel microwave radiometer in the Chinese Chang'E lunar project will be the first onboard the lunar exploration satellite.

All four channels are estimated to have a sensitivity of 0.5 K in brightness temperature. The spatial resolutions on the lunar surface are in the range of 35–50 km. CE-1's microwave radiometry data set is the first obtained from a passive sensor in lunar orbit and covers the entire Moon in unprecedented spatial resolution and temporal span. CE-2 attempts to change the height of the orbit and better coverage.

CE-2's microwave radiometer (MRM) data preprocessing flow was described in Table 9.2. Our work in this chapter is based on the MRM level 2C dataset, which are stored in PDS format. Measurements in each individual track are stored in a single file. Each file contains a header and a table of measured data. The header provides the background data information and describes the meaning of every table column. The MRM level 2C data include solar incident and azimuth angles, which describe the position of the Sun. Information on the lunar surface latitude and longitude, data sampling time to calculate the corresponding time of sampling points in the lunar are now available. It is now publicly available on the Web site http://moon.bao.ac.cn.

9.2 Methodology and Model

9.2.1 Hour Angle Correction

To correct Chang'E2 microwave radiometer data, we need to introduce the concept of the hour angle. In astronomy and celestial navigation, the hour angle is one of the coordinates used in the equatorial coordinate system to

TABLE 9.1

Major Technical Specifications of CE-2 Microwave Radiometer

Frequencies (GHz)	3	7.8	19.35	37
Bandwidth (MHz)	100	200	500	500
Integration time (ms)	200	200	200	200
Temperature sensitivity (K)	≤0.5	≤0.5	≤0.5	≤0.5
Linearity	≥0.99	≥0.99	≥0.99	≥0.99
3 dB beam width	E:15+2° H:12+2°	E:9+2° H:9+2°	E:9+2 H:10+2°	E:10+2°H:10+2°
Footprint (from 200 km orbital altitude)	30	30	30	30

Source: Zheng, Y.C., et al. *Planet. [J] Space Sci,* 56, 881–886, 2008.

TABLE 9.2

Definition of MRM Data at Various Level of Preprocessing

Levels	Data Preprocessing Description	Physical Meaning
Raw data	Raw data is baseband data received from data transmission receiver in the two ground stations	Bit-stream data
Level 0A	Produced after the procedure of frame synchronization, descrambling, and decompression. The coordinated universal time was added	MRM source package data
Level 0B	Processed after sequencing, removing duplication, and optimized stitching of Level 0A data, which were received from the two ground stations. The detached, format reorganized, stitched data, which were acquired in one calibrating and measurement period, were put into one MRM data block	MRM data block
Level 1	Several level 0B data block were incorporated into one track data file. The science data (sampled by observation antenna), temperature data (sampled at different part of the instrument), calibration data (sampled by cold space calibration antenna), and auxiliary data were separated. MRM voltage data were produced after physical quantities conversion. After adding the description about the data and orbit parameter, level 1 data were produced	MRM voltage data
Level 2A	Firstly, antenna temperature was retrieved from level 1 MRM voltage data. Level 2A data were produced after the justification of data validity and rectification	MRM antenna temperature data
Level 2B	Level 2B data were produced after geometric positioning of MRM antenna temperature data	Positioned MRM antenna temperature data
Level 2C	Brightness temperature of the Moon, level 2C data, were retrieved from antenna temperature of level 2B data	Brightness temperature of the Moon

give direction of a point on the celestial sphere. The hour angle of a point is the angle between two planes. The angle may be measured in either degrees or time, which are mutual convertible 24^h precisely equals to 360°.

In the MRM level 2C data, the local time for each observation record is characterized by two coordinates in the horizon coordinate system (HCS), the solar incidence and azimuth angle, which describe the position of the Sun from observer's prospective. However, since the local time should be latitude independent, it is more formal to characterize the position of the Sun using the lunar equatorial coordinate system (LECS) instead.

The transformation between the HCS and the LECS leads to the following expression of the solar hour angle in terms of the solar incidence and azimuth angle (Zheng et al. 2012). The following equations can be derived from projecting a unit vector along with the direction of the Sun and different axes in the observer's local frame:

$$\sin \theta \sin \phi = \sin i \sin a \tag{9.1}$$

$$\cos \theta = \sin \lambda \cos i + \cos \lambda \sin i \cos a \tag{9.2}$$

$$\sin \theta \cos \phi = \cos \lambda \cos i - \sin \lambda \sin i \cos a \tag{9.3}$$

where i = the incident angle (zenith distance) $[0,\pi]$, a = the azimuth angle (clockwise from north) $[0,2\pi]$, h = the hour angle (negative before noon, positive after noon) $[-\pi,\pi]$, θ = the polar angle (polar distance) $[0.\pi]$, ϕ = the negative hour angle (positive if the Sun is on the west side) = $-h$ $[-\pi,\pi]$, λ = the latitude $[-\pi/2,\pi/2]$.

Dividing the first equation by the third one, one obtains

$$\tan \phi = \sin a \tan i / (\cos\lambda - \sin \lambda \cos a \tan i) \tag{9.4}$$

Given the latitude (λ), incident angle (i), and azimuth (a), the function returns the value of the hour angle.

9.2.2 Solar Solar Altitude Angle

For a site on Earth, the altitude of the Sun refers to the angle between sunlight incident direction and the horizon, which is represented by h. It is equal to the horizontal height of Sun in the celestial horizon coordinate system. Following sunup to sundown, the solar altitude at the same place is constantly changing within a day and the maximum occurs at noon. Furthermore, the largest solar altitude changes with the change of solar declination. δ stands for solar declination and φ represents geographic latitude of observation point. The following formula is used:

$$\sin H = \sin\varphi\sin\delta + \sin\varphi\cos\delta \tag{9.5}$$

So, $H = 90° - (\varphi - \delta)$ for the northern hemisphere and $H = 90° - (\delta - \varphi)$ for the southern hemisphere. The largest solar altitude of different places at the same latitude is changing with the distance between one place and direct sunlight point.

9.2.3 Terrain Screen Angle

As the sunlight can be shielded by the slope of target or surrounding terrain, the direct solar radiation cannot reach the surface of Moon. ξ is stands of this character, which is a binary factor. The target can be lighted when ξ = 1 and the target cannot be lighted when ξ = 0.

When the solar altitude is less than the slope angle of solar incident direction, the sunlight will be shielded by slope and cannot reach the target. When the solar altitude is less than terrain shading angle in all directions, no matter how the Moon orbits, the Sun will be shielded by the surrounding terrain and also cannot reach the target pixels. So only the sunlight is not shielded by the target itself. Only the target's terrain and the surrounding terrain can receive light.

As shown in Figure 9.1, on the solar azimuth direction BC, we set a large enough search length, d which is 10° in this chapter. Combined with the DEM data, we can calculate maximum slope angle in all directions and the slope angle computation formula as follows, taking AB as an example.

$$\tan \alpha = \frac{AB}{AG} \tag{9.6}$$

Respectively calculate the angle between every intersection angle and the point B, and the maximum angle is the terrain screen angle.

9.2.4 Microwave Radiative Transfer Simulation

Lately, Shkuratov (2001), Ya-Qiu Jin (Jin, Yan, and Liang 2003), and Wang et al. (2010) showed that the lunar regolith layer model is a two-layer model of the lunar regolith and lunar rocks (two-layer model). The lunar regolith layer model should consider the temperature distribution in the lunar regolith layer, which divided by the dielectric constant gives the effective reflectivity of the lunar surface (Figure 9.2).

According to microwave radiative transfer theory and second-rate approximation theory (Ulaby, Moore, and Fung 1981), the brightness temperature

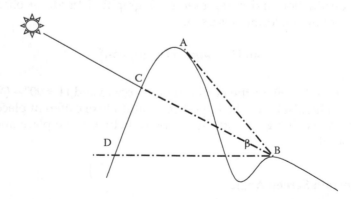

FIGURE 9.1
Terrain screen angle.

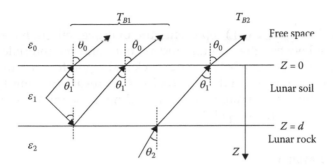

FIGURE 9.2
Radiative transfer model of passive microwave in lunar regolith.

(T_R) is made up of the brightness temperature T_{B1}, which is from the radiation in the dense medium, and the brightness temperature T_{B2}, which is from the radiation in the underlying surface under microwave radiometer observations direction angle θ.

$$T_B = T_{B1} + T_{B2} = T_{1up} + T_{1dn} + T_{2up} \tag{9.7}$$

$$T_{1up} = \int_0^d \frac{1-r_{p1}}{1-L} k_{a1}(z)T(z)\sec\theta_1 e^{-\int_0^z k_{a1}(z')\sec\theta_1 dz'} dz \tag{9.8}$$

$$T_{1dn} = \int_0^d \frac{(1-r_{p1})r_{p2}}{1-L} k_{a1}(z)T(z)\sec\theta_1 e^{-\left(\int_z^d k_{a1}(z')\sec(\theta_1)dz' + \int_0^d k_{a1}(z')\sec\theta_1 dz'\right)} dz \tag{9.9}$$

$$T_{2up} = \int_d^\infty \frac{(1-r_{p1})(1-r_{p2})}{1-L} k_{a2}(z)T(z)\sec\theta_2 e^{-\int_d^z k_{a2}(z')\sec\theta_2 dz'} dz \cdot e^{-\int_0^d k_{a1}(z')\sec\theta_1 dz'} \tag{9.10}$$

Here θ_1 and θ_2 are the exit angles of the microwave in the lunar soil and lunar rocks, which meet the refraction theorem. d is the thickness of the lunar regolith layer.

$k = 2\pi\varepsilon'' / (c / \sqrt{\varepsilon_i})$, $r_{ij} = |R_{ij}|^2$ $(i = 0,1,2; j = 1,2,3)$ is the absorption coefficient of lunar soil ($i = 1$) and the absorption coefficient of lunar rocks. f represents the frequency. μ1, μ2 indicates the magnetic permeability of the lunar soil and lunar rocks, respectively. If the Moon is assumed to be a smooth spherical object, rp1, rp2 is the reflectance of the vacuum and the regolith interface and the reflectance of the rock and the regolith interface. And

$$L = r_{p1}r_{p2}e^{-2\int_0^d k_{a1}(z)\sec\theta_1 dz}$$, $1/(1\text{-}L)$ represents the multiple reflection coefficient of the interface between the vacuum and regolith and the multiple reflection coefficient of the interface between the rock and regolith. $\varepsilon1$, $\varepsilon2$ are dielectric constants of the regolith and lunar rocks. As a result, the simulated lunar surface brightness temperature varies according to the thickness and dielectric constant of the lunar regolith.

9.2.5   SVD Model

The singular value decomposition (SVD) is a diagnosis method that is used to study the relational structure of a two-variable field. It unfolds in two covariance based on maximum, and calculates separately cross-covariance matrix of a singular value and orthogonal, left and right singular vector and time factor, and then pairing of singular vectors constitute a SVD spatial mode. Heterogeneous is the correlation between the time factor of the left (right) singular vectors and the time series of the right (left) observation, also known as nonhomogeneous. Heterogeneous shows a certain mode time variation of a field effect on another field conditions (namely heterogeneous correlation field), and its significant correlation zone represents another field affected by the impact of the most critical modal area. Homogeneous is the correlation between the time factor of the left (right) singular vectors and the time series of the left (right) observation.

Homogeneous shows a certain mode time variation of a field effect on their own field conditions (namely homogeneous correlation field), and its significant correlation zone represents the most critical modal area of their own changes. Most researchers have used the anisotropic correlation coefficient to analyze the relationship between left field and right field. This chapter also discusses time and spatial distribution characteristics of the anisotropic correlation coefficient.

9.3   Results and Analysis

9.3.1   The Permanently Shadowed Regions

The constant shadow regions at the lunar poles are extracted using the DEM data of Chang'E-1. If the maximum altitude angle is greater than solar altitude at noon and the terrain elevations in calculation area are below the line of sight, the grid I is obscured and regarded as constant shadow area. Light rate of each position in polar region can be obtained, which shows that permanently shadowed regions mainly exist in high latitudes and are

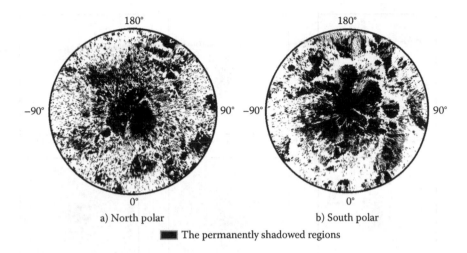

a) North polar   b) South polar

■ The permanently shadowed regions

FIGURE 9.3
The result of the permanently shadowed regions for the North Pole a) and the South Pole b).

distributed in the relatively large and deep impact craters, with the quantity in South Pole being more than that in the North Pole (Figure 9.3).

9.3.2 Brightness Temperature Map of Different Hour Angle

Based on hour angle correction and kriging interpolation, different time and 19 and 37 GHz frequency microwave brightness temperature values can be obtained. CE-2 microwave radiometer data were divided into different time periods according to the hour angle. Then, mapping is done on the microwave brightness temperature map of the Moon for different frequency and time in 1°*1° grid using kriging interpolation.

Figure 9.4 is a map of brightness temperature distribution of the lunar South Pole in 37 GHz, for example. It is shown that the lunar brightness temperature generally reduces with the rise of latitude, which is because the lunar physical surface temperature that impacts brightness temperature reduces gradually from the equator to the poles on account of varying solar radiation distribution. Generally, brightness temperature distribution is obviously related to the latitude and brightness temperature is not the same even on the same latitude, which is because the local characteristics of microwave radiation are influenced by the lunar terrain and reflectivity of the lunar surface.

9.3.3 Simulation of Detecting Depth

Based on radiation transfer model, the microwave detecting brightness temperature is under the influence of some factors like temperature distribution in the soil layer, dielectric constant, and lunar soil thickness. Assuming that lunar surface temperature is 71.28 K, deep soil (below 1 m) temperature

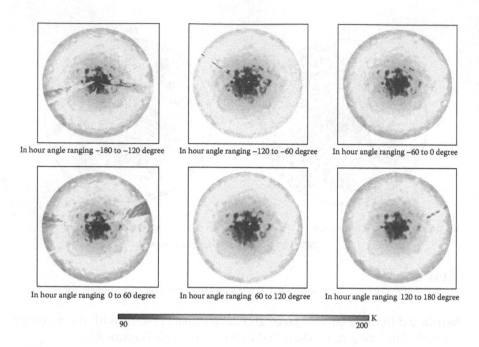

In hour angle ranging −180 to −120 degree   In hour angle ranging −120 to −60 degree   In hour angle ranging −60 to 0 degree

In hour angle ranging 0 to 60 degree   In hour angle ranging 60 to 120 degree   In hour angle ranging 120 to 180 degree

90   200   K

FIGURE 9.4
Brightness temperature at South Pole for the 37 GHz channel on the full sphere by kriging interpolation method.

is 134 K, instrument observation angle is 0°, lunar soil dielectric constant e1 = 2.42(1+0.00486i), and lunar rock dielectric constant e2 = 7.8(1+0.056i), the brightness temperature is calculated using 37 and 19 GHz frequency at thickness of the soil from 0.1 to 4 m. It can be seen from Figure 9.2 that detecting depth of 37 GHz is 0.6 m, and detecting depth of 19 GHz is 1.4 m. Because water ice is buried 40 cm below the lunar surface, 37 GHz can only detect the shallow parts of the water ice, and 19 GHz can detect the presence of deep water ice. This chapter will research on abnormal brightness temperature changes of 19 and 37 GHz, as the abnormal brightness temperature is likely caused by water ice (Figure 9.5).

9.3.4. Regional Temperature Anomalies of the Moon by SVD Model

Seeking the clues of water ice is one of the main goals of current lunar explorations, and water ice exists in constant shadowed areas. Using the SVD method, the brightness temperature temporal variation of 19 and 37 GHz in constant shadow areas are analyzed to research brightness temperature variation characteristics and interrelation of two detection channels and variation difference of depth of brightness temperature changes.

Based on the previous analysis, Figure 9.6 shows the correlation between the distribution of 19 GHz and the distribution of 37 GHz brightness

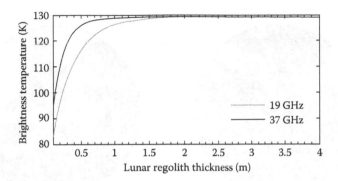

FIGURE 9.5
Change of the microwave brightness temperature with lunar regolith thickness.

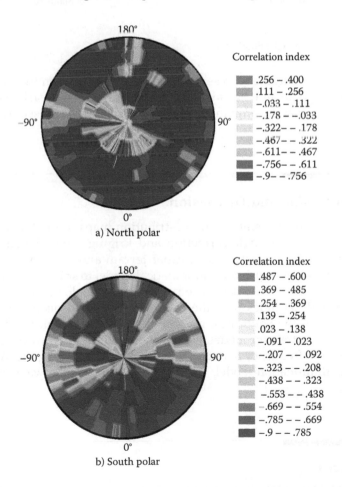

FIGURE 9.6
SVD analysis between brightness temperature in 37 and 19 GHz.

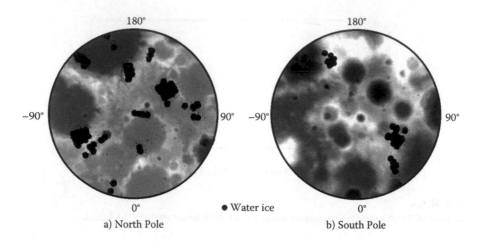

a) North Pole   ● Water ice   b) South Pole

FIGURE 9.7
The area of water ice in the South Pole.

temperature. The figure represents the differences between the radiation of 1.5 m lunar regolith and the radiation of 0.5 m lunar regolith in behavior over time. We think that the brightness temperature change is caused by water ice (Figure 9.7).

9.4 Conclusion and Discussion

The lunar brightness temperature distribution maps of the lunar poles are obtained using hour angle correction and kriging interpolation methods, which reflects the characteristic of lunar terrain and reflectivity very well. Furthermore, we establish the mathematical model to study the polar illumination condition by DEM data. The permanently shadowed regions mainly located in high latitudes with the relatively large and deep impact crater. The quantity at the South Pole is more than that at the North Pole. Based on the radiative transfer model, we get the depth of microwave radiometer and choose the frequency to detect the water ice. Combining the SVD model and lunar illumination model, the position with possible water ice can be obtained.

References

Butler BJ, Muhleman DO, Slade MA. Mercury: Full-disk radar images and the detection and stability of ice at the North Pole. *J Geophys Res*, 1993, 98(E8): 15003–15023.

Dobson MC, Ulaby FT, Hallikainen MT, et al. Microwave dielectric behavior of wet soil: Part II: dielectric mixing models. *IEEE Trans Geosci Remote Sens*, 1985, 23(l): 35–46.

Feldman WC, Lawrence DJ, Elphic RC, et al. Chemical information content of lunar thermal and epithermal neutrons. *J Geophys Res*, 2000, 105(E8): 20347–20363.

Feldman WC, Maurice S, Binder AB, et al. Fluxes of fast and epithermal neutrons from lunar prospector: Evidence for water ice at the lunar poles. *Science*, 1998, 281(5382): 1496–1500.

Jin Y, Yan F, Liang Z. Simulation of brightness temperature from the Lunar surface using multi-channel microwave radiometers. *Chin J Radio Sci*, 2003, 18: 477–486.

Jin SG, Arivazhagan S, Araki H. New results and questions of lunar exploration from SELENE, Chang'E-1, Chandrayaan-1 and LRO/LCROSS. *Adv Space Res*, 2013, 52(2): 285–305.

Meng Z. Lunar regolith parameters retrieval using radiative transfer simulation and look-up table. Doctoral dissertation. Changchun: Jilin University, 2008.

Neal CR. The moon 35 years after Apollo: What's left to learn? *Chem der Erde*, 2009, 69(1): 3–43.

Nozette S, Lichtenberg CL, Spudis P, et al. The Clementine biostatic radar experiment. *Science*, 1996, 274(5292): 1495–1498.

Ulaby FT, Moore RK, Fung A. *Microwave Remote Sensing*. Reading, MA: Addison-Wesley-Longman, 1981.

Vasavada AR, Paige DA, Wood SE. Near-surface temperatures on Mercury and the Moon and the stability of polar ice deposits. *Icarus*, 1999, 141(2): 179–193.

Wang ZZ, Li Y, Jiang JS, Li DH. Lunar surface dielectric constant, regolith thickness, and 3He abundance distributions retrieved from the microwave brightness temperatures of CE-1 Lunar Microwave Sounder. *Sci China Ser D*, 2010, 53(9): 1365–1378.

Wei E, Yan W, Jin SG, Liu J, Cai J. Improvement of Earth orientation parameters estimate with Chang'E-1 VLBI observations. *J Geodyn*, 2013, 72, 46–52.

Zhang WG, Jiang JS, Liu HG, et al. Distribution and anomaly of microwave emission at lunar South Pole (in Chinese). *Sci China Ser D-Earth Sci*, 2009, 39(8): 1059–1068.

Zheng YC, Tsang KT, Chan KL, Zou YL, Zhang F, Ouyang ZY. First microwave map of the Moon with Chang'E-1 data: The role of local time in global imaging [J]. *Icarus*, 2012, 219(1): 194–210.

Zheng YC, Ouyang ZY, Li CL, Liu JZ, Zou YL. China' Lunar Exploration Program: Present and future. *Planet. [J] Space Sci*, 2008, 56(7): 881–886.

10

Mercury's Magnetic Field in
the MESSENGER Era

Johannes Wicht and Daniel Heyner

CONTENTS

10.1 Introduction 223
10.2 Mercury's Internal Structure 225
10.3 Mercury's External Magnetic Field 231
10.4 Mercury's Internal Magnetic Field 235
10.5 Numerical Models of Mercury's Internal Dynamo.....................................241
    10.5.1 Dynamo Theory...241
    10.5.2 Standard Earth-Like Dynamo Models 246
    10.5.3 Dynamos with a Stably Stratified Outer Layer.....................................248
    10.5.4 Inhomogeneous Boundary Conditions..252
    10.5.5 Alternatives 255
10.6 Conclusion 257
Acknowledgments..258
References...259

10.1 Introduction

In 1974, the three flybys of the Mariner 10 spacecraft revealed that Mercury has a global magnetic field. This was a surprise for many scientists since an internal dynamo process was deemed unlikely because of the planet's relative small size and its old inactive surface (Solomon 1976). Either the iron core would have already solidified completely or the heat flux through the core–mantle boundary (CMB) would be too small to support dynamo action. The Mariner 10 measurements also indicated that Mercury's magnetic field is special (Ness et al. 1974). Being 100 times smaller than the geomagnetic field, it seems too weak to be supported by an Earth-like core dynamo. And though the data were scarce, they nevertheless allowed to constrain that the internal field is generally large scale and dominated by a dipole but possibly also a sizable quadrupole contribution. Both the Hermean field amplitude and its geometry are unique in our solar system.

Mercury is the closest planet to the Sun and therefore subject to a particularly strong and dynamic solar wind. Since Mercury's magnetic field is so weak, the solar wind plasma can come extremely close to the planet and may even reach the surface. Mariner 10 data showed that Mercury's magnetosphere is not only much smaller than its terrestrial counterpart but also much more dynamic. Adapted models originally developed for Earth failed to adequately describe the Hermean magnetosphere, which therefore remained little understood in the Mariner 10 era (Slavin et al. 2007).

Knowing a planet's internal structure is crucial for understanding the core dynamo process. Mercury's large mean density pointed toward an extraordinary huge iron core and a relatively thin silicate mantle covering only about the outer 25% in radius. Since little more data were available in the Mariner era, the planet's interior properties and dynamics remained poorly understood.

Solving the enigmas about Mercury's magnetic field and interior were major incentives for NASA's MESSENGER mission (Solomon et al. 2007). After the launch in August 2004 and a first Mercury flyby in January 2008, the spacecraft went into orbit around the planet in March 2011. At the date of writing, more than 2800 orbits have been completed. MESSENGER's orbit is highly eccentric with a periapsis between 200 and 600 km at 60–70° northern latitude and an apoapsis of about 15,000 km altitude. This has the advantage that the spacecraft passes through the magnetosphere on each orbit but complicates the extraction of the internal field component because of a strong covariance of equatorially symmetric and antisymmetric contributions (Anderson et al. 2012; Johnson et al. 2012). The trade-off between the dipole and quadrupole field harmonics, which was already a problem with *Mariner 10* data, therefore remains an issue in the MESSENGER era. The situation is further complicated by the fact that the classical separation of external and internal field contributions developed by Gauss (Olsen, Glassmeier, and Jia 2010) does not directly apply at Mercury. It assumes that the measurements are taken in a source-free region with negligible electric currents, an assumption not necessarily fulfilled in such a small and dynamic magnetosphere.

In order to nevertheless extract information on the internal magnetic field, the MESSENGER team analyzed the location of the magnetic equator (MEQ) where B_ρ, the magnetic field component perpendicular to the planet's rotation axis, passes through zero (Anderson et al. 2011, 2012). Since the internal field changes on a much slower time scale than the magnetosphere, the time-averaged location should basically not be affected by the magnetospheric dynamics. The analysis not only confirmed that the Hermean field is exceptionally weak with an axial dipole of only 190 nT but also suggested that the internal field is best described by an axial dipole that is offset by 480 km to the north of the planet's equator (Anderson et al. 2012). This configuration, which we will refer to as the MESSENGER offset dipole model (MODM) in the following, requires a strong axial quadrupole and a very low dipole tilt, a combination that is unique in our solar system.

This chapter attempts to summarize the new understanding of Mercury's magnetic field in the MESSENGER era at the date of writing. MESSENGER is still orbiting its target and continues to deliver outstanding data that will further improve our knowledge of this unique planet. Section 10.3 briefly reviews the current knowledge of Mercury's magnetosphere. Section 10.4 describes recent models for the planet's interior, focusing in particular on the possible core dynamics. The MEQ analysis and the offset dipole model MODM are then discussed in Section 10.4. Explaining the weakness of Mercury's magnetic field already challenged classical dynamo theory and the peculiar field geometry further raises the bar. Section 10.5 reanalyzes several dynamo model candidates in the light of the new MESSENGER data. Some concluding remarks in Section 10.6 close the chapter.

10.2 Mercury's Internal Structure

MESSENGER observations of Mercury's gravity field (Smith et al. 2012) and Earth-based observations of the planet's spin state (Margot et al. 2012) provide valuable information on the interior structure. The fact that Mercury is in a special rotational state (Cassini state 1) allows to deduce the polar moment of inertia C from the degree two gravity moments and the planet's obliquity, the tilt of the spin axis to the orbital normal (Peale 1969). The moment of inertia factor $C/(MR_M^2)$, where M is the planet's total mass and R_M its mean radius, contains the interior mass distribution. The factor is 0.4 for uniform density and decreases when the mass is increasingly concentrated toward the center. The Hermean value of $C/(MR_M^2) = 0.346 \pm 0.014$ (Margot et al. 2012) indicates a significant degree of differentiation.

The observation of the planet's 88-day libration amplitude g_{88}, a periodic spin variation in response to the solar gravitational torques on the asymmetrically shaped planet, allows to also deduce the moment of inertia of the rigid outer part C_m. If the iron core is at least partially liquid, C_m is the moment of the silicate shell and thus smaller than C. The Herman value of $C_m/C = 0.431 \pm 0.025$ (Margot et al. 2012) confirms that the core remains at least partially liquid.

In addition to M and R_M, the ratios $C/(MR_M^2)$ and C_m/C provide the main constraints for models of Mercury's interior (Smith et al. 2012; Hauck et al. 2013). Note that Rivoldini and Van Hoolst (2013) follow a somewhat different approach, taking into account the possible coupling between the core and the silicate shell. The coupling has the effect that C_m cannot be determined independently of the interior model and Rivoldini and Van Hoolst (2013) therefore directly use g_{88} rather than C_m as a constraint. The updated interior modeling indicates that the core radius is relatively well constrained at 2020 ± 30 km (Hauck et al. 2013) or 2004 ± 39 km (Rivoldini and Van Hoolst

2013). This leaves only the outer 16–19% of the mean planetary radius R_M = 2440 km to the mantle.

Hauck et al. (2013) found a mean mantle density (including the crust) of $3380 \pm 200 \, kg/m^3$. Measurements of MESSENGER's X-Ray Spectrometer (XRS) show that the volcanic surface rocks have a low content of iron and other heavier elements (Nittler et al. 2011). Smith et al. (2012) and Hauck et al. (2013) therefore speculate that a solid FeS outer core layer may be required to explain the mean mantle density. Rivoldini and Van Hoolst (2013), however, argue that the mantle density is not particularly well constrained. Compositions compatible with XRS measurements are well within the allowed solutions and a denser lower mantle layer is not required by the data.

Naturally, information about the core is of particular interest for the planetary dynamo. There is a rough consensus on the mean core density with Hauck et al. (2013) and Rivoldini and Van Hoolst (2013) suggesting 6980 ± 280 km/m^3 and $7233 \pm 267 \, km/m^3$, respectively. However, the core composition and the radius of a potential inner core are not well constrained. Admissable interior models cover all inner core radii from zero to very large values with an aspect ratio of about $a = r_i/r_o = 0.9$ (Rivoldini and Van Hoolst 2013) where r_i and r_o are the inner and outer core radii, respectively.

An additional constraint on the inner core size relies on the observations of the so-called lobate scarps on the planet's surface, which are likely caused by global contraction. MESSENGER data based on 21% of the surface suggested a contraction between 1 and 3 km (Di Achille et al. 2012). This sets severe bounds on the amount of solid iron in Mercury's core because of the density decrease associated with the phase transition of the liquid core alloy. Several thermal evolution models therefore favor a completely liquid core or only a very small inner core (Grott, Breuer, and Laneuville 2011; Tosi et al. 2014). Recent more comprehensive MESSENGER observations, however, allow for a contraction of up to 7 km. This somewhat releases the constraints (Solomon et al. 2014), though very large inner cores may still be unlikely.

Sulfur has been found in many iron–nickel meteorites and is therefore a prime candidate for the light constituent in Mercury's core. Rivoldini and Van Hoolst (2013) consider iron–sulfur core alloys and find a likely bulk sulfur concentration of 4.5 ± 1.8 wt%. Since this composition lies on the iron-rich side of the eutectic, iron crystalizes out of the liquid when the temperature drops below the melting point. Where this happens first depends on the form of the melting curve and the adiabat describing core conditions.

Since Mercury's mantle is so thin, it has likely cooled to a point where mantle convection is very sluggish or may have stopped altogether (Grott, Breuer, and Laneuville 2011; Michel et al. 2013; Tosi et al. 2014). The heat flux through the CMB is thus likely subadiabatic and therefore too low to support a core dynamo driven by thermal convection alone. The required additional driving power may then either be provided by a growing inner core or by an iron snow zone. The solid inner core starts to grow as soon as the adiabat crosses the melting curve in the planetary center. Since the solid iron phase can incorporate only a relatively small sulfur fraction, most of the sulfur is expelled at the inner core front and drives compositional convection. The latent heat released upon iron solidification provides additional thermal driving power. Contrary to the situation for Earth, freezing could also start at the CMB because of the lower pressures in Mercury's core. The iron crystals would then precipitate or snow into the center and remelt when encountering temperatures above the melting point at a depth r_m. This process leaves a sulfur-enriched lighter residuum in the layer $r > r_m$. As the planet cools, r_m decreases and a stabilizing sulfur gradient is established that follows the liquidus curve and covers the whole snow zone $r > r_m$ (Hauck, Aurnou, and Dombard 2006). Since the heat flux through the CMB is likely subadiabatic today, thermal effects will also suppress rather than promote convection in the outer part of Mercury's core. A stably stratified layer underneath the planet's core mantle boundary and probably extending over the whole iron snow region therefore seems likely. The liquid iron entering the layer below r_m serves as a compositional buoyancy source. The latent heat being released in the iron snow zone diffuses to the core mantle boundary. Today's low CMB heat flux implies that this can be achieved by a relatively mild temperature gradient.

The possible core scenarios are illustrated in Figure 10.1 with melting curves for different sulfur concentrations and core adiabats with CMB temperatures in the range between 1600 and 2000 K suggested by interior (Rivoldini and Van Hoolst 2013) and thermal evolution models (Grott, Breuer, and Laneuville 2011; Michel et al. 2013; Tosi et al. 2014). Data on the melting behavior of iron–sulfur alloys are few and the melting curves shown in Figure 10.1 therefore rely on simple parameterizations (Rivoldini et al. 2011). The adiabats have been calculated by Rivoldini and Van Hoolst (2013). Mercury's core pressure is only grossly constrained, with CMB pressures in the range 4–7 GPa and central pressures in the range 30–45 GPa (Hauck et al. 2013). We adopt a central pressure of 40 GPa here. Figure 10.1 suggests that iron starts to solidify in the center for initial sulfur concentrations below about 4 wt%. Sulfur released from the inner core boundary increases the concentration in the liquid core over time and thereby slows down the inner core growth and delays the onset of iron snow. For an initial sulfur concentration beyond 4 wt% iron solidification starts with the CMB snow regime. A convective layer that is enclosed by a solid inner core and a stably stratified outer iron snow layer seems possible for sulfur concentrations between about 2.5 and 7 wt%. For sulfur concentration beyond 7 wt% an inner core would only grow when the snow zones extends through the whole core and the snow starts to accumulate in the center.

The adiabats and thin red lines in Figure 10.1 illustrate the evolution for an initial sulfur concentration of 3 wt%. For the hot (red) adiabat with $T_{cmb} =$ 2000 K neither inner core growth nor iron snow would have started and there would be no dynamo. When the temperature drops, iron starts to solidify first at the center. For a CMB temperature of $T_{cmb} = 1910$ K (solid green adiabat),

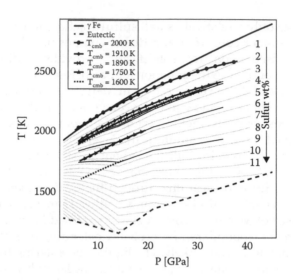

FIGURE 10.1

Melting curves for different initial sulfur concentrations and possible Mercury adiabats for different temperatures shown as $\rightarrow$, $\rightarrow$, $\rightarrow$, $\rightarrow$, ... — from top to bottom show the melting curves for the convecting part of the core for an initial sulfur concentration of 3 wt% and a core state described by the $\rightarrow$, $\rightarrow$, —, and ..... — shows the melting curve for pure iron while the ._ shows the eutectic temperature. The figure, provided by Attilio Rivoldini, has been adapted from Rivoldini et al. (2011) to include the Mercury core adiabats calculated in Rivoldini and Van Hoolst (2013). A central pressure of 40 GPa is assumed for Mercury but the adiabats are only drawn in the liquid part of the core. (Adapted from Rivoldini A., et al. 2011. Geodesy constraints on the interior structure and composition of Mars. *Icarus*, 213, 451.)

the inner core has already grown to a radius of about 600 km while the outer snow layer is only about 160 km thick. The sulfur released upon inner core growth has increased the bulk concentration in the liquid part of the core to 3.4 wt% (first thin red line from the top). The decrease in the sulfur abundance due to the remelting of iron snow has not been taken into account in this model. When the CMB temperature has dropped to $T_{cmb} = 1890$ K (dashed green adiabat), the inner core and snow layer have grown by a comparable amount while the sulfur concentration has increased to 4.4 wt% (second thin red line from the top). At $T_{cmb} = 1750$ K (gray) there remains only a relatively thin convective layer between the inner core boundary at $r_i = 1440$ km and the lower boundary of the outer snow layer at $r_m = 1650$ km. For the coldest adiabat shown in Figure 10.1 with $T_{cmb} = 1890$ K (blue) only the outer 300 km of the core remain liquid but belong to the iron snow zone so that no dynamo seems possible.

Additional, sometimes complex scenarios have been discussed in the context of Ganymede by Hauck et al. (2006) and may also apply to Mercury since the iron cores of both bodies cover similar pressure ranges. For example, Figure 10.1 illustrates a kink in the melting curve for pressures around 21 GPa and compositions larger than 5 wt% sulfur. This could lead to a double snow regime where not only the very outer part of the core precipitates iron but also an intermediate layer around 21 GPa. This possibility has been explored in a dynamo model by Vilim, Stanley, and Hauck (2010), which we will discuss in Section 10.2. Since the kink is not very pronounced, however, such a double snow dynamo would not be very long lived.

Another interesting scenario unfolds when the light element concentration lies on the S-rich side of the eutectic. Under these conditions, FeS rather than Fe would crystallize out when the temperature drops below the FeS melting curve. Since FeS is lighter than the residuum fluid, the crystals would rise toward the CMB. However, eutectic or even higher sulfur concentrations cannot represent bulk conditions since it would be difficult to match Mercury's total mass (Rivoldini et al. 2011). Inner core growth would increase the sulfur concentration in the remaining fluid over time but never beyond the eutectic point. This has likely not been reached in Mercury because the eutectic temperature of 1200–1300 K (Rivoldini et al. 2011) is significantly lower than today's CMB temperature suggested by thermal evolution (Grott, Breuer, and Laneuville 2011, Tosi et al. 2014) and interior models (Rivoldini and Van Hoolst 2013).

An alternative explanation for a locally high sulfur concentration was suggested by the XRS observations. The low Fe but large S abundance in surface rocks indicates that Mercury's core could have formed at strongly reducing conditions. This promotes a stronger partitioning of Si into the liquid iron phase leading to a ternary Fe-Si-S core alloy (Malavergne et al. 2010). Experiments have shown that Si and S are immiscible for pressures below 15 GPa (Morard and Katsura 2010), which is the pressure range in the outer part of Mercury's core. However, the immiscibility only happens for sizable Si and S concentrations. Experiments by Morard and Katsura (2010), for example, demonstrate that at 4 GPa and 1900 K abundances of 6 wt% S and 6 wt% Si are required to trigger the immiscibility and lead to the formation of a sulfur rich phase with a composition of about 25 wt% S. For FeS crystallization to play a role at today's CMB temperatures, the sulfur-rich phase should lie significantly to the right of the eutectic where the FeS melting temperature increases with light element abundance. Thus, even higher S and Si contributions are required but seem once more difficult to reconcile with the planet's total mass (Rivoldini and Van Hoolst 2013). Since Si partitions much more easily into the solid iron phase than sulfur, its contribution to compositional convection and the stabilization of the snow zone is significantly weaker.

Several numerical studies in the context of Earth and Mars have shown that the CMB heat flux pattern can have a strong effect on the dynamo mechanism (see e.g., Wicht et al. (2011) and Dietrich and Wicht (2013) for overviews).

Like the mean heat flux out of the core, this pattern is controlled by the lower mantle structure. The Martian dynamo ceased about 4 Gyr ago but has left its trace in the form of a strongly magnetized crust. The fact that the magnetization is much stronger in the southern than in the northern hemisphere could reflect a special configuration of the planet's ancient dynamo. Impacts or large degree mantle convection may have significantly decreased the heat flux through the northern CMB and therefore weakened dynamo action in this hemisphere (Stanley et al. 2008; Amit, Christensen, and Langlais 2011; Dietrich and Wicht 2013). Mercury's magnetic field is distinctively stronger in the northern than in the southern hemisphere and it seems attractive to invoke an increased northern CMB heat flux as a possible explanation.

Clues about the possible pattern may once more come from MESSENGER observations. A combination of gravity and altimeter data allow to estimate the crustal thickness in the northern hemisphere. On average, the crust is about 50 km thicker around the equator than around the pole (Smith et al. 2012), which points toward more lava production and thus a hotter mantle at lower latitudes. This is consistent with the fact that the northern lowlands are filled by younger flood basalts since melts more easily penetrate a thinner crust (Denevi et al. 2013). Missing altimeter data and the degraded precision of gravity measurements does not allow to constrain the crustal thickness in the southern hemisphere. The lack of younger flood basalts, however, could indicate a thicker crust and hotter mantle. Since a hotter mantle would reduce the CMB heat flux, these ideas indeed translate into a pattern with increased flux at higher northern latitudes. However, Mercury's volcanism ceased more than 3.5 Gyr ago and today's thermal mantle structure may look completely different. Even simple thermal diffusion should have eroded any asymmetry over such a long time span. Thermal evolution simulations show that at least the lower part of the mantle may still convect today (Smith et al. 2012; Tosi et al. 2014), which would change the structure on much shorter time scales. Since the active shell is so thin, the pattern would be rather small scale without any distinct north/south asymmetry.

Because of Mercury's 3:2 spin-orbit resonance, the high eccentricity of the orbit, and the very small obliquity the time-averaged insolation pattern shows strong latitudinal and longitudinal variations. Williams et al. (2011) calculates that the mean polar temperature can be 200 K lower than the equatorial. Longitudinal variations show two maxima that are about 100 K hotter than the minima at the equator. If Mercury's mantle convection has ceased long ago, the respective pattern may have diffused into the mantle and could determine the CMB heat flux variation. Higher than average flux at the poles and a somewhat weaker longitudinal variation would be the consequence. We discuss the impact of the CMB heat flux pattern on the dynamo process in Section 10.2.

10.3 Mercury's External Magnetic Field

Planetary magnetospheres are the result of the interaction between the planetary magnetic field and the impinging solar wind plasma. Because of Mercury's weak and asymmetric magnetic field and the position close to the Sun, the Hermean and terrestrial magnetospheres differ fundamentally. Mercury experiences the most intense solar wind of all solar system planets. Under average conditions, the ratio of the solar wind speed and the Alfvén velocity, called the Alfvénic Mach number, is comparable to the terrestrial one. With values of 6.6 for Mercury (Winslow et al. 2013) and 8 for Earth, the solar wind plasma is super-magnetosonic at both planets, that is, the medium propagates faster than magnetic disturbances and a bow shock (BS) therefore forms in front of the magnetosphere. Because of the weak Hermean magnetic field, the subsolar point of the BS is located rather close to the planet at an average position of only 1.96 planetary radii (Winslow et al. 2013) compared to 14 planetary radii for Earth.

Behind the BS, the cold solar wind plasma is heated up and interacts with the planetary magnetic field, thereby creating the magnetosphere. To first order, the planetary field lines form closed loops within the dayside magnetosphere and a long tail on the nightside. The outer boundary of the magnetosphere, the magnetopause (MP), is located where the pressure of the shocked solar wind and the pressure of the planetary magnetic field balance. The solar wind ram pressure, on average 14.3 nPa at Mercury (Winslow et al. 2013), is an order of magnitude higher than at Earth while the magnetic field is two orders of magnitude weaker. Like the BS, the MP is therefore located much closer to the planet at Mercury than at Earth with mean standoff distances of about 1.45 (Winslow et al. 2013) and 10 planetary radii, respectively. Both the Hermean magnetosphere and magnetosheath, the region between BS and MP, are thus much smaller than the terrestrial equivalents in relative and absolute terms.

Figure 10.2 shows the current density in a numerical hybrid simulation that models the solar wind interaction with the planet (Müller et al. 2012). The location of the BS and the magnetosphere can be identified via the related current systems. Along a spacecraft trajectory these boundaries can be identified by the related magnetic field changes. Figure 10.3 shows MESSENGER magnetic field measurements for a relatively quiet orbit (orbit number 14) where both the BS and the MP can be clearly classified on both sides of the planet.

Another important element of the magnetosphere is the neutral current sheet, which is responsible for the elongated nightside magnetotail and separates the northern and southern magnetotail lobes. Johnson et al. (2012) report that the sheet starts at $1.41R_M$, where R_M is the mean Hermean radius, which is approximately the standoff distance of the dayside MP. Roughly the same proportion is also found at Earth.

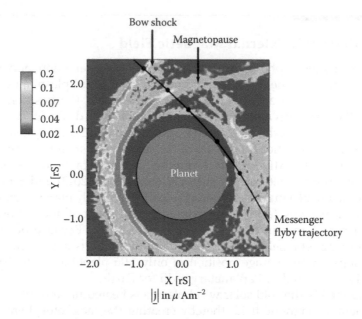

FIGURE 10.2
Electrical currents in a numerical simulation of the Hermean magnetosphere. The amplitude of the current density j is color-coded. An equatorial cross section is shown in a coordinate system where X points towards the Sun (negative solar wind direction) and the Y-axis lies in the Hermean ecliptic. The bow shock standing in front of the planet slows down the solar wind. The MP is the outer boundary of the magnetosphere. The neutral current sheet is located in the nightside of the planet. An arc of electrical current visible close to the flyby trajectory (January 14, 2008) could be interpreted as a partial ring current. This figure is a snapshot from a solar wind hybrid simulation. (Adapted from Müller J., et al. 2012. Origin of Mercury's double magnetopause: 3D hybrid simulation study with AIKEF. *Icarus*, 218, 666.)

The locations of BS, MP and neutral current sheet are not stationary but vary with time. The density of the average solar wind decreases with distance r_S to the Sun like $1/r_S^2$. Since Mercury orbits the Sun on a highly elliptical orbit (ellipticity: 0.21), the local solar wind pressure varies significantly on the orbital time scale of 88 days. The solar wind characteristics also changes constantly on much shorter time scales because of spatial inhomogeneities due to, for example, coronal-mass ejections. As a result, the Hermean magnetosphere is very dynamic. And since the magnetosphere is so small, the magnetic disturbance also propagates deep into the magnetosphere and impedes the separation of the field into internal and external contributions (Glassmeier et al. 2010). Reconnection processes in the magnetotail are another source for variations in the Hermean magnetosphere (Slavin et al. 2012).

The Hermean and terrestrial magnetospheres differ in several additional aspects. Mercury's surface temperature can reach several hundred Kelvin, which means that the planet's gravitational escape velocity of 4.3 km/s can

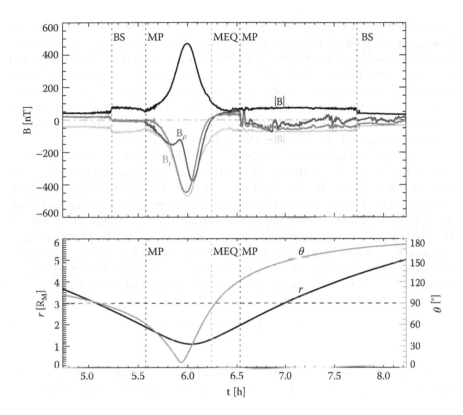

FIGURE 10.3

Magnetic field data recorded by the MESSENGER magnetometer (10s average) during orbit 14 on the DOY 84 in 2011. The *upper panel* shows the time series of the absolute magnetic field $|B|$, the negative absolute field, the radial component B_r and the component B_ρ perpendicular to the rotation axis. Time is measured in hours since the last apocenter passage. The plasma boundaries are marked with vertical dashed lines (BS: bow shock, MP: magnetopause). The location where B_ρ vanishes define the MEQ. The *lower panel* shows the planetocentric distance r and the co-latitude θ. (Data are taken from the Planetary Data System/Planetary Plasma Interactions Node.)

easily be reached thermally. The thermal escape rate is therefore significant and the remaining atmosphere too thin to form an ionosphere. At Earth, the ionosphere hosts substantial current systems that significantly affect the magnetospheric dynamics, for example magnetic substorms. Field-aligned currents that close via the ionosphere at Earth must close within the magnetospheric plasma or the planetary body at Mercury (Janhunen and Kallio 2004).

When the planetary magnetic field on the inside of the MP is nearly anti-parallel to the magnetosheath field on the outside, the respective field lines can reconnect. This typically happens when the interplanetary magnetic field has a component parallel to the planetary field. The reconnected field lines are advected tailward by the solar wind, which drives a global scale magnetospheric convection loop that ultimately replenishes the dayside field (Dungey cycle). Due to the small size of the Hermean magnetosphere, the typical time scale of this plasma circulation is only about 1–2 min compared to 1 h at Earth (Slavin et al. 2012), which demonstrates that the Hermean magnetosphere can adapt much faster to changing solar wind conditions. The rate of reconnection, measured by the relative amplitude of the magnetic field component perpendicular to the MP, is about 0.15 at Mercury and thus three times higher than that at Earth (Dibraccio et al. 2013).

Charged particles that are trapped inside the magnetosphere and drift around the planet in azimuthal direction form a major magnetospheric current system at Earth, the so-called ring current. The drift is directed along isocontours of the magnetic field strength. However, since internal and magnetospheric field can reach comparable values, these contours close via the MP at Mercury, as is illustrated in Figure 10.4. At Earth, the planetocentric distance $R_{rc,E}$ of the ring current is about four times the terrestrial radius. When assuming that the position scales linearly with the planetary dipole moment, the distance can be rescaled to the Hermean situation by

$$R_{rc,M} = R_{rc,E}\,\frac{m_M}{m_E} \approx 820 \text{ km} \qquad (10.1)$$

where m_M and m_E are the dipole moments of Mercury and Earth, respectively. The ring current would thus clearly lie below Mercury's surface. Hybrid

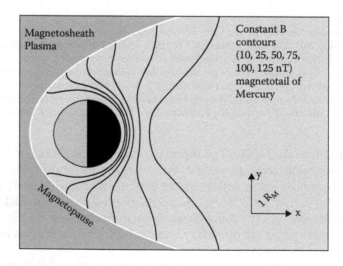

FIGURE 10.4
Equatorial isocontours of the total magnetic field in a Hermean model magnetosphere. The MP is shown as a white line and the planet as a sphere. (From Baumjohann W., et al. 2010. Current systems in planetary magnetospheres and ionospheres. *Space Sci. Rev.* 152, 99–134.)

simulations by Müller et al. (2012) indicate that the solar wind protons entering the magnetosphere can drift roughly half-way around the planet before being lost to the MP, as is illustrated in Figure 10.2. This could be interpreted as a partial ring current. The protons create a diamagnetic current that locally decreases the magnetic field.

Because MESSENGER delivers only data from one location at a time inside a very dynamic magnetosphere, it is not only challenging to separate internal from external field contributions but also temporal from spatial variations. Numerical simulations for the solar wind interaction with the planetary magnetic field, like the hybrid simulation used to investigate the partial ring current (see Figure 10.2), can improve the situation by constraining the possible spatial structure for a given solar wind condition. However, as these codes are numerically very demanding, it becomes impractical to perform simulation for all the different conditions possibly encountered by MESSENGER. A more practical approach is to use simplified models where a few critical properties like the shape of the MP and the strength and shape of the neutral current sheet are described with a few free parameters. Johnson et al. (2012) demonstrate how the parameters can be fitted to MESSENGER's accumulated magnetic field data to derive a model for the time-averaged magnetosphere.

The offset of Mercury's magnetic field by 20% of the planetary radius to the north can cause an equatorial asymmetry of the planet's exosphere. Ground-based observations of sodium emission lines suggest that there is more sodium released from the southern than the northern planetary surface. Mangano et al. (2013) argue that precipitating solar wind protons are the main player in the sodium release and more likely reach the southern surface where the magnetic field is weaker.

The Hermean magnetosphere resembles its terrestrial counterpart in several aspects but there are also huge differences. Mercury's magnetosphere is much smaller and significantly more dynamic, responding much faster to changing solar wind conditions. While the external field contributions are orders of magnitude smaller than internal contributions at Earth, they can become comparable at Mercury (see Figure 10.5). This lead Glassmeier, Auster, and Motschmann (2007) to investigate the long-term effect of the external field on the internal dynamo process, as we will further discuss in Section 10.5.

10.4   Mercury's Internal Magnetic Field

The difficulties in separating internal and external field and the strong covariance of different spherical harmonic contributions caused by the highly elliptical orbit complicate a classical field modeling with Gaussian

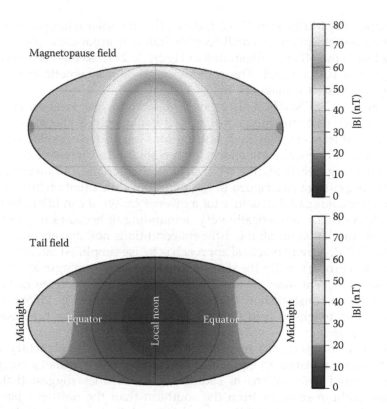

FIGURE 10.5
External fields from the paraboloid model based on measurements of the MESSENGER mission at the planetary surface. *Top panel:* amplitude of the magnetopause field. *Bottom panel:* amplitude of the neutral sheet magnetic field. (From Johnson C.L., et al. 2012. MESSENGER observations of Mercury's magnetic field structure. *J. Geophys. Res.* 117, 0.)

coefficients for Mercury (Anderson et al. 2011). Instead, the MESSENGER magnetometer team analyzed the location of the MEQ to indirectly deduce the internal field. The MEQ is the point where the magnetic field component B_ρ perpendicular to the planetary rotation axis vanishes. Changing solar wind conditions lead to variations in the equator location on different time scales from seconds to months but should average out over time, at least as long as the planetary body itself has no first order impact on the magnetospheric current system. The mean location of the MEQ is then primarily determined by the internal field.

Anderson et al. (2012) analyzed the MEQ for 531 descending orbits with altitudes between 1000 and 1500 km and 120 ascending orbits with altitudes between 3500 and 5000 km. They find that the equator crossings are confined to a relatively thin band offset by about $Z = 480$ km to the north of the planet's equator. We adopt a planet-centered cylindrical coordinate system here where ρ and z are the coordinates perpendicular to and along the rotation axis, respectively, and Φ is the longitude. Anderson et al. (2012) minimized the effects of solar wind related magnetic field variations by considering a mean where each equator location is weighted with the inverse of the individual standard error σ_e. This procedure yields a mean offset of $\bar{Z}_d = 479$ km with a standard deviation of $\Delta\bar{Z}_d = 46$ km for the descending orbits. The mean three standard error in determining the individual equator crossings is $3\bar{\sigma}_e = 24$ km. Because of the increased solar wind influence and the closer proximity to the magnetosphere, the MEQ is less well defined for the ascending orbits with $\bar{Z}_a = 486$ km, $\Delta\bar{Z}_a = 270$ km, and $3\bar{\sigma}_a = 86$ km (see Table 10.1 in Anderson et al. (2012)).

These observations suggest that the offset of the MEQ has a constant value of 480 km independent of the distance to the planet. Such a configuration can readily be explained by an internal axial dipole that is offset by 480 km to the north of the equatorial plane. This translates into an infinite sum of axisymmetric Gaussian field coefficients g_ℓ in the classical planet centered representation with

$$g_{\ell 0} = \ell\, g_{10} Z^{\ell-1}, \tag{10.2}$$

TABLE 10.1

Comparison of Magnetic Field Models

Quantity	Mercury: MODM (Anderson et al. 2012)	Earth: Grimm (Lesur et al. 2012)	Jupiter: VIP4 (Connerney et al. 1998)	Saturn (Cao et al. 2012)	Uranus (Holme & Bloxham, 1996; up to degree $\ell = 4$)
g_{10} [nT]	-190 ± 10	-29560	420500	21191	11855
tilt [°]	<0.8	10.2	9.5	<0.06	58.8
g_{20}/g_{10}	0.392 ± 0.010	0.079	-0.012	0.075	-0.496
g_{30}/g_{10}	0.116 ± 0.009	-0.045	-0.004	0.112	0.353
g_{40}/g_{10}	0.030 ± 0.005	-0.031	-0.040	0.003	0.034
H	0.20	0.017	0.045	0.050	0.251
$\bar{Z}$	2.0×10^{-1}	2.6×10^{-2}	3.5×10^{-3}	3.8×10^{-2}	5.3×10^{-2}
$\bar{Z}_d$	2.0×10^{-1}	3.6×10^{-3}	2.8×10^{-2}	4.0×10^{-2}	1.6×10^{-1}
$\Delta\bar{Z}$	$1.7 \times 10^{-2}(1.1 \times 10^{-1})$	2.2×10^{-1}	1.6×10^{-1}	2.9×10^{-3}	1.0
$\Delta\bar{Z}_d$	$7.5 \times 10^{-3}(1.9 \times 10^{-2})$	1.3×10^{-1}	9.3×10^{-2}	8.5×10^{-4}	6.6×10^{-1}

Note: The Neptunian magnetic field is similar to the field of Uranus and has therefore not been included. The last four lines list mean offset values $\bar{Z}$ for all spherical surfaces up to $4R$ and the mean offset $\bar{Z}_d$ for the distances between $1.3R$ and $1.5R$ covered by MESSENGER's descending orbits. R refers to the planetary radius (1 bar level for gas planets). $\Delta\bar{Z}$ and $\Delta\bar{Z}_d$ are the related standard deviations. For Mercury, we list the deviation caused by a $0.8°$ tilt and also the observed standard deviations in brackets.

where $\mathcal{Z} = Z/R_M$ is the normalized offset and ℓ the spherical harmonic degree (Bartels 1936; Alexeev et al. 2010). Note that all contributions have the same sign. In the Gaussian representation the planetary surface field is expanded into spherical surface harmonics $Y_{\ell m}$ of degree ℓ and order m (Olsen et al. 2010). The coefficients $g_{\ell m}$ and $h_{\ell m}$ express the $\cos(m\Phi)$ and $\sin(m\Phi)$ dependence for a given degree ℓ. Only coefficients $g_{\ell 0}$ contribute to an axisymmetric field.

Anderson et al. (2012) report that coefficients up to $\ell = 4$ suffice to explain the mean MEQ locations in the MODM. To illustrate the characteristics of MODM, we experiment with different combinations of the spherical harmonic contributions and perform a numerical search for the MEQ on a dense longitude/latitude grid for spherical surfaces with radii up to $4R_M$. Figure 10.6a illustrates how the different axisymmetric contributions in the MODM team up to yield an offset that is nearly independent of the distance to the planet. A large axial quadrupole contribution, which amounts to nearly 40% of the axial dipole, guarantees a realistic offset for $\rho > 2R_M$. Additional higher harmonic contributions are required to achieve a consistent offset at closer distances. Figure 10.6b demonstrates that already the relative axial octupole g_{30}/g_{10} is not particularly well constrained and values between 0.05 and 0.12 seem acceptable. Anderson et al. (2012), however, suggest a surprisingly tight range of 0.116 ± 0.009. Contributions beyond $\ell = 3$ cannot be particularly large to retain a nearly constant offset value in the observed range. Constraining them further, however, would require data closer to the planet than presently available. The analysis shows that the mean offset $\mathcal{Z}$ further away from the planet can serve as a proxy for the ratio of the axial quadrupole to axial dipole contribution while the dependence of $\mathcal{Z}$ on the distance closer to the planet provides information on higher order axial contributions.

Anderson et al. (2012) estimate an upper limit for the dipole tilt of $\Theta = 0.8°$. A tilt of the planetary centered dipole causes a longitudinal variation of the MEQ location that increases with distance to the planet, as is demonstrated in Figure 10.6b and c. A tilt as large as 2° seems still compatible with the data but the more complex longitudinal dependence of the offset (Anderson et al. 2012) indicates that either higher order harmonics or more likely the solar wind interaction contributes to the variation around the mean offset. A tilt below <0.8° is also consistent with a more complete field analysis by Johnson et al. (2012) that includes a parameterized magnetospheric model.

Table 10.1 compares primary magnetic field characteristics of the MODM with models for other planets and Figure 10.7 shows the respective radial magnetic surface fields. MODM's large quadrupole contribution is comparable to that inferred for Uranus or Neptune. Unlike the fields of the ice giants, however, Mercury's field is also very axisymmetric, a property it shares with Saturn. The seemingly perfect axisymmetry of Saturn's field is also the reason for the small spread $\Delta\bar{\mathcal{Z}}$ of MEQ locations for this planet. Saturn's relative quadrupole contribution, however, and thus the relative offset is much smaller than at Mercury.

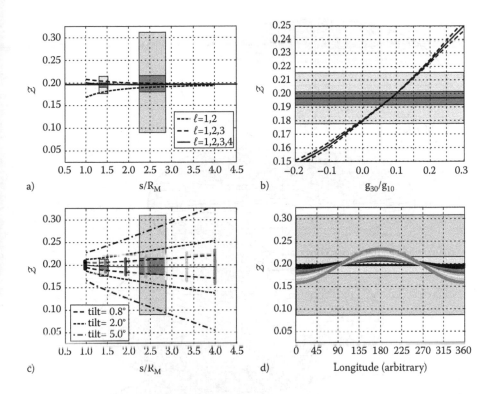

FIGURE 10.6

Illustration of the offset dipole model by Anderson et al. (2012). Panel a) demonstrates how the location of the magnetic equator for the descending (left box) and ascending (right box) orbits is explained by combining axial Gauss coefficients up to degree $\ell = 4$. Light gray boxes illustrate the standard deviation, middle grey boxes the mean three sigma error (see text), and the horizontal black line corresponds to the mean offset. Panel b) illustrates the impact of different relative octupole amplitudes g_{30}/g_{10}. Colored dots in panels c) and d) show the equator locations found on a dense spherical longitude/latitude grid when an equatorial dipole component g_{11} has been added that corresponds to a dipole tilt of 0.8°. In panel c) the dashed, dotted, and dash-dotted lines show the mean equator offset for each spherical surface of radius s/R_M plus and minus the standard deviation. (From Anderson B.J., et al. 2012. Low-degree structure in Mercury's planetary magnetic field. *J. Geophys. Res.* (1991–2012), 117.)

Magnetic harmonics where the sum of degree ℓ and order m is odd (even) represent equatorially antisymmetric (symmetric) field contributions. The axial dipole field is thus equatorially antisymmetric while the axial quadrupole field is symmetric. Mercury's field has a significant equatorially symmetric contribution because of the strong axial quadrupole. Another measure related to the equatorial symmetry breaking is the hemisphericity.

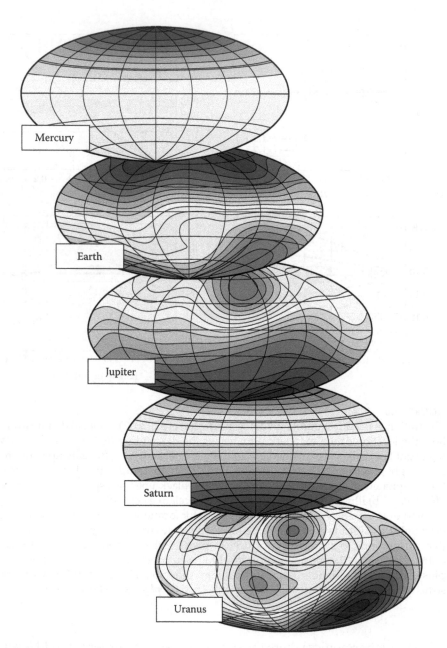

FIGURE 10.7
Comparison of different radial magnetic fields at planetary surface. Blue (red and yellow) indicates radially inward (outward) field. See Table 10.1 for information on the different field models.

$$H = \frac{B_N - B_S}{B_N + B_S} \tag{10.3}$$

where B_N and B_S are the rms surface field amplitudes in the northern and southern hemispheres, respectively. Due to the offset dipole geometry, the Hermean magnetic field is significantly stronger in the northern than in the southern hemisphere so that the hemisphericity reaches a relatively large value of 0.2. In conclusion, Mercury's magnetic field is not only very weak but also has a peculiar geometry unlike any other planet in our solar system that combines a relatively large axial quadrupole contribution with a very small dipole tilt.

The time-averaged residual field after subtracting the internal and external field models by Johnson et al. (2012) from the observational data is surprisingly strong with amplitudes of up to 45 nT at 300 km altitude above Mercury's surface (Purucker et al. 2012). The fact that the residual field is concentrated at high northern latitudes, is relatively small scale, and correlates with the boundary of the northern volcanic plains to a fair degree points toward crustal remanent magnetization though, an internal field contribution can also not be excluded. A crustal origin would suggest that Mercury's dynamo is long-lived and probably older than 3.5 Gyr. Since the residual field opposes the current dipole direction, the dynamo must have reversed its polarity at least once. This would put valuable constraints on thermal evolution models and dynamo simulations for Mercury.

10.5 Numerical Models of Mercury's Internal Dynamo

10.5.1 Dynamo Theory

Numerical dynamo simulations solve for convection and magnetic field generation in a viscous, electrically conducting, and rotating fluid. Since the solutions are very small disturbances around an adiabatic, well-mixed, nonmagnetic, and hydrostatic background state, only first-order terms are taken into account. For terrestrial planets, the mild density and temperature variations of the background state are typically neglected in the so-called Boussinesq approximation (Braginsky and Roberts 1995). The mathematical formulation of the dynamo problem is then given by the Navier–Stokes equation,

$$E\frac{d\mathbf{U}}{dt} = -\nabla P - 2\hat{\mathbf{z}} \times \mathbf{U} + \mathrm{Ra}\frac{r}{r_o}C\hat{\mathbf{r}} + \frac{1}{\mathrm{Pm}}(\nabla \times \mathbf{B}) \times \mathbf{B} + E\nabla^2\mathbf{U}, \tag{10.4}$$

the induction or dynamo equation

$$\frac{\partial \mathbf{B}}{\partial t} = (\mathbf{B} \cdot \nabla)\mathbf{U} + \frac{1}{Pm}\nabla^2 \mathbf{B}, \tag{10.5}$$

the codensity evolution equation

$$\frac{dC}{dt} = \frac{1}{Pr}\nabla^2 C + q, \tag{10.6}$$

the flow continuity equation

$$\nabla \cdot \mathbf{U} = 0, \tag{10.7}$$

and the magnetic continuity equation

$$\nabla \cdot \mathbf{B} = 0. \tag{10.8}$$

Here, d/dt stands for the substantial time derivative $\partial/\partial t + \mathbf{U} \cdot \nabla$, $\mathbf{U}$ is the convective flow, $\mathbf{B}$ the magnetic field, P is a modified pressure that also contains centrifugal effects, and C is the codensity.

The equations are given in a nondimensional form that uses the thickness of the fluid shell $d = r_o - r_i$ as a length scale, the viscous diffusion time d^2/ν as a time scale, the codensity difference ΔC across the shell as the codensity scale, and $(\bar{\rho}\mu\lambda\Omega)^{1/2}$ as the magnetic scale. Here, r_i and r_o are the radii of the inner and outer boundary, respectively, ν is the kinematic viscosity, $\bar{\rho}$ the reference state core density, μ the magnetic permeability, λ the magnetic diffusivity, and Ω the rotation rate.

The problem is controlled by five dimensionless parameters: the Ekman number

$$E = \frac{\nu}{\Omega d^2}, \tag{10.9}$$

the Rayleigh number

$$Ra = \frac{\bar{g}_o \alpha \Delta c \, d^3}{\kappa \nu} \tag{10.10}$$

the Prandtl number

$$Pr = \frac{\nu}{\kappa}, \tag{10.11}$$

the magnetic Prandtl number

$$Pm = \frac{\nu}{\lambda},$$ (10.12)

and the aspect ratio

$$a = \frac{r_i}{r_o}.$$ (10.13)

These five dimensionless parameters replace the much larger number of physical properties of which the thermal and/or compositional diffusivity κ, the thermal and/or compositional expansivity α, and the outer boundary reference gravity $\bar{g}_0$ have not been defined so far.

Convection is driven by density variations due to superadiabatic temperature gradients—only this component contributes to convection—or due to deviations from a homogeneous background composition. Possible sources for thermal convection are secular cooling, latent heat, and radiogenic heating. Possible sources for compositional convection are the light elements released from a growing inner core and iron from an iron snow zone. To simplify computations, both types of density variation are often combined into one variable called codensity C despite the fact that the molecular diffusivities of heat and chemical elements differ by orders of magnitude. The approach is often justified with the argument that the small-scale turbulent mixing, which cannot be resolved in the numerical simulation, should result in larger effective turbulent diffusivities that are of comparable magnitude (Braginsky and Roberts 1995). This has the additional consequence that the "turbulent" Prandtl number and magnetic Prandtl number would become of order one (Braginsky and Roberts 1995). The codensity evolution equation 10.6 contains a volumetric source/sink term q that can serve different purposes depending on the assumed buoyancy sources. For convection driven by light elements released from the inner core, q acts as a sink that compensates the respective source. When modeling secular cooling, the outer boundary is the sink and q the balancing volumetric source (Kutzner and Christensen 2000). For iron snow that remelts at depth q should be positive in the snow zone but negative in the convective zone underneath.

Typically, no-slip boundary conditions are assumed for the flow. For the condensity, either fixed codensity or fixed flux boundary conditions are used. The latter translates to a fixed radial gradient and requires a modification of the Rayleigh number (10) where ΔC then stands for the imposed gradient times the length scale d. For terrestrial planets, the much slower evolving mantle controls how much heat is allowed to leave the core, so that a heat flux condition is more appropriate. Lateral variations on the thermal lower mantle structure translate into an inhomogeneous CMB heat flux (Aubert et al. 2008). Since the electrical conductivity of the rocky mantle in terrestrial planets is orders of magnitudes lower than that of the core, the magnetic field can be assumed to match a potential field at the interface $r = r_o$. This matching condition can be formulated as a magnetic boundary condition for the individual spherical harmonic field contributions (Christensen and Wicht 2007). A simplified induction equation 10.5 must be solved for the magnetic field in a conducting inner core, which has to match the outer core field at r_i. We refer to Christensen and Wicht (2007) for a more detailed discussion of dynamo theory and the numerical methods employed to solve the system of equations.

Explaining the weakness of Mercury's magnetic field proved a challenge for classical dynamo theory. In convectively driven core dynamos, the Lorentz force and thus the magnetic field need to be sufficiently strong to influence the flow and thereby saturate magnetic field growth. The impact of the Lorentz force is often expressed via the Elsasser number

$$\Lambda = \frac{B^2}{\rho\mu\lambda\Omega} \tag{10.14}$$

where B is the typical magnetic field strength. The Elsasser number estimates the ratio of the Lorentz to the Coriolis force, which is known to enter the leading order convective force balance. For Earth, Λ is of order one, which suggests that the Lorentz force is indeed significant. For Mercury, however, extrapolating the measured surface field strength to the planet's core mantle boundary yields $\Lambda_{cmb} \approx 10^{-5}$, a value much too low to be compatible with an Earth-like convectively driven core dynamo (Wicht et al. 2007). Several authors therefore pursued alternative theories like crustal magnetization or a thermo-electric dynamo (see Wicht et al. (2007) for an overview).

However, convectively driven core dynamos remain the preferred explanation since different modifications of the numerical models originally developed to explain the geodynamo successfully reduced the surface field strength toward more Mercury-like values (for recent overviews see Wicht et al. (2007); Stanley and Glatzmaier (2010); Schubert and Soderlund (2011)). We revisit several of these models in the following and test whether they are consistent with MESSENGER magnetic field data.

All numerical dynamo simulations have the problem that numerical limitations do not allow to use realistic diffusivities. For example, the viscous diffusivity is many orders of magnitude too large to damp the very small-scale convection motions that cannot be resolved with the available computer power. Dynamo modelers typically fix the Ekman number E, the ratio of viscous to Coriolis forces, to the smallest value accessible with the numerical resources. The most advanced computer simulations reach down to $E = 10^{-7}$, which is still many orders of magnitude larger than the planetary value of $E \approx 10^{-12}$ (see Table 10.2). The Prandtl number Pr can assume realistic values but the magnetic Prandtl number Pm has to be set to a value that guarantees dynamo action. Because of the increased viscous diffusivity, Pm is also orders of magnitudes too large. The Rayleigh

TABLE 10.2

Comparison of Dynamo Models

Model	MODM	E5R6	E5R36	E5R45	CW2	CW3	CW4	Y_{10} BD	Y_{10} ID	Y_{20}
Ra	—	2×10^7	1.2×10^8	1.5×10^8	2×10^8	4×10^8	6×10^8	4×10^7	4×10^7	4×10^7
E	10^{-13}	3×10^{-5}	3×10^{-5}	3×10^{-5}	10^{-4}	10^{-4}	10^{-4}	10^{-4}	10^{-4}	10^{-4}
Pm	10^{-6}	1	1	1	3	3	3	2	2	2
Pr	0.1	1	1	1	1	1	1	1	1	1
a	—	0.35	0.35	0.35	0.50	0.50	0.50	0.35	0.35	0.20
Ro_t	8	0.02	0.10	0.18	0.42	2.7	3.6	0.11	0.06	0.05
Λ_{cmb}	10^{-5}	2.2×10^{-2}	1.8×10^{-1}	1.6×10^{-2}	4.9×10^{-4}	1.9×10^{-4}	4.7×10^{-5}	1.8×10^{-1}	1.9×10^{-1}	4.0×10^{-2}
$|g_{10}|$ [nT]	190	8.8×10^3	2.6×10^4	1.4×10^3	1.4×10^3	924	432	2.1×10^4	1.7×10^4	8.3×10^3
tilt [°]	<0.8	0	2.5	38.1	3.5	4.2	8.6	3.4	10.7	8.2
$|g_{20}/g_{10}|$	0.39	0	0.34	3.8	0.08	0.16	0.31	0.06	0.25	0.52
H	0.20	0	0.02	0.11	0.05	0.09	0.09	0.04	0.20	0.23
$|\bar{Z}|$	2.0×10^{-1}	0	2.2×10^{-2}	2.1×10^{-1}	4.0×10^{-2}	8.1×10^{-2}	8.4×10^{-2}	3.8×10^{-2}	1.5×10^{-1}	2.0×10^{-1}
$\Delta\bar{Z}$	1.7×10^{-2} (1.1×10^{-1})	0	6.9×10^{-2}	8.6×10^{-1}	8.3×10^{-2}	2.4×10^{-1}	1.1×10^{-1}	2.2×10^{-1}	2.3×10^{-1}	

Note: For Mercury we list the MODM while time averaged local Rossby numbers Ro_t and magnetic field properties are listed for the numerical simulations. Mercury's core Rayleigh number and aspect ratio are basically unconstrained but we have assumed an Earth-like value of $a = 0.35$ to calculate the Ekman number. Other Mercury parameters follow Schubert and Soderlund (2011), and Olson and Christensen (2006). Two models with a Y_{10} CMB heat flux pattern are listed, one is BD while the other is ID. The model with a $Y20$ pattern is ID.

number is then adjusted to a value that yields the desired dynamics. The fact that numerical dynamo simulations are very successful in reproducing many aspects of planetary dynamos suggest that at least the large-scale dynamics responsible for producing the observable magnetic field is captured correctly.

The simulation results must be rescaled to the planetary situation. For simplicity, we will rescale the magnetic field strength by assuming that the Elsasser number would not change when pushing the parameter toward realistic values. Assuming Mercury's rotation rate, mean core density, magnetic permeability, and magnetic diffusivity then allows us to deduce the dimensional magnetic field strength via Equation 10.14. Note, however, that other scalings have been proposed (Christensen 2010) and may lead to somewhat different answers.

10.5.2   Standard Earth-Like Dynamo Models

To highlight the difficulties of classical dynamo simulations to reproduce the Hermean magnetic field, we start with analyzing three models that have been explored in the geomagnetic context by Wicht, Stellmach, and Harder (2011). All have the same Ekman ($E = 3 \times 10^{-5}$), Prandtl number ($Pr = 1$), magnetic Prandtl number ($Pm = 1$), aspect ratio ($a = 0.35$), use rigid and fixed codensity boundary conditions, and are driven by a growing inner core. They differ only in the Rayleigh number: Model E5R6 has the lowest Rayleigh number of $Ra = 2 \times 10^7$, 6 times the critical value for onset of convection. Model E5R36 has an intermediate Rayleigh number of $Ra = 1.2 \times 10^8$ while model E5R45 has the largest Rayleigh number at $Ra = 1.5 \times 10^8$. All model parameters are listed in Table 10.2.

Figure 10.8 shows the time evolution of the axial dipole strength, the dipole tilt, the mean MEQ offset $\bar{Z}$ for up to four planetary radii (assuming Mercury's thin crust), and the related standard deviation $\Delta\bar{Z}$. At the lowest Rayleigh number, convective driving is too small to break the equatorial symmetry. The magnetic field is therefore perfectly equatorially antisymmetric and very much dominated by the axial dipole contribution. Dipole tilt, offset, and standard deviation therefore vanish. At the intermediate Rayleigh number, the solution is sufficiently dynamic and asymmetric to be considered very Earth-like (Wicht, Stellmach, and Harder 2011; Christensen, Aubert, and Hulot 2010). While the axial quadrupole and other equatorially symmetric field contributions have grown, the strong axial dipole still clearly dominates. The mean offset $\bar{Z}$ therefore remains small but oscillates around zero since neither the northern nor the southern hemisphere are preferred in the dynamo setup. The spread $\Delta\bar{Z}$ is of the same order as the offset itself mainly because of the Earth-like dipole tilt. The inertial contributions in the flow force balance have increased to a point where magnetic field reversals can be expected (Christensen and Aubert 2006; Wicht, Stellmach, and Harder 2011).

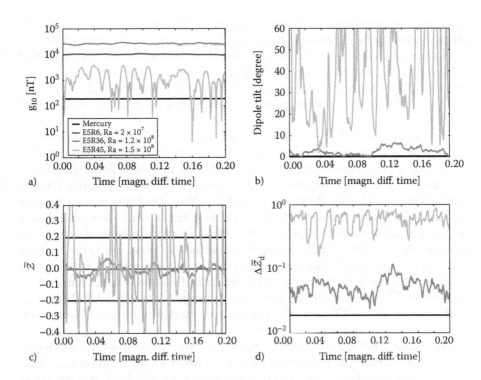

FIGURE 10.8
Time evolution of three standard dynamo models with different Rayleigh numbers. The thick black horizontal lines indicate the MESSENGER offset dipole model. Panel a) shows the axial dipole coefficient, panel b) the dipole tilt, panel c) the mean offset $\overline{Z}$ averaged over all radii up to $4R_M$, and panel d) shows the standard deviation for the offset in the distance range of the descending orbits $\Delta\overline{Z}_d$. For the numerical simulations time is given in units of the magnetic diffusion time $\tau_\lambda = d^2/\lambda$. When assuming an Earth-like aspect ratio of 0.35 and a magnetic diffusivity of $\lambda = 1$, the Hermean magnetic diffusion time amounts to $\tau_\lambda \approx 54$ kyr.

Christensen and Aubert (2006) introduced the local Rossby number

$$\text{Ro}_\ell = \frac{U}{L\Omega} \tag{10.15}$$

to quantify the ratio of inertial to Coriolis forces. Here, U is the rms flow amplitude and L is a typical flow length scale defined by

$$L = d\,\pi\,\frac{\sum U_\ell}{\sum \ell U_\ell}. \tag{10.16}$$

U_ℓ is the rms flow amplitude of spherical harmonic contributions with degree ℓ. The Coriolis force is responsible for organizing the flow into quasi-two-dimensional convective columns, which tend to produce the larger scale dipole dominated magnetic field. Inertia and in particular the nonlinear advective term, on the other hand, is responsible for the mixing of different scales and therefore the braking of flow symmetries. At $Ro_\ell = 0.10$ inertia is likely large enough in model E5R36 to trigger reversals though no such event has been observed in the relatively short period we could afford to simulate.

Christensen and Aubert (2006) report that this typically happens for $Ro_\ell = 0.08$, a limit clearly exceeded at $Ro_\ell = 0.18$ in model E5R45. Smaller scale contributions dominate the now multipolar magnetic field, which also becomes very variable in time and constantly changes its polarity (Wicht, Stellmach, and Harder 2011). Consequently, the offset also varies rapidly and may even exceed Mercury's offset value at times where the axial dipole is particularly low (See Figure 10.8b). While axial dipole and offset can assume Mercury-like values during brief periods in time, this is not true for the dipole tilt and $\Delta \overline{Z}$. The larger Rayleigh number promotes not only the axial quadrupole but also higher harmonics and nonaxial field contribution in general. The tilt is therefore typically rather large and the MEQ covers a wide latitude range. Closer to the planet, even two or more closed lines with $B_\rho = 0$ can be found at a given radius.

Olson and Christensen (2006) estimate a large local Rossby number of $Ro_\ell \approx 8$ for Mercury, mainly because of the planet's slow rotation rate. This suggests that the dynamo produces a multipolar field at least as complex as in the large Rayleigh number model E5R45. This is at odds with the observations unless we could add a physical mechanism to the model that would filter out smaller scale field contributions while retaining the strong axial quadrupole. As we discuss in the following, the stably stratified layer underneath Mercury's CMB (see Section 10.2) may meet these requirements.

10.5.3 Dynamos with a Stably Stratified Outer Layer

The idea of a stably stratified layer in the outer part of a dynamo region was first proposed by Stevenson (1980) to explain Saturn's very axisymmetric magnetic field. The immiscibility of helium and hydrogen in Saturn's metallic envelope (Lorenzen, Holst, and Redmer 2009) may cause helium to precipitate into the deeper interior. Similar to the iron snow scenario discussed in Section 10.2, this process may establish a stabilizing helium gradient in the rain zone.

Christensen (2006) and Christensen and Wicht (2008) adopt this idea for Mercury. They propose that the subadiabatic heat flux through the CMB leads to the stable stratification but since they use a condensity approach the model is not able to distinguish between thermal and compositional effects. The magnetic field that is produced in the convecting deeper core region has to diffuse through the largely stable outer layer so that the magnetic skin effect applies here. The time variability of the magnetic field increases with spatial complexity (Christensen and Tilgner 2004; Lhuillier et al. 2011). The higher harmonic field contributions are therefore more significantly damped by the skin effect than for example dipole or quadrupole. Zonal motions that may still penetrate the stable layer cannot lead to significant dynamo action but further increase the skin effect for nonaxisymmetric field contributions. Thanks to this filtering effect, the multipolar field of a high Ro_ℓ dynamo should look more Mercury-like when reaching the planetary surface.

Testing different dynamo setups, Christensen (2006) and Christensen and Wicht (2008) demonstrate that the surface field is indeed weaker and less complex when a sizable stable layer is included. We reanalyze the models 2, 3, and 4 published in Christensen and Wicht (2008) to test whether they are consistent with the new MESSENGER data. All three models, which we will refer to as CW2, CW3, and CW4 in the following, have a solid inner core that occupies the inner 50% in radius and a stable region that occupies the outer 28%. Thus, only a relatively thin region is left to host the active dynamo. Like for the standard models explored earlier, all three cases have the same Ekman number ($E = 10^{-4}$), Prandtl number ($Pr = 1$), and magnetic Prandtl number ($Pm = 3$) but differ in Rayleigh number. Once more, they use rigid flow boundary conditions and are driven by a growing inner core. The model parameters are listed in Table 10.2.

Figure 10.9 shows the time evolution of the axial dipole contribution, the dipole tilt, the mean offset $\overline{Z}$, and of its standard deviation $\Delta \overline{Z}$. At the lowest Rayleigh number of $Ra = 2 \times 10^8$ in CW2, the magnetic field strength is already significantly weaker and the dipole tilt and offset standard deviation can actually reach Mercury-like small values. The axial dipole component, however, is still somewhat strong and dominant and the offset value therefore too small. Increasing the Rayleigh number to $Ra = 4 \times 10^8$ in model CW3 decreases the axial dipole in absolute and relative terms. The mean tilt, offset, and spread increase, but there are times when Mercury-like field geometries are approached. The axial dipole is still by a factor four too strong. At $Ra = 4 \times 10^8$ in model CW4, however, the axial dipole can even become smaller than at Mercury. The field is very time-dependent during these episodes and is characterized by large dipole tilts and $\Delta \overline{Z}$ values since higher harmonic and nonaxisymmetric field contributions dominate. Very Mercury-like fields, which combine small dipole tilts with larger offset values but small offset standard deviations, can be found during brief periods when the axial dipole is somewhat stronger than the Mercury value.

Figure 10.10 illustrates the location of the MEQ for two particularly Mercury-like snapshots in the two larger Rayleigh number models CW3 and CW4. Figure 10.11 directly compares the respective radial magnetic fields with the MESSENGER model. Both figures demonstrate that solutions very similar to the offset dipole field proposed for Mercury can be found with a stably stratified outer core layer and a sufficiently high Rayleigh number. However, the magnetic field varies considerably in time and since neither hemisphere is preferred, the offset can switch from north to south and back.

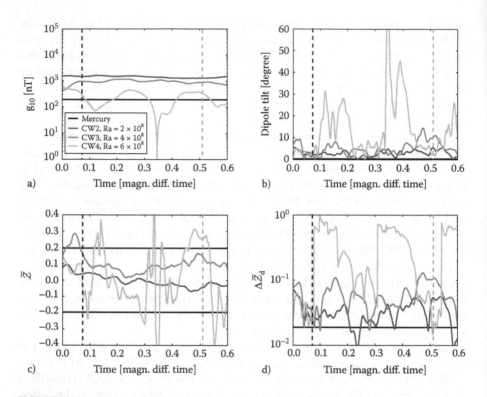

FIGURE 10.9
Time evolution of three dynamo models with a stably stratified layer. See Figure 10.8 for more explanation. The dashed vertical black and grey lines mark the times for the snapshots illustrated in Figure 10.10 and Figure 10.11.

The particular offset dipole configuration encountered by MESSENGER would thus only be transient and representative for only a few percent of the time at best.

Figure 10.12 compares the time-averaged spherical harmonics surface spectrum of models CW3 and CW4 with MODM, confirming that the relative quadrupole contribution and thus the equator offset is typically too low. The relative energy in spherical harmonic degrees $\ell = 3$ and 4, however, agrees quite well with MESSENGER observations.

Manglik, Wicht, and Christensen (2010) explore what happens to the stable layer when giving up the codensity formulation. They use a so-called double-diffusive approach where two equations of the form of Equation 10.6 separately describe the evolution of temperature and composition. When assuming a compositional diffusivity that is one order of magnitude lower than the thermal diffusivity, the compositional plumes that rise from the inner core boundary already stay significantly narrower than their thermal counterparts. This allows them to more easily penetrate and destroy the stable outer layer. The desirable filtering effect

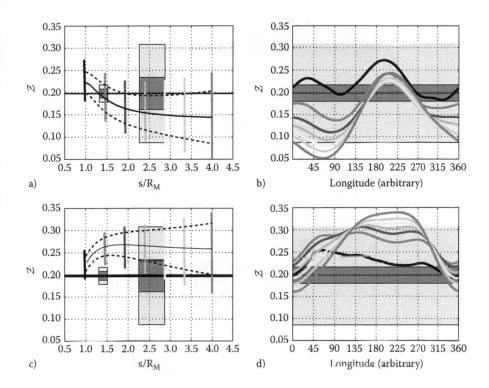

FIGURE 10.10

Magnetic equator location for two snapshots in dynamo models CW3 (top panels) and CW4 (bottom panels). The respective snapshot times have been marked by the vertical dashed lines in Figure 10.9. Colored dots show the equator locations found on a dense spherical longitude/latitude grid. The curved solid lines in panels a) and b) show the mean equator offset for each spherical surface with radius s/R_M, the dashed lines show the mean offset plus and minus the standard deviation. Thick horizontal lines illustrate the mean offset measured by the MESSENGER magnetometer while mid gray and light gray boxes show mean three sigma error and standard deviation for descending (left) and ascending orbits (right), respectively.

is greatly lost unless the sulfur concentration is below 1 wt% where compositional convection starts to play an inferior role. For such a low light element concentration, however, Mercury's core would likely be completely solid today.

The iron snow mechanism discussed in Section 10.2 offers an alternative scenario where the stable stratified layer is likely to persist even in a double-diffusive approach. The sulfur gradient that develops in the iron snow zone is potentially much more stabilizing than the subadiabatic thermal gradient assumed by Christensen and Wicht (2008) and Manglik, Wicht, and Christensen (2010). Furthermore, the additional convective driving source represented by the remelting snow would counteract the effects of the more sulfur-rich plumes rising from a growing inner core.

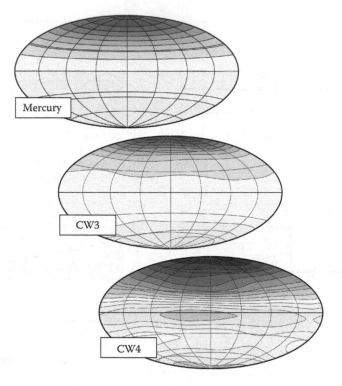

FIGURE 10.11
Comparison of the MODM radial magnetic field for Mercury with the two particularly
Mercury-like snapshots in models CW3 and CW4 already depicted in Figure 10.10. Blue (red
and yellow) indicates radially inward (outward) field.

10.5.4 Inhomogeneous Boundary Conditions

As already discussed in Section 10.1, an inhomogeneous heat flux through
the CMB is an obvious way to break the north/south symmetry and enforce
a more permanent offset of the MEQ. To explain the stronger magnetization
of the southern crust on Mars, several authors explored a variation following
a spherical harmonic function Y_{10} of degree $\ell = 1$ and order $m = 0$ (Stanley
et al. 2008; Amit, Christensen, and Langlais 2011; Dietrich and Wicht 2013).
The total CMB heat flux is then given by

$$q = q_0 \left(1 - q_{10}^* \cos\theta\right) \tag{10.17}$$

where q_0 is the mean heat flux, q_{10}^* the relative amplitude of the lateral
variation, θ the colatitude. Positive values of q_{10}^* are required at Mars and
negative values should enforce the stronger northern magnetic field observed
on Mercury.

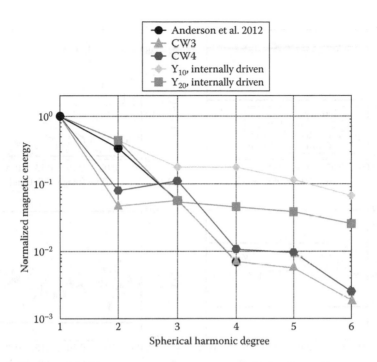

FIGURE 10.12

Comparison of the normalized MODM surface spectrum by Anderson et al. (2012) with time-averaged spectra for four different dynamo models: the dynamo models CW3 and CW4 that incorporate a stably stratified outer layer and models with an inhomogeneous CMB heat flux following a spherical harmonic Y_{10} or Y_{20} pattern, respectively. (From Anderson B.J., et al. 2012. Low degree structure in Mercury's planetary magnetic field. *J. Geophys. Res.* (1991–2012), 117 and Christensen U.R., Wicht J. 2008. Models of magnetic field generation. In partly stable planetary cores: Applications to Mercury and Saturn. *Icarus*, 196, 16.)

To explore the impact of the CMB heat flux pattern, we use dynamo simulations in the parameter range discussed by Dietrich and Wicht (2013) and Cao et al. (2014). The parameters are $E = 10^{-4}$, $Ra = 4 \times 10^7$, $Pr = 1$, $Pm = 2$, and $a = 0.35$. Once more, rigid boundary conditions are used and we impose the heat flux at the outer boundary. The Rayleigh number is then defined based on the mean CMB heat flux (Dietrich and Wicht 2013). Figure 10.13 demonstrates that a relative variation amplitude of $q_{10} = -0.10$ is nearly sufficient to enforce the observed offset when the dynamo is driven by homogeneously distributed internal sources (red line). These may model secular cooling, radioactive heating, or the remelting of iron snow. We have used a codensity formulation here and set the codensity flux from the inner core boundary to zero. Significantly larger heat flux variations are required for the other end member when the dynamo is driven by bottom sources that mimic a growing inner core. This is consistent with the findings by (Hori, Wicht, and Christensen 2012) who report that the impact of thermal CMB

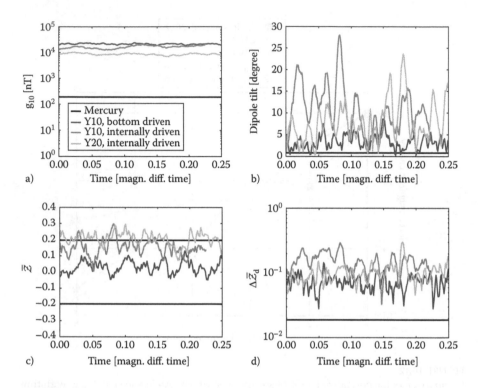

FIGURE 10.13

Time evolution of three dynamo models with inhomogeneous CMB heat flux. A Y_{10} pattern with increased heat flux through the northern hemisphere but also a Y_{20} pattern with a larger heat flux in the equatorial region promotes a Mercury-like offset. See text and Figure 10.8 for more explanation.

boundary conditions is generally larger for internally driven (ID) than bottom driven (BD) simulations.

Another not so obvious method to promote a north/south asymmetry is to increase the heat flux through the equatorial region. Cao et al. (2014) explore a Y_{20} pattern, which means that the total CMB flux is given by

$$q = q_0 (1 - q_{20}^*) \frac{1}{2} (3\cos\theta - 1). \tag{10.18}$$

The green line in Figure 10.13 illustrates that a variation amplitude of $q_{20} = 1/3$ causes a more or less persistent Mercury-like offset value. This translates into an increase of the equatorial flux by 17% and a decrease to polar flux by 33%. Cases with an increased heat flux at the poles, that is, negative values of q_{20}, did not yield the desired result. Except for the CMB heat flux pattern and a smaller inner core that only occupies 20% of the radius, the models are identical to the Y_{10} cases explored above. Though the Y_{20} pattern is equatorially symmetric, it promotes an equatorially asymmetric flow and therefore an asymmetric magnetic field production. A preliminary analysis of the system suggests that the Y_{20} pattern significantly decreases the critical Rayleigh number for the onset of equatorially antisymmetric convection modes, which is very large when the CMB heat flux is homogenous (Landeau and Aubert 2011).

The inhomogeneous CMB heat flux mainly helps to promote a Mercury-like mean offset of the MEQ while other important field characteristics seem not consistent with the observations. The field is generally much too strong and the often large dipole tilt and offset spread $\Delta \bar{Z}$ testify that higher harmonic and nonaxisymmetric field contributions remain too significant. This is confirmed by the time-averaged spectra shown in Figure 10.12. Adding a stably stratified outer layer, probably in combination with a larger Rayleigh number to bring down the too strong axial dipole contribution, seems like an obvious solution to this problem. This was confirmed by the first results presented by Tian, Zuber, and Stanley (2013) who explore the combination of the stable layer with the Y_{10} heat flux pattern.

10.5.5 Alternatives

Several authors varied the inner core size to explore its impact on the dynamo process. Heimpel et al. (2005) analyze models with aspect ratios between $a = 0.65$ and $a = 0.15$ that are all driven by a growing inner core. They report that the smallest inner core yields a particularly weak magnetic field with a CMB Elsasser number of $\Lambda_{cmb} = 10^{-2}$ when the Rayleigh number is close to onset for dynamo action. This is still more than two orders of magnitude too large for Mercury. Convection and dynamo action are mainly concentrated at only one convective column attached to the inner core. Such localized magnetic field production is not very conducive to maintaining a large-scale magnetic field, which is confirmed by the magnetic surface spectrum of a model snapshot shown in Figure 10.14. The relative quadrupole contribution nearly matches the MODM value but the higher harmonic contributions can reach a similar level and are thus too strong. This is also true for the dipole tilt that has a mean value of 8° for this model.

Takahashi and Matsushima (2006) find that the magnetic field strength is also reduced when using a large inner core with $a = 0.7$ in combination with a large Rayleigh number. Once more, the field is still too strong for Mercury with an Elsasser number around $\Lambda_{cmb} = 10^{-2}$ and is also much too small in scale with $\ell = 3$ and 5 contributions dominating the spectrum (see Figure 10.14). Stanley et al. (2005) explore even larger inner cores with aspect ratios up to $a = 0.9$ and report particularly weak fields at rather low Rayleigh numbers. The use of stress-free flow boundary conditions set this dynamo model apart from all the other cases discussed here. Field strength, dipole tilt, and offset are highly variable but seem to assume Mercury-like values at times. Little

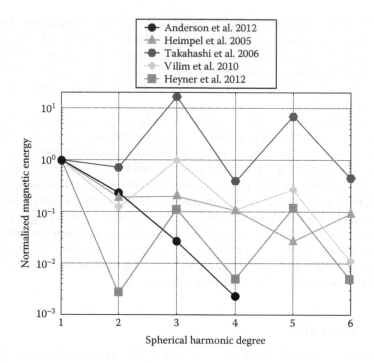

FIGURE 10.14

Comparison of the normalized MODM surface spectrum by Anderson et al. (2012) with spectra for different dynamo models. A time-averaged spectrum is shown for the models by Vilim et al. (2010) and Heyner et al. (2011b) while the spectra for Heimpel et al. (2005), Takahashi and Matsushima (2006) represent snapshots.

more is published about the field geometry and it seems worth to explore these models further.

Vilim, Stanley, and Hauck (2010) explore the double snow zone regime that may develop when the sulfur content in Mercury's core exceeds 10 wt%, as briefly discussed in Section 10.4. They consider a thin outer snow zone and a thicker zone in the middle of the liquid core in addition to a growing inner core. Since both snow zones are stably stratified, the dynamo action is concentrated in the two remaining shells. The magnetic fields that are produced in these two dynamo regions tend to oppose each other, which leads to a reduced overall field strength that matches the MESSENGER observation. However, the octupole component is generally too strong while the quadrupole contribution is too weak, as is demonstrated in Figure 10.14.

Since internal and external magnetic field can reach similar magnitudes at Mercury, the latter may actually play a role in the core dynamo. The idea of a feedback between internal and external dynamo processes was first proposed by Glassmeier et al. (2007) for Mercury and further developed in a series of papers (Heyner et al. 2010, 2011b, 2011a). Because the internal dynamo process operates on time scales of decades to centuries, only the long time-averaged magnetospheric field needs to be considered. This can be approximated by an external axial dipole that opposes the direction of the internal axial dipole within the core. The ratio of the external to internal dipole field depends on the distance of the MP to the planet and thus on the intensity of the internal field. Heyner et al. (2011b) find that the feedback quenches the dynamo field to Mercury-like intensities when the simulation is started off with an already weak field and the Rayleigh number is not too high. These conditions can, for example, be met when dynamo action is initiated with the beginning of iron snow or inner core growth at a period in the planetary evolution where mantle convection is already sluggish and the CMB heat flux therefore low. The feedback process modifies the dipole dominated field by concentrating the flux at higher latitudes. The result is a spectrum where the relative quadrupole (and other equatorially antisymmetric contributions) is too weak while the octupole (and other equatorially symmetric contributions) is too strong (see Figure 10.14).

10.6 Conclusions

The MESSENGER data have shown that Mercury has an exceptional magnetic field (Anderson et al. 2012; Johnson et al. 2012). The internal field is very weak and has a simple but surprising geometry that is consistent with an axial dipole offset by 20% of the planetary radius to the North. This implies a very strong axial quadrupole but at the same time also small higher harmonic and nonaxial contributions, a unique combination in our solar system.

Numerical dynamo models have a hard time to explain these observations. Strong axial quadrupole contributions and thus a significant mean offset of the MEQ can be promoted by different measures. Very small and very large inner cores or strong inertial forces are three possibilities that lead to a sizable but also very time-dependent axial quadrupole contribution.

A more persistent Mercury-like mean offset can be enforced by imposing lateral variations in the CMB heat flux. Pattern with either an increased heat flux in the northern hemisphere or in the equatorial region yield the desired result. They are particularly effective when the dynamo is not driven by a growing inner core but by homogeneously distributed buoyancy sources (Cao et al. 2014). New models for Mercury's interior, however, suggest that neither pattern is likely to persist today.

Unfortunately, the measures that promote a stronger axial quadrupole also tend to promote nondipolar and nonaxisymmetric field contributions in general. The offset of the MEQ therefore strongly depends on longitude and distance to the planet, which is at odds with the MESSENGER observations. Dynamo simulations by Christensen (2006) and Christensen and Wicht (2008) have shown that a stably stratified outer core layer helps to solve this problem. The magnetic field that is produced in the deeper core regions has to diffusive through this largely passive layer to reach the planetary surface. And since the magnetic field varies in time, it is damped by the magnetic skin effect during this process. Higher harmonic and nonaxisymmetric contributions are damped more effectively than axial dipole or quadrupole because the variation time scale decreases with increasing spatial complexity. When reaching the surface, the field is therefore not only more Mercury-like in geometry but also similarly weak.

Recent interior models for Mercury suggest that a stable outer core layer may indeed exist. Because of the low pressures in Mercury's outer core, an outer iron snow zone should develop underneath the CMB for mean core sulfur concentration beyond about 2 wt%. As the planet cools, the snow zone extends deeper into the core and a stably stratifying sulfur gradient develops. Since the mean heat flux out of the Hermean core is likely subadiabatic today, thermal effects would further contribute to stabilizing the outer core region. Such a layer is also likely to persist when double-diffusive effects are taken into account (Manglik, Wicht, and Christensen 2010). Additional work on the FeS melting behavior, on Mercury's interior properties, and the planet's thermal evolution is required to better understand and establish this scenario. The possible presence of Si in the Hermean core could further complicate matters (Malavergne et al. 2010).

Dynamo simulations that more realistically model the iron snow stratification and the convective driving in the presence of an iron snow zone and possibly also a growing inner core seem a logical next step. Lateral variations in the CMB heat flux and a feedback with the magnetospheric field are two other features that may play an important role in Mercury's dynamo process.

The Hermean magnetospheric field remains a challenging puzzle despite the wealth of data delivered by the MESSENGER magnetometer. Its small size and high variability complicates the separation of internal and external field contributions, of temporal and spatial variations, and of solar wind dynamics and Mercury's genuine field dynamics. The BepiColombo mission, scheduled for launch in 2016, will significantly improve the situation since two spacecrafts will orbit the planet at the same time, a planetary orbiter build by ESA and a magnetospheric orbiter build by JAXA.

Acknowledgments

Johannes Wicht was supported by the Helmholz Alliance "Planetary Evolution and Live" and by the Special Priority Program 1488 "Planetary Magnetism" of the German Science Foundation. D. Heyner was supported by the German Ministerium für Wirtschaft und Technologie and the German Zentrum für Luft- und Raumfahrt under contract 50 QW 1101. We thank

Attilio Rivoldini, Tina Rückriehmen, Wieland Dietrich, Hao Cao, Brian Anderson, Karl-Heinz Glassmeier, and Ulrich R. Christensen for helpful discussions. Attilio Rivoldini also kindly provided Figure 10.1.

References

Alexeev I.I., Belenkaya E.S., Slavin J.A., et al. 2010. Mercury's magnetospheric magnetic field after the first two MESSENGER flybys. *Icarus* 209, 23.

Amit H., Christensen U.R., Langlais B. 2011. The influence of degree-1 mantle heterogeneity on the past dynamo of Mars. *Phys. Earth Planet. Inter.* 189, 63.

Anderson B.J., Johnson C.L., Korth H., et al. 2011. The global magnetic field of Mercury from MESSENGER orbital observations. *Science*, 333, 1859.

Anderson B.J., Johnson C.L., Korth H., et al. 2012. Low-degree structure in Mercury's planetary magnetic field. *J. Geophys. Res.* (1991–2012), 117.

Aubert J., Amit H., Hulot G., Olson P. 2008. Thermochemical flows couple the Earth's inner core growth to mantle heterogeneity. *Nature* 454, 758.

Bartels J. 1936. The eccentric dipole approximating the earth's magnetic field. *J. Geophys. Res.* 41, 225.

Baumjohann W., Blanc M., Fedorov A., Glassmeier K.H. 2010. Current systems in planetary magnetospheres and ionospheres. *Space Sci. Rev.* 152, 99.

Braginsky S., Roberts P. 1995. Equations governing convection in Earth's core and the geodynamo. *Geophys. Astrophys. Fluid Dyn.* 79, 1.

Cao H., Russell C.T., Wicht J., Christensen U.R., Dougherty M.K. 2012. Saturn's high degree magnetic moments: Evidence for a unique planetary dynamo. *Icarus* 221, 388.

Cao H., Aurnou J., Wicht J., et al. 2014. A dynamo explanation for the large offset of Mercury's MEQ, to appear in *Geophys. Res. Lett.*

Christensen U., Aubert J. 2006. Scaling properties of convection-driven dynamos in rotating spherical shells and application to planetary magnetic fields. *Geophys. J. Int.* 116, 97.

Christensen U., Tilgner A. 2004 Power requirement of the geodynamo from ohmic losses in numerical and laboratory dynamos. *Nature* 429, 169.

Christensen U.R., 2006. A deep dynamo generating Mercury's magnetic field. *Nature* 444, 1056.

Christensen U., Wicht J. 2007. In: P. O. (ed.) *Core Dynamics*, vol. 8 of Treatise on Geophysics, 245–282, Elsevier.

Christensen U.R. 2010. Dynamo scaling laws and application to the planets. *Space Sci. Rev.*, 152, 565.

Christensen U.R., Wicht J. 2008. Models of magnetic field generation in partly stable planetary cores: Applications to Mercury and Saturn. *Icarus*, 196, 16.

Christensen U.R., Aubert J., Hulot G. 2010. Conditions for Earth-like geodynamo models. *Earth Planet. Sci. Lett.*, 296, 487.

Connerney J.E.P., Acuña M.H., Ness N.F., Satoh T. 1998. New models of Jupiter's magnetic field constrained by the Io flux tube footprint. *J. Geophys. Res.*, 103, 11929.

Denevi B.W., Ernst C.M., Meyer H.M., et al. 2013. The distribution and origin of smooth plains on Mercury. *J. Geophys. Res.*, 118, 891.

Di Achille G., Popa C., Massironi M., et al. 2012. Mercury's radius change estimates revisited using MESSENGER data. *Icarus*, 221, 456.

Dibraccio G.A., Slavin J.A., Boardsen S.A., et al. 2013. MESSENGER observations of magnetopause structure and dynamics at Mercury. *J. Geophys. Res.* 118, 997.

Dietrich W., Wicht J. 2013. A hemispherical dynamo model: Implications for the Martian crustal magnetization. *Phys. Earth Planet. Inter.*, 217, 10.

Glassmeier K., Auster H., Motschmann U., 2007. A feedback dynamo generating Mercury's magnetic field. *Geophys. Res. Lett.*, 34, 22201.

Glassmeier K.H., Auster H.U., Heyner D., et al. 2010. The fluxgate magnetometer of the BepiColombo Mercury planetary orbiter. *Planet. Space Sci.* 58, 287.

Grott M., Breuer D., Laneuville M. 2011. Thermo-chemical evolution and global contraction of Mercury. *Earth Planet. Sci. Lett.*, 307, 135.

Hauck S.A., Aurnou J.M., Dombard A.J. 2006. Sulfur's impact on core evolution and magnetic field generation on Ganymede. *J. Geophys. Res.*, 111, 9008.

Hauck S.A., Margot J.L., Solomon S.C., et al. 2013. The curious case of Mercury's internal structure. *J. Geophys. Res.* 118, 1204.

Heimpel M.H., Aurnou J.M., Al-Shamali F.M., et al. 2005. A numerical study of dynamo action as a function of spherical shell geometry. *Earth Planet. Sci. Lett.* 236, 542.

Heyner D., Schmitt D., Wicht J., et al. 2010. The initial temporal evolution of a feedback dynamo for Mercury. *Geophys. Astrophys. Fluid Dyn.*, 104, 419.

Heyner D., Schmitt D., Glassmeier K.H., Wicht J. 2011a. Dynamo action in an ambient field. *Astronom. Nach.*, 332, 36.

Heyner D., Wicht J., Gómez-Pérez N., et al. 2011b. Evidence from numerical experiments for a feedback dynamo generating Mercury's magnetic field. *Science*, 334, 1690.

Holme R., Bloxham J., 1996. The magnetic fields of Uranus and Neptune: Methods and models. *J. Geophys. Res.*, 101, 2177.

Hori K., Wicht J., Christensen U.R. 2012. The influence of thermo-compositional boundary conditions on convection and dynamos in a rotating spherical shell. *Phys. Earth Planet. Inter.*, 196, 32.

Janhunen P., Kallio E., 2004. Modelling the solar wind interaction with Mercury by a quasi-neutral hybrid model. *Ann. Geophys.*, 22, 1829.

Johnson C.L., Purucker M.E., Korth H., et al. 2012. MESSENGER observations of Mercury's magnetic field structure. *J. Geophys. Res.* 117, 0.

Kutzner C., Christensen U., 2000. Effects of driving mechanisms in geodynamo models. *Geophys. Res. Lett.*, 27, 29.

Landeau M., Aubert J. 2011. Equatorially asymmetric convection inducing a hemispherical magnetic field in rotating spheres and implications for the past Martian dynamo. *Phys. Earth Planet. Inter.*, 185, 61.

Lesur V., Kunagu P., Asari S., et al. 2012. GRIMM–3, Third version of the GFZ reference internal magnetic model, In preparation.

Lhuillier F., Fournier A., Hulot G., Aubert J. 2011. The geomagnetic secular-variation timescale in observations and numerical dynamo models. *Geophys. Res. Lett.* 38, L09306.

Lorenzen W., Holst B., Redmer R. 2009. Demixing of hydrogen and helium at megabar pressures. *Phys. Rev. Lett.*, 102, 115701.

Malavergne V., Toplis M.J., Berthet S., Jones J. 2010. Highly reducing conditions during core formation on Mercury: Implications for internal structure and the origin of a magnetic field. *Icarus*, 206, 199.

Mangano V., Massetti S., Milillo A., et al. 2013. Dynamical evolution of sodium anisotropies in the exosphere of Mercury. *Planet. Space Sci.*, 82, 1.

Manglik A., Wicht J., Christensen U.R. 2010. A dynamo model with double diffusive convection for Mercury's core. *Earth Planet. Sci. Lett.* 289, 619.

Margot J.L., Peale S.J., Solomon S.C., et al. 2012. Mercury's moment of inertia from spin and gravity data. *J. Geophys. Res.* 117, 0.

Michel N.C., Hauck S.A., Solomon S.C., et al. 2013. Thermal evolution of Mercury as constrained by MESSENGER observations. *J. Geophys. Res.*, 118, 1033.

Morard G., Katsura T. 2010. Pressure–temperature cartography of Fe–S–Si immiscible system. *Geoch. Cosmoch. Acta* 74, 3659.

Müller J., Simon S., Wang Y.C., et al. 2012. Origin of Mercury's double magnetopause: 3D hybrid simulation study with AIKEF. *Icarus*, 218, 666.

Ness N.F., Behannon K.W., Lepping R.P., Whang Y.C., Schatten K.H. 1974. Magnetic field observations near Mercury: Preliminary results from Mariner 10. *Science*, 185, 151.

Nittler L.R., Starr R.D., Weider S.Z., et al. 2011. The major-element composition of Mercury's surface from MESSENGER x-ray spectrometry. *Science* 333, 1847.

Olsen N., Glassmeier K.H., Jia X. 2010. Separation of the magnetic field into external and internal parts. *Space Sci. Rev.* 152, 135.

Olson P., Christensen U. 2006. Dipole moment scaling for convection-driven planetary dynamos. *Earth Planet. Sci. Lett.*, 250, 561.

Peale S.J. 1969. Generalized Cassini's laws. *Astronom. J.*, 74, 483.

Purucker M.E., Johnson C.L., Winslow R.M., et al. 2012. In: *Lunar and Planetary Institute Science Conference Abstracts*, vol. 43 of Lunar and Planetary Institute Science Conference Abstracts, 1297.

Rivoldini A., Van Hoolst T. 2013. The interior structure of Mercury constrained by the low-degree gravity field and the rotation of Mercury. *Earth Planet. Sci. Lett.*, 377, 62.

Rivoldini A., Van Hoolst T., Verhoeven O., Mocquet A., Dehant V. 2011. Geodesy constraints on the interior structure and composition of Mars. *Icarus*, 213, 451.

Schubert G., Soderlund K.M. 2011. Planetary magnetic fields: Observations and models. *Phys. Earth Planet. Inter.*, 187, 92.

Slavin J.A., Krimigis S.M., Acuña M.H., et al. 2007. MESSENGER: Exploring Mercury's magnetosphere. *Space Sci. Rev.*, 131, 133.

Slavin J.A., Anderson B.J., Baker D.N., et al. 2012. MESSENGER and Mariner 10 flyby observations of magnetotail structure and dynamics at Mercury. *J. Geophys. Res.*, 117, 1215.

Smith D.E., Zuber M.T., Phillips R.J., et al. 2012. Gravity field and internal structure of mercury from MESSENGER. *Science*, 336, 214.

Solomon S.C. 1976. Some aspects of core formation in Mercury. *Icarus*, 28, 509.

Solomon S.C., McNutt R.L., Gold R.E., Domingue D.L. 2007. MESSENGER mission overview. *Space Sci. Rev.*, 131, 3.

Solomon S.C., Byrne P.K., Klimczak C., et al. 2014. Geological evidence that Mercury contracted by more than previously recognized, presented at AGU, Dec. 2013.

Stanley S., Glatzmaier G.A. 2010. Dynamo models for planets other than Earth. *Space Sci. Rev.*, 152, 617.

Stanley S., Bloxham J., Hutchison W., Zuber M. 2005. Thin shell dynamo models consistent with Mercury's weak observed magnetic field. *Earth Planet. Sci. Lett.*, 234, 341.

Stanley S., Elkins-Tanton L., Zuber M.T., Parmentier E.M. 2008. Mars' paleomagnetic field as the result of a single-hemisphere dynamo. *Science*, 321, 1822.

Stevenson D.J. 1980. Saturn's luminosity and magnetism. *Science*, 208, 746.

Takahashi F., Matsushima M. 2006. Dipolar and non-dipolar dynamos in a thin shell geometry with implications for the magnetic field of Mercury. *Geophys. Res. Lett.*, 33, L10202.

Tian Z., Zuber M.T., Stanley S., 2013. Explaining Mercurys magnetic field observables using dynamo models with stable layers and laterally variable heat flux, presented at AGU, Dec. 2013.

Tosi N., Grott M., Plesa A., Breuer D. 2013. Thermochemical evolution of Mercury's interior. *Geophys. Res. Lett.*, 118, 2474–2487.

Vilim R., Stanley S., Hauck S.A. 2010. Iron snow zones as a mechanism for generating Mercury's weak observed magnetic field. *J. Geophys. Res.*, 115, 11003.

Wicht J., Mandea M., Takahashi F. et al. 2007. Mercury's internal magnetic field. *Space Sci. Rev.*, 132, 261.

Wicht J., Stellmach S., Harder H. 2011. Numerical dynamo simulations: From basic concepts to realistic models. In: Freeden W., Nashed M., Sonar T. (eds.) *Handbook of Geomathematics*, Springer, Berlin – Heidelberg, New York, 459–502.

Williams J.P., Ruiz J., Rosenburg M.A., Aharonson O., Phillips R.J. 2011. Insolation driven variations of Mercury's lithospheric strength. *J. Geophys. Res.* 116, 1008.

Winslow R.M., Anderson B.J., Johnson C.L., et al. 2013. Mercury's magnetopause and bow shock from MESSENGER magnetometer observations. *J. Geophys. Res.*, 118, 2213.

11

Lunar Gravity Field Determination from Chang'E-1 and Other Missions' Data

Jianguo Yan, Fei Li, and Koji Matsumoto

CONTENTS

11.1 Status of Lunar Gravity Field 263
11.2 Theory of Lunar Gravity Field Model Determination 265
11.3 Chang'E-1 Precision Orbit Determination and Lunar Gravity
Field Recovery 270
    11.3.1 Chang'E-1 Precision Orbit Determination 270
    11.3.2 Lunar Gravity Field Model (CEGM-01) Solution 272
    11.3.3 Integrated Solution from Chang'E-1, SELENE, and
    Historical Tracking Data...277
11.4 Gravity Field Models from the GRAIL Mission 284
11.5 Conclusions..287
Acknowledgments..289
References...289

11.1 Status of Lunar Gravity Field

The gravity field of the Moon is a key quantity in lunar science. It substantially contributes to improve our knowledge about the mass distribution in the interior of the Moon, crustal properties, deep tectonic features, and the lunar evolution history (e.g., Andrews-Hanna et al. 2013; Wieczorek et al. 2013; Zuber et al. 2013b), especially when coupled with geologic mapping information (Wilhelms, Howard, and Wilshire 1979). Geophysical parameters, such as the selenopotential Love number k_2, can be estimated together with the gravity field from satellite tracking data. Due to the lack of appropriate seismic data and other in situ surface investigations, these geophysical quantities act as critical constraints to determine the state of the lunar core (Goossens and Matsumoto (2008), Williams, Boggs, and Ratcliff 2009, Yan et al. 2013a, 2013b).

Numerous efforts have been made to develop lunar gravity field models since the first lunar orbiter Russian Luna 10 mission in 1966 (Jin, Arivazhagan, and Araki 2013). Typical early representative models included a 16 × 16 degree and order model proposed by Bill and Ferrari (1980). During the same period, a 16 × 16 order gravity field model was proposed by Sagitov from the Soviet Union. Although the details of its implementation were not disclosed, it performed similarly to Bill and Ferrari's 16 × 16 order model.

In the 1990s lunar exploration missions Clementine and Lunar Prospector (LP) were launched. Based on the tracking data of previous missions and Clementine, GLGM-2 with degree and order of 70 was developed by Goddard Space Flight Center (GSFC; Lemoine et al. 1997). After including orbital tracking data of LP, high-resolution gravity field models LP165P and LP150Q were developed by Konopliv (2001). LP was a nearly circular polar satellite with an orbital height of 100 km in normal phase and 30 km in extended phase. The tracking data of LP significantly improved lunar nearside gravity.

After years of development, the method using lunar satellite orbital tracking data to determine lunar gravity field model matured, and the available lunar gravity field models had played significant roles in lunar spacecraft orbit determination and navigation. However, the tracking data of lunar spacecraft were limited to the nearside. The Kaula constraint had to be introduced because of the data gap and it decreased the gravity accuracy on the lunar far side.

The SELENE mission was put forward to tackle the data gap on the lunar far side by high-low satellite-to-satellite tracking mode for the first time. The SELENE mission was successfully launched by Japan in September 2007 and obtained direct far-side measurement. The latest lunar gravity field models, such as SGM90d, SGM100h, and SGM100i (Namiki et al. 2009; Matsumoto et al. 2010; Goossens et al. 2011), were developed using the tracking data of SELENE as well as other historical spacecraft. Compared with LP series model LP100K, they showed a significant improvement of precision of orbit determination on the lunar far side, and discovered new gravity anomaly features on the lunar far side.

Figures 11.1 and 11.2 show gravity anomaly maps from LP150Q and SGM09d. It is clear that SGM90d is better in resolution than LP150Q in the

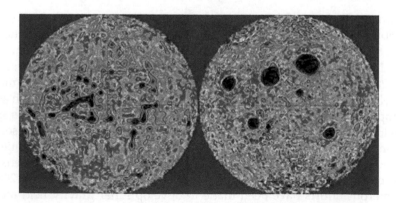

FIGURE 11.1
Lunar gravity anomaly based on LPQ150Q. (From Konopliv, A.S., et al. Recent gravity models as a result of the Lunar Prospector mission. *Icarus* 150(1), 1–18, 2001.)

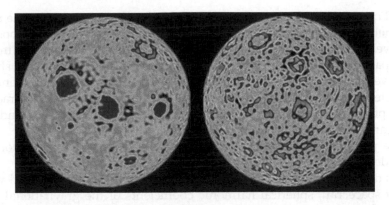

FIGURE 11.2
Lunar gravity anomaly based on SGM90d. (From Namiki, N., et al. Farside gravity field of the Moon from four-way Doppler measurements of SELENE (Kaguya), *Science* 323(5916), 900 905, 2009.)

lunar far side and can discern more gravity features. In the maria mascon nearside area, negative gravity anomaly is surrounded by positive gravity anomaly. In the typical area of lunar far side, just like Korevev, an annularly surrounded form of gravity anomaly is discovered, in other words, a negative gravity anomaly exists in the center area surrounded by positive and negative gravity anomaly alternately. This gravity anomaly feature provides a new basis to study the crust mantle structure, isostatic compensation status, lunar dichotomy, and other scientific issues.

Currently, lunar gravity field determination is greatly improved by the GRAIL mission, and the detailed introduction about the achievements from GRAIL will be described in Section 11.4.

11.2 Theory of Lunar Gravity Field Model Determination

The dynamic method is used in lunar spacecraft precision orbit determination and gravity field modeling. The precision orbit determination is implemented by parameters' differential correction using estimation algorithms such as least squares and Kalman filter (Tapley et al. 2004b). The goal is to obtain a minimum sum of weighted squares of differences between the measurement and the theoretical observation from the model. Two basic equations, observables equation and differential equation of motion, are involved in the precision orbit determination. Here is a brief description of the lunar spacecraft precision orbit determination.

The observables equation in time t establishes in the following form:

$$O_c = f_0[\vec{r}(t,p), \dot{\vec{r}}(t,p)] + b + RF_c \tag{11.1}$$

In Equation 11.1, O_c is an observable calculated from model in time t; f_o is geometric relationship determined by observation data type; $\vec{r}, \dot{\vec{r}}$ are position and velocity of spacecraft in time t. For the planetary orbit determination the spacecraft position and velocity used in observable computation should be given in solar barycenter frame. b is constant bias of measurement; RF_c is systematic correction of measurement caused by atmospheric refraction, ionospheric delay, transponder delay, antenna pointing errors, etc., and usually can be removed in data preprocessing.

The parameters of observation model to be estimated in Equation 11.1 include:

p is dynamic parameters vector that consists of initial position and velocity of spacecraft, spherical harmonic coefficients of the gravitational field, atmospheric drag coefficient, solar pressure coefficient and other force model parameters such as empirical acceleration parameters;

b is constant bias of measurement and station;

Based on known a priori information of parameters, an estimation model is built by linearization:

$$O_o - O_c = \frac{\partial O_c}{\partial \bar{q}} \Delta \bar{q} + n \tag{11.2}$$

where O_o is an actual observation in time t and O_c is a theoretical value obtained from observation model. n is measurement noise and $\Delta \bar{q}$ is correction of parameters vector which includes dynamic parameter p and measurement bias b.

Equation 11.2 can be expanded by differential correction as:

$$O_o - O_c = (\frac{\partial O_c}{\partial \bar{p}}) \Delta \bar{p} + (\frac{\partial O_c}{\partial b}) \Delta b + n \tag{11.3}$$

or

$$v = \left[\frac{\partial O_c}{\partial \bar{p}} \vdots \frac{\partial O_c}{\partial b} \right] \begin{bmatrix} \Delta \bar{p} \\ \cdots \\ \Delta b \end{bmatrix} + n \tag{11.4}$$

By setting the partial derivative coefficient matrix as F and weight matrix of observables as W, the parameters can be computed based on the theory of least squares estimation:

$$\Delta q = (F^T W F)^{-1} (F^T W l) \tag{11.5}$$

In Equation 11.5 the critical point is to determine the partial derivative coefficient matrix F. The part of F, $\dfrac{\partial O_c}{\partial \bar{p}}$, can be computed by the following equation:

$$\frac{\partial O_c}{\partial \bar{p}} = \frac{\partial f_0}{\partial \bar{r}}\frac{\partial \bar{r}}{\partial \bar{p}} + \frac{\partial f_0}{\partial \dot{\bar{r}}}\frac{\partial \dot{\bar{r}}}{\partial \bar{p}} \tag{11.6}$$

where $\dfrac{\partial f_0}{\partial \bar{r}}$ and $\dfrac{\partial f_0}{\partial \dot{\bar{r}}}$ can be computed directly from observables in Equation 11.1;

$\dfrac{\partial \bar{r}}{\partial \bar{p}}$ and $\dfrac{\partial \dot{\bar{r}}}{\partial \bar{p}}$ should be computed by the motion equation. The form of the motion equation of spacecraft is

$$\ddot{\bar{r}} = f(\bar{r}, \dot{\bar{r}}, \bar{p}, t) \tag{11.7}$$

In lunar inertial frame, the motion equation of spacecraft is

$$\ddot{r} = -\frac{GM}{r^2}(\frac{\vec{r}}{r}) + F_\varepsilon \tag{11.8}$$

$$F_\varepsilon = \sum_{j=1}^{4} F_j \tag{11.9}$$

The first term of the equation is center of gravity of lunar, and F_ε is perturbation acceleration. The perturbation forces applied to lunar spacecraft include nonspherical perturbation force, three-body perturbation of Sun and Earth, lunar tide perturbation, solar radiation and albedo force.

We list four dominant perturbation forces as follows. The first is nonspherical perturbation force. Its spherical harmonic function form is (Kaula 1966)

$$\vec{F_1} = \nabla V$$

$$V = \frac{\mu}{r}\sum_{l=2}^{N} \bar{C}_{l0}(\frac{a_m}{r})^l \bar{P}_l(\sin \varphi) + \sum_{l=2}^{N}\sum_{m=1}^{l} (\frac{a_m}{r})^l \bar{P}_{lm}(\sin \varphi)[\bar{C}_{lm}\cos m\lambda + \bar{S}_{lm}\sin m\lambda]$$

$$\tag{11.10}$$

The first term of the second equation in Equation 11.10 is zonal harmonic terms and the second term is tesseral harmonic terms. In this equation, a_m is the average equatorial radius of lunar reference ellipsoid and r, ϕ, λ is lunar geodetic coordinates.

The second perturbation force is three-body perturbations between satellite and Sun and Earth and can be written as:

$$\vec{F_2} = -\sum_{j=1}^{2} Gm_j(\frac{\vec{r_j}}{r_j^3} - \frac{\vec{\Delta_j}}{\Delta_j^3}) \tag{11.11}$$

$$\vec{\Delta_j} = \vec{r} - \vec{r_j} \tag{11.12}$$

In Equations 11.11 and 11.12, $\vec{r_j}$ and m_j ($j = 1,2$) are the position vectors relative to lunar and mass of Sun and Earth, and $\vec{r}$ is position vector of satellite.

The third is lunar tidal perturbation. It can be expressed as gradient form of force function (Rowlands et al. 2000):

$$\vec{F_4} = \nabla V_{OT}$$

$$V_{OT} = k_2(\frac{Gm_l}{r})(\frac{a_m}{r})^2(\frac{a_m}{r_l})^3 \bar{P_2}(\sin \psi) \tag{11.13}$$

In Equation 11.13, $k_2 = 0.0240$ is lunar's Love number (Konopliv et al. 2013), and ψ is an angle between the satellite's position vector and Earth position vector in lunar geodetic coordinate system, and can be expressed as the inner product between two unit vectors:

$$\cos \psi = (\frac{\vec{r}}{r}, \frac{\vec{r_j}}{r_j}) = (\frac{\vec{r}}{r})(\frac{\vec{r_j}}{r_j}) \tag{11.14}$$

Solar radiation pressure and albedo force are dissipative forces. By using the image data of Clementine and measuring the absolute albedo, scientists of Delft Institute for Earth-Oriented Space Research (DEOS) obtained lunar global albedo model in the form of spherical harmonics (Floberghangen, Visser, and Weischede 1999). The solar radiation acceleration of unit facet is:

$$\vec{da_i} = C_R(\tau a E_s \cos \theta_s + cM_B)\frac{A_c}{mc\pi r_i^2} \cos a dA \vec{r_i} \tag{11.15}$$

where C_R is solar radiation pressure coefficient, τ is shadow factor $(0<\tau<1)$, a is reflected solar radiation pressure value calculated by expanding spherical harmonic coefficients, E_s is stream of photons from Sun and the Moon, θ_s is incident angle of solar photons to the i-th unit area, a is scattering angle of this unit area, M_B is ideal black body radiation of the Moon, A_c is cross-sectional area of the satellite, m is mass of the satellite, c is light speed, r_i is the distance of the satellite and unit area in lunar surface, dA is the lunar surface unit area, and $\vec{r_i}$ is a unit vector from the lunar surface unit area to the satellite. The solar pressure and albedo force of the satellite can be computed by integrating Equation 11.15.

Considering that dynamic parameters vectors $\vec{P}$ and time variables t are independent, the derivative operator and the partial derivative operator can be exchanged in the left-side of Equation 11.7 after taking derivative to dynamic parameters vector $\vec{P}$ of both side. Thus, we can get:

$$\frac{d^2}{dt^2}\left(\frac{\partial \vec{r}}{\partial \vec{p}}\right) = \frac{\partial f}{\partial \vec{r}}\frac{\partial \vec{r}}{\partial \vec{p}} + \frac{\partial f}{\partial \dot{\vec{r}}}\frac{\partial \dot{\vec{r}}}{\partial \vec{p}} + \frac{\partial f}{\partial \vec{p}} \tag{11.16}$$

Let $A(t) = \left.\dfrac{\partial f}{\partial \vec{r}}\right|_{3\times3}$, $B(t) = \left.\dfrac{\partial f}{\partial \dot{\vec{r}}}\right|_{3\times3}$, $C(t) = \left.\dfrac{\partial f}{\partial \vec{p}}\right|_{3\times l}$ and $Y(t) = \left.\dfrac{\partial \vec{r}}{\partial \vec{p}}\right|_{3\times l}$, where l is the number of dynamic parameters. Then Equation 11.16 can be simplified as:

$$\ddot{Y} = A(t)Y + B(t)\dot{Y} + C(t) \tag{11.17}$$

In Equation 11.17, $A(t)$, $B(t)$ and $C(t)$ can be computed by partial derivative of force model on parameters, and Y is an unknown variable. Partial derivative of state vector on dynamic parameters vector can be solved by numerical integration of variational Equation 11.17. Taking it back to Equation 11.6, the partial derivative of observables on dynamic parameters can be obtained and partial derivative coefficient matrix F can also be determined. Correction values can be obtained by solving the estimation Equation 11.5. Iterations are necessary for the estimation because the whole process is based on linear expansion of perturbation theory. Therefore, in order to obtain the unknown values of parameters, we need to set convergence criterion to determine iteration number. The gravity field will be developed by introducing potential coefficients sequence $\{C_{nm}, S_{nm}\}$ to dynamic parameter vector $\vec{P}$. If the degree and order of gravity field is high, specific algorithm of matrix inversion should be considered to ensure the precision of inversion (Ulman, 1994; Konopliv et al. 2013; Lemoine et al. 2013). In Figure 11.3 we take an example of Chang'E-1 to show the flow chart of precision orbit determination and lunar gravity field recovery.

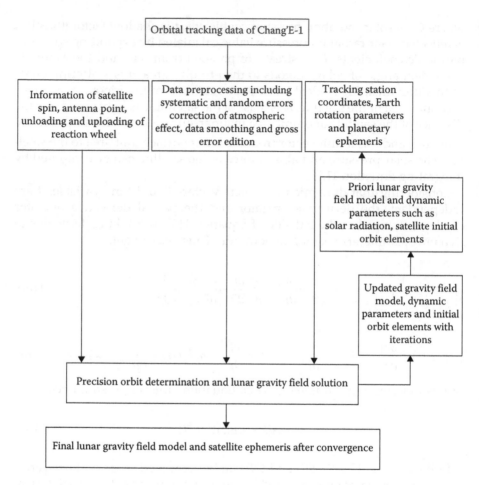

FIGURE 11.3
The flow chart of precision orbit determination and gravity field recovery.

11.3 Chang'E-1 Precision Orbit Determination and Lunar Gravity Field Recovery

11.3.1 Chang'E-1 Precision Orbit Determination

Chang'E-1 precision orbit determination was mainly based on the tracking data from the in-orbit test phase of the mission (Yan et al. 2010; Wei et al. 2013). The in-orbit test phase spanned from November through December 2007. During this period, intense track was carried out by unified S-band (USB) system and S/X dual frequency very long baseline interferometry (VLBI) network with a maximum bandwidth of 16 MHz. The four VLBI stations involved in the VLBI network are Sheshan (31.09N, 121.19E), Miyun

(40.55N, 116.97E), Kunming (25.03N, 102.42E), and Nanshan (43.47N, 87.17E). The Sheshan and Nanshan stations are conventional VLBI stations, while the Miyun and Kunming stations were newly built for the Chang'E-1 mission. The close loop S-band round-trip signal was received and coherently transponded by RF phase-lock-loop and a low-gain or high-gain antenna with a diameter of 0.6 m. The X-band down-link VLBI beacon signal was transmitted by a X-band antenna (Yan et al. 2010). The tracking accuracies of USB range and range rate were 2 m and 1 mm/s, while for VLBI delay and delay rate the accuracies were 3 ns and 1 ps/s, respectively.

The dynamical method was used for Chang'E-1 precision orbit determination. Dynamic models used in orbit determination included nonspherical gravitational perturbation of the Moon, three-body perturbation of Earth and Sun, and the solar radiation pressure perturbation. The maneuvers of treaction-wheel unloading and uploading were accommodated by estimating three-axis empirical accelerations along the radial, along-track, and cross-track directions during the time interval of the maneuvers.

Regarding the data processing, Earth-fixed coordinate system and locations of the tracking stations were consistent with ITRF2000. The lunar-fixed coordinate system was chosen to be consistent with the orientation parameters of JPL DE403 planetary ephemeris (Standish 1998) and so were the

TABLE 11.1

Statistical Information of Overlap Difference in Different Orbit Determination Modes for Selected Arcs

Arc Date	Data Type	Radial (m)	Along-Track (m)	Cross-Track (m)
20/21	RR&R+VLBI	8.920	28.312	16.319
	RR&R	3.606	273.565	515.121
21/22	RR&R+VLBI	22.252	99.666	112.136
	RR&R	19.364	156.875	413.145
22/23	RR&R+VLBI	3.530	114.503	13.351
	RR&R	3.696	104.996	90.009
23/24	RR&R+VLBI	3.714	156.883	8.720
	RR&R	6.249	155.941	19.684
24/25	RR&R+VLBI	6.146	18.723	13.292
	RR&R	9.904	32.155	334.610
25/26	RR&R+VLBI	50.421	114.465	447.329
	RR&R	55.074	127.560	644.588
26/27	RR&R+VLBI	7.545	78.499	9.459
	RR&R	6.120	61.707	364.677
27/28	RR&R+VLBI	4.382	107.623	15.511
	RR&R	2.967	90.132	288.501
28/29	RR&R+VLBI	0.880	10.024	5.753
	RR&R	2.166	7.354	127.649

ephemeredes of Sun, Earth, and other planets. The inertial system applied in orbit integration was lunar-centered inertia coordinate system J2000. The estimated parameters included satellite initial orbital parameters, the pass-dependent measurement biases, solar radiation pressure coefficient and empirical accelerations.

Overlap differences of arcs from November 20 to 29 were computed to evaluate the contribution of VLBI tracking data in Chang'E-1 precision orbit determination. During this period, mitigatory reaction-wheel unloading and uploading maneuver was done only once between every two consecutive orbital arcs. However, during other periods such maneuvers were performed more than once, and broke the orbital arc into shorter pieces, thus were not suitable for overlap analysis.

Table 11.1 gave the statistics of overlap difference of each arc. The overlap difference was computed by extrapolating one arc for 12 hours, and compared with the reconstructed orbit in the same time of neighboring arc. R meant range data and RR meant range-rate data. From Table 11.1, it was clear that after adding VLBI data, the overlap difference decreased significantly, especially in the normal direction of the orbital plane. The overlap errors presented here were not an absolute way to judge the accuracy of POD, however, they still could reflect the degree of consistency between neighboring arcs.

11.3.2 Lunar Gravity Field Model (CEGM-01) Solution

The Chang'E-1 tracking data used for gravity field solution were two-way range and range rate (R&RR) data from Qingdao and Kashi stations. The data sampling rate was 1 second. During Chang'E-1 normal mission, the daily observation time length was 6 hours for December 2007, 3 hours for January and February 2008, and about 3 hours for May to July 2008. There was no tracking data in March and April 2008. Nadir tracking data covering the lunar surface are shown in Figure 11.4. The left region in Figure 11.4 was the lunar far side. The total length of the observed arc used in the gravity

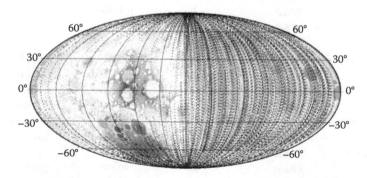

FIGURE 11.4
Picture of orbit tracking data covering.

TABLE 11.2

Tracking Data Length of Chang'E-1

Month	Total Period of Observation Stations (h)			
	Qingdao Station Ranging	Qingdao Station Ranging Rate	Kashi Station Ranging	Kashi Station Ranging Rate
07/12	111.95	113.81	49.19	49.44
08/01	52.31	51.76	53.10	52.54
08/02	48.54	48.68	30.16	29.95
08/05	55.62	55.68	40.09	40.09
08/06	49.71	49.57	29.41	29.41
08/07	76.22	77.25	4.81	4.82
Total	394.35	396.75	206.76	206.25

model recovery is given in Table 11.2. Because of the lunar physical libration effects, tracking data could cover partial polar regions in the lunar far side.

The data were preprocessed to reduce the effects of ionosphere and neutral atmosphere. The dynamic models used in gravity field recovery included nonspherical gravitational perturbation of the Moon, three-body perturbation of Earth and Sun, indirect perturbation for Earth's oblateness, and the solar radiation pressure. GLGM-2 was used as a priori model (Lemoine et al. 1997). The estimated parameters included local and global parameters. Local parameters were similar to the parameters estimated in POD, and the global parameters were the lunar gravitational field coefficients with degree and order from 2 to 50. Due to the limitation of Chang'E-1 orbital altitude and lack of no direct observation data on the lunar far side, Kaula constraints must be introduced in inversion. In order to extract the gravity field information by iteration, the Kaula constraint was gradually weakened until convergence.

The lunar gravity field power spectrum was used to evaluate the solution model. The spectrum includes the sigma and error sigma of the coefficients per degree computed as follows (Heiskanen & Moritz 1967):

$$\sigma_n = \sqrt{\frac{\sum_{m=0}^{n}(\bar{C}_{nm}^2 + \bar{S}_{nm}^2)}{2n+1}} \tag{11.18}$$

$$\delta_n = \sqrt{\frac{\sum_{m=0}^{n}(\sigma_{\bar{C}_{nm}}^2 + \sigma_{\bar{S}_{nm}}^2)}{2n+1}} \tag{11.19}$$

in which $\bar{C}_{nm}$, $\bar{S}_{nm}$ are the regularized gravity field coefficients, $\sigma_{\bar{C}_{nm}}$ and $\sigma_{\bar{S}_{nm}}$ are the sigmas of $\bar{C}_{nm}$ and $\bar{S}_{nm}$. The sigma per degree stands for the RMS

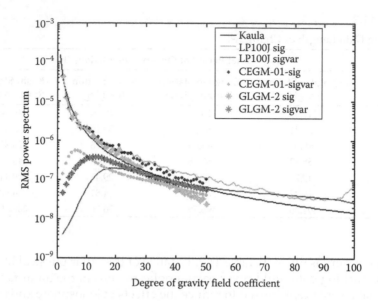

FIGURE 11.5
The power spectra of GLGM-2, CEGM-01 and LP100J.

magnitude of normalized coefficients, which shows the power of gravity field in the frequency domain. The error sigma per degree stands for the RMS magnitude of normalized coefficient error, and it is a standard measure for formal error of the gravity field model.

Figure 11.5 shows the power spectrum of solution model CEGM-01, LP100J, and GLGM-2. In Figure 11.5, "sig" and "sigvar" stand for sigma and error sigma, respectively. As shown in Figure 11.5, the degree-wise sigma of GLGM-2 was less than error degree-wise sigma after 25 degree, which indicated that the coefficients of GLGM-2 beyond this degree were obtained mainly through mathematical constraint. The tracking data of Chang'E-1 with the orbital height of 200 km would mainly contribute to the medium wavelength of lunar gravity field. Beyond degree 10, the accuracy of CEGM-01 model showed a significant improvement over GLGM-2, indicating that the tracking data of Chang'E-1 could contribute to the lunar gravity field solution with certain performance.

In order to evaluate the accuracy and reliability of CEGM-01, precision orbit determination of Chang'E-1 was done by using GLGM-2, LP100J, and CEGM-01. Table 11.3 gives the statistic information of residuals of 6 months. It shows that CEGM-01 and LP100J had same level residuals and are more accurate than GLGM-2 with one order of magnitude. It affirmed the improvement of CEGM-01 compared with GLGM-2 to some extent.

To illustrate the reliability CEGM-01 model further, the tracking data of LP in April 1998 were reprocessed using various gravity field models. It could

TABLE 11.3

Residuals RMS of R&RR for Chang'E-1 Based on the Three Models for All of the Tracing Data

Month	Model	Range		Range Rate	
		Mean	RMS	Mean	RMS
07/12	LP100J	1.11891	0.21724	0.0041	0.00098
	CEGM-01	1.20817	0.37885	0.0048	0.00153
	GLGM-2	30.3257	41.28487	0.12492	0.26518
08/01	LP100J	1.071	0.281	0.0037	0.0043
	CEGM-01	1.000	0.226	0.00096	0.001
	GLGM-2	12.408	17.103	0.075	0.068
08/02	LP100J	2.16091	2.75503	0.01493	0.03005
	CEGM-01	2.00399	2.58758	0.0143	0.02976
	GLGM-2	5.11534	6.47749	0.02596	0.03421
08/05	LP100J	1.8318	2.84827	0.00507	0.00247
	CEGM-01	1.04157	0.21393	0.00469	0.00107
	GLGM-2	3.40165	5.5905	0.01362	0.01286
08/06	LP100J	1.11549	0.35523	0.00442	0.00094
	CEGM-01	0.95959	0.14025	0.00456	0.00085
	GLGM-2	3.81151	4.68373	0.01972	0.0261
08/07	LP100J	1.02552	0.23316	0.00462	0.00114
	CEGM-01	0.98703	0.37878	0.00481	0.00117
	GLGM-2	8.21857	16.96503	0.02623	0.0436

be used as independent data to test CEGM-01 and GLGM02. As can be seen from Figures 11.6 and 11.7, the residuals of R&RR for LP using CEGM-01 model are significantly superior to GLGM-2 model. The precision could be improved by a factor of 4. Both these models showed worse residuals than LP100J. The reason was the divergence and error introduced by the downward continuation of the gravity field model when it was introduced in the precision orbit determination for a spacecraft with a lower orbital height.

The lunar surface gravity anomaly based on a different lunar gravity field model was also analyzed to assess the model. Lunar surface gravity anomaly is an important geophysical quantity for investigating lunar-specific features and internal structures such as mascons and the crustal dichotomy (Yan et al. 2013a). The lunar surface gravity anomaly is defined as (Heiskanen & Moritz 1967):

$$\Delta g = \frac{GM}{r^2} \left[\sum_{n=0}^{N_{max}} \sum_{m=0}^{n} (\frac{R}{r})^n (n-1)\bar{P}_{nm}(\sin\varphi)(\bar{C}_{nm}\cos m\lambda + \bar{S}_{nm}\sin m\lambda) \right] \quad (11.20)$$

In which R is the mean lunar radius (assumed to be 1738 km), $\bar{C}_{nm}$ and $\bar{S}_{nm}$ are the regularized gravity field coefficients, $\bar{P}_{nm}$ is the normalized associated

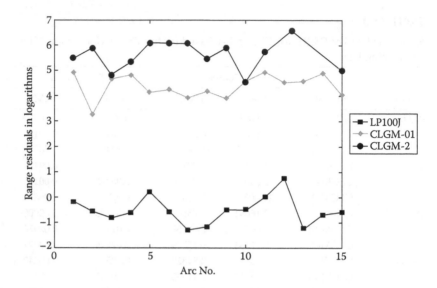

FIGURE 11.6
Residuals of Rang for LP based on the three models (LP100J, CEGM-01 and GLGM-2) (Natural logarithm, 0 corresponds to 1 m, 2.30 corresponding to 10 m, 4.60 corresponding to 100 m).

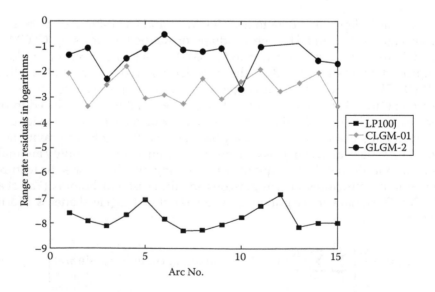

FIGURE 11.7
Residuals of Rang for LP based on the three models (LP100J, CEGM-01 and GLGM-2) (Natural logarithm, 0 corresponds to 1 m/s, 4.6 to correspond to 1 cm/s, 6.9 corresponds to 1 mm/s).

Legendre function, (r, ϕ, λ) are the radius, latitude, and longitude in lunar-fixed coordinates, GM is the lunar gravity constant, and N_{max} is the highest degree of the gravity field model in the computation. The highest degree of LP100J, GLGM-2, and CEGM-01 were 100, 50, and 50 respectively in term of the gravity anomaly computation.

Figures 11.8 to 11.10 show the gravity anomaly maps of various models, in which the center section shows the lunar far side and the left and right sides show the lunar nearside. From the figures, we could see that CEGM-01 could distinguish more details than GLGM-2 in terms of gravity anomaly on the lunar far side. The resolution of LP100J was superior to CEGM-01 due to inclusion of tracking data of LP. The lunar gravity anomaly presented a certain degree of stripe effect based on CEGM-01, which was mainly due to the limited coverage of the polar orbiter. Figure 11.11 showed the difference of gravity anomaly between CEGM-01 and LP100J, and it was clear that there was a much bigger difference on the lunar far side than the nearside, which might indicate that the tracking data of Chang'E-1 could make only a minor contribution to the lunar gravity field in the far side.

11.3.3 Integrated Solution from Chang'E-1, SELENE, and Historical Tracking Data

On the basis of Chang'E-1 data processing, we continued to solve lunar gravity field model by combining orbital tracking data of other spacecrafts, especially of SELENE. SELENE was launched by Japan in September 2007, and contributed to lunar gravity and selenodetic research by first measuring lunar far side directly with high-low satellite-to-satellite tracking mode. Using 3 months four-way tracking, Namiki et al. (2009) presented a model SGM90d and showed significant improvement on lunar far-side gravity compared to previous LP series models. From this model, the gravity distribution of lunar far-side basins were found to be different from the nearside basins, which had multiring structure with negative gravity anomaly on the center. Using all the tracking data of SELENE, Matsumoto et al. (2010) solved a gravity field model SGM100h with a higher resolution and gave a modified classification of lunar far-side basins.

We had developed a gravity field model CEGM-01 using Chang'E-1 data (Yan et al. 2010), and it showed that Chang'E-1 can contribute more to determining gravity field than Clementine. To further investigate the potential of the Chang'E-1 tracking data in lunar gravity field modeling, we solved a model CEGM02 based on the tracking data of Chang'E-1, SELENE, LP, and other historical spacecraft by a combination of normal matrix from Chang'E-1 and SGM100h (Yan et al., 2012). The techniques used for Chang'E-1 data processing had been mentioned earlier. The arc length, data type, and weight information of SELENE and other tracking data were described in Table 11.2 of Matsumoto et al. (2010).

Figures 11.12 and 11.13 show the power spectrum of CEGM02, as well as SGM100h and LP100K for comparison. LP100K was chosen here because

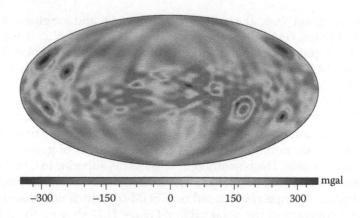

FIGURE 11.8
Lunar surface gravity anomaly of GLGM-2.

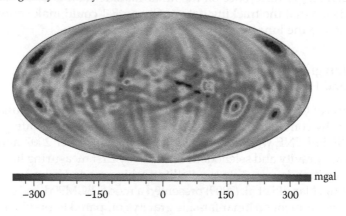

FIGURE 11.9
Lunar surface gravity anomaly of CEGM-01.

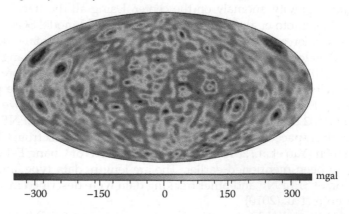

FIGURE 11.10
Lunar surface gravity anomaly of LP100J.

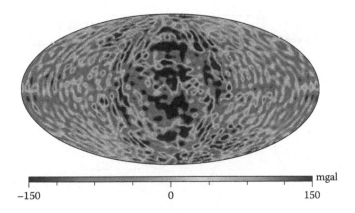

FIGURE 11.11
Lunar surface gravity anomaly differences between CEGM-01 and LP100J.

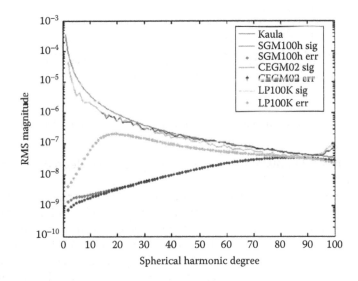

FIGURE 11.12
RMS degree variance and error degree variances of gravity field model coefficients.

of the similar data source used in SGM100h and CEGM02, especially of LP tracking data. From Figure 11.12 we could see that CEGM02 showed significant improvement of low degree coefficients below degree 25 compared with SGM100h, especially for coefficients lower than degree 5, and the formal error was reduced by a factor of 2. The improvement of CEGM02 was due to the combination of Chang'E-1 tracking data with the orbital height of 200 km. From Figure 11.13 we could see that the coefficient differences below 5 were larger than the formal error. For the higher degrees, the coefficient difference was smaller than the formal error, which meant the gravity

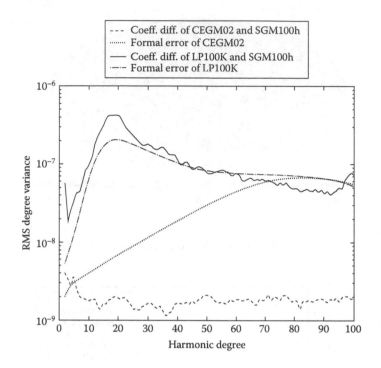

FIGURE 11.13
RMS degree variance of coefficient differences of model CEGM02, SGM100h and LP100K.

information of CEGM02 with coefficients higher than degree 5 were mainly determined by SGM100h.

In order to verify the improvements of the low-degree coefficients, the gravity anomaly errors of various models are presented in Figure 11.14. The gravity anomaly error was computed from the covariance matrix and truncated to degree and order 25. The RMS errors of CEGM02, SGM100h, and LP100K were 0.032, 0.108, and 7.814 mgal, respectively. The improvements in gravity anomaly errors were corresponding to the formal error improvements of low-degree coefficients in Figure 11.12.

The performance of CEGM02 was further assessed by spacecraft orbit determination. In Table 11.4 we showed the orbital overlap differences of LP. The data of LP spanned from February 16, 1998, to June 16, 1998, and the arc length was 2 days. The overlap differences were computed by extrapolating the 2-day arc by 2 hours to compare with the neighboring arc. From Table 11.4 we could see there were marginal improvements of overlap differences of SGM100h compared with LP100K, which could be due to the fact that the LP orbital errors were dominated by high-degree coefficients. The minor differences between SGM100h and CEGM02 indicated that the LP orbit was not sensitive to the improvement of CEGM02 in low-degree coefficients.

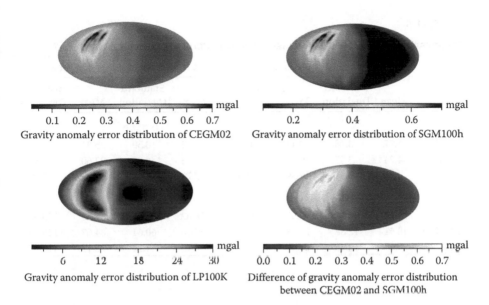

Gravity anomaly error distribution of CEGM02

Gravity anomaly error distribution of SGM100h

Gravity anomaly error distribution of LP100K

Difference of gravity anomaly error distribution between CEGM02 and SGM100h

FIGURE 11.14

Gravity anomaly error up to degree and order 25 of various models. The maps are centered at 270°E and with Hammer projection and show the nearside on the right and far side on the left.

TABLE 11.4

Statistical Information of Orbital Overlap Differences in Three Directions of 2-Day Arc for LP, from February 16, 1998 to June 16, 1998

	Radial (m)		Along-Track (m)		Cross-Track (m)	
Models	Mean	Standard Deviation	Mean	Standard Deviation	Mean	Standard Deviation
LP100K	0.83	0.60	16.20	13.74	16.83	21.45
SGM100h	0.96	0.67	13.20	10.02	13.47	17.06
CEGM02	0.95	0.64	13.56	10.26	13.66	17.26

Table 11.5 and Figure 11.15 gave the overlap differences of SELENE subsatellites Rstar and Vstar. These two spacecraft were in higher, elliptical orbits and were more sensitive to the low-degree coefficients. During orbital determination of Rstar and Vstar, we chose arc length of one week, and the overlap difference was obtained by extending the arc by one day to compare with same time span of the neighboring arc. From Table 11.5 and Figure 11.15 we could see that there were small improvements of overlap differences, especially for Rstar, which could be attributed to the improvement of low-degree gravity field coefficients. In order to consider the effect of low-degree coefficients on orbit determination more significantly, we computed the overlap difference of LP with a long time gap. During the LP normal phase, there

TABLE 11.5

Statistical Information of the Average Orbital Differences of Rstar during 2008

	Radial (m)		Along-Track (m)		Cross-Track (m)		Position (m)	
	Mean	Standard Deviation	Mean	Standard Deviation	Mean	Standard Deviation	Mean	Standard Deviation
Rstar								
LP100K	4.77	3.78	19.44	12.89	12.32	10.50	25.31	14.17
SGM100h	4.74	3.55	19.24	10.33	8.58	6.03	22.60	10.50
CEGM02	4.56	3.41	17.89	9.61	7.79	5.46	20.94	9.80
Vstar								
LP100K	9.95	10.66	70.67	106.45	36.62	57.88	82.54	120.04
SGM100h	10.51	12.68	73.78	112.35	33.40	60.61	84.07	126.69
CEGM02	9.73	12.46	68.58	104.33	30.93	54.14	78.15	116.67

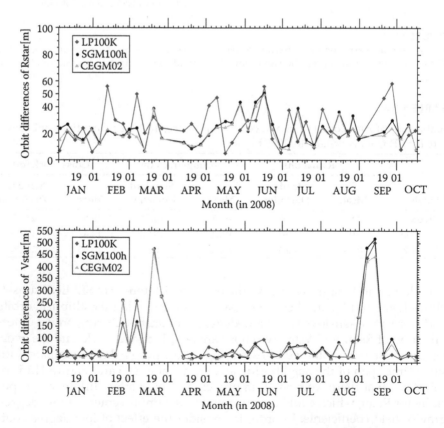

FIGURE 11.15
Average orbital differences of Rstar (upper) and Vstar (lower) during year 2008.

TABLE 11.6

Orbital Overlap Differences between May 21, 1998 and June 23, 1998

Model	Radial (m)	Along-Track (m)	Cross-Track (m)	Total (m)
LP100K	3.16	1563.25	50.80	1564.07
SGM100h	9.41	1558.23	15.44	1558.33
CEGM02	9.64	1462.57	17.04	1462.70

were only a few times of orbital maneuvers, and it was possible to predict orbit in long time. We chose two arcs with the date of May 21, 1998 and June 23, 1998. For the first arc, we did the precise orbit determination with 2-day arc length, and extended the ephemeris to June 23, 1998, to compare the reconstructed orbit·of the second arc. The overlap differences in three directions were given in Table 11.6. As the long-arc extrapolated orbit error was mainly caused by the errors of low-degree coefficients, the reduction of the differences from CEGM02 could be caused by the improvements of its low-degree coefficients.

Lunar moment of inertia is an important quantity to investigate lunar internal structure, and it can be deduced from degree 2 coefficients and lunar libration parameters β and γ (Konopliv et al. 1998). The low-degree coefficients of various models including GLGM-3 (Mazarico et al. 2010) and LP150Q were presented in Figure 11.16. The relationships between moment of inertia, libration parameters and degree 2 coefficients are:

$$\beta = \frac{C-A}{B}, \gamma = \frac{B-A}{C}$$

$$C_{20} = \frac{A+B-2C}{2MR^2}, C_{22} = \frac{B-A}{4MR^2}$$

$$I = \frac{A+B+C}{3} \tag{11.21}$$

where A, B, and C are the principal moments of inertia with $A<B<C$. They are defined in lunar principal axis system, and the axis associated with the moment A is oriented toward Earth, C is coincident with the rotation axis, and B is nearly aligned with the orbital velocity vector (Bills 1995). I is mean moment of inertia. As there are four equations compared with three parameters, it was overdetermined and could be solved by choosing three equations each time.

Table 11.7 gives the moment of inertia computed from different constraints. As the libration parameters had same precision in each computation, the higher accurate coefficients C_{20} and C_{22} will obtain better self-consistency among the four estimates of the moments of inertia. As expected, CEGM02 had minimum discrepancy.

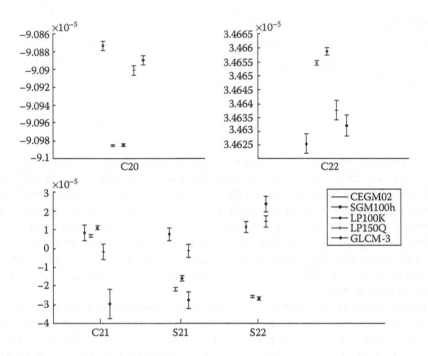

FIGURE 11.16
Low degree coefficients and their standard deviation of various models.

TABLE 11.7

Mean Moment of Inertia of Various Models

	CEGM02	SGM100h	LP100K
Omitting γ	0.393469 ± 0.000058	0.393476 ± 0.000064	0.393043 ± 0.00011
Omitting β	0.393462 ± 0.000067	0.393510 ± 0.000090	0.393216 ± 0.00021
Omitting C_{20}	0.393470 ± 0.000071	0.393468 ± 0.000077	0.393004 ± 0.00013
Omitting C_{22}	0.393462 ± 0.000065	0.393510 ± 0.000090	0.393216 ± 0.00021
Mean	0.393466 ± 0.000065	0.393491 ± 0.000080	0.393120 ± 0.00016
Maximum discrepancy	0.00001	0.00005	0.00021

11.4   Gravity Field Models from the GRAIL Mission

As the SELENE mission realized high-low satellite-to-satellite tracking measurement on the Moon for the first time, the Gravity Recovery And Interior Laboratory (GRAIL) realized low-low satellite-to-satellite tracking measurement and obtained significant improvement on lunar gravity field modeling (Yan et al. 2013b). The GRAIL mission was launched in September 2011. Scientific data were collected from March to May 2012 in the normal phase (with the average orbital height of 50 km) and September to December 2012 in the extension phase (with the average orbital height of 30 km). The main scientific objective of the GRAIL mission was to gain improved knowledge of the lunar gravity field in order to learn more about the subsurface and interior structure of the Moon, and hence to reveal its formation and thermal evolution (Zuber et al. 2013a). The first GRAIL lunar gravity field solutions using tracking data in normal phase have been published (Konopliv et al. 2013; Lemoine et al. 2013; Zuber et al. 2013b).

GRAIL was the lunar analog of the terrestrial Gravity Recovery and Climate Experiment (GRACE) (Tapley et al. 2004a). Compared to GRACE, the GRAIL mission was simplified by omitting the K-band intersatellite link (24 GHz) as well as the accelerometer; the latter was due to the fact that the Moon orbiters do not have to contend with atmosphere drag. In addition, for GRAIL a S-band (2 GHz) time transfer system was added. It acted as the global positioning system (GPS) timing functionality used for GRACE, which was because of the lack of GPS availability at the distance of the Moon (Asmar et al. 2013). The GRAIL twin orbits were determined by two-way S-band and one-way X-band (8 GHz) Doppler tracking data from the deep space network (DSN). High-precision gravity field recovery was accomplished by the analysis of measurements provided by the Ka-band (32 GHz) lunar gravity ranging system (LGRS; Zuber et al. 2013a).

Using GRAIL tracking data in the normal phase, Zuber presented a primary gravity field model GL0420A. The DSN two way S-band Doppler tracking data and the LGRS Ka-band range-rate tracking data were included in the process. The postfit residual of range-rate tracking data was of the order of 0.02 to 0.05 μm s^{-1}, which was better than mission expectation (Zuber et al. 2013b). Such high-accuracy tracking data, especially the satellite-to-satellite range-rate tracking data improved the gravity field significantly. As shown in Figure 11.17, the improvement was two to four orders of magnitude compared with model SGM150j from the SELENE mission (Goossens et al. 2011). From the coherence between gravity and LRO topography, we could see the nearly complete positive correlation from degree 100 to 320, which was mainly due to the prominent improvement of short wavelength of lunar gravity field model. The decrease after 350 was caused by the gravity error, and it could be refined after the gravity field model is expanded with higher degree and order.

In Figure 11.18 the free-air gravity anomaly and Bouguer gravity anomaly of model GL0420A were presented. It was for the first time to give a high-resolution presentation of lunar far-side gravity. Compared with previous gravity field models from SELENE mission, GL0420A revealed many newly found distinctive gravitational signature, such as impact basin rings, central peaks of complex craters, volcanic landforms, and a large amount of simple bowl-shaped craters (Zuber et al. 2013b). By combination with LRO topography and image data, it should be possible to find more potential lunar basins with the help of GRAIL gravity data (Frey 2011).

As shown in Figure 11.17, there was an obvious jump of GRAIL power spectrum beyond degree 400, which showed that there was still gravity signature

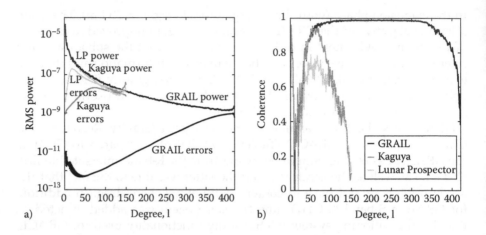

FIGURE 11.17

a) RMS power and b) Coherence versus harmonic degree for the gravity field model GL0420A, SGM150j and LP150Q. (From Zuber, M.T., et al. Gravity field of the moon from the Gravity Recovery and Interior Laboratory (GRAIL) mission. *Science* 339(6120), 668–671, 2013b.)

in the tracking data after the gravity field model expanded to degree and order 420. JPL and Goddard group reprocessed the GRAIL tracking data of the normal phase and obtained gravity field models with a degree and order 660 (Konopliv et al. 2013; Lemoine et al. 2013). The models developed by these two groups were named GL0660B and GRGM660PRIM. The accuracy and postfit measurement residuals of two models were close, while there were some differences on the process of orbit determination and gravity inversion. For example, JPL group solved the gravity field using SRIF filter directly, while Goddard group used the method of variance-component estimation to adjust the a priori data weights.

In the upper part of Figure 11.19 the power spectrum of GL0660B and GRGM660PRIM, and the power spectrum of difference of GL0660B and GRGM660PRIM are presented. In the lower part the differences of GL0660B and GRGM0660PRIM with SGM100h are added. Here σ means power spectrum, δ means error power spectrum, and Δ means power spectrum of model difference. We could see that GL0660B and GRGM660PRIM had close formal error, especially beyond degree 200. The small deviation of long wavelength part may by caused by the different solution strategy employed in the development of the model. The more striking point was the big differences between SGM100h and GRAIL models. For the coefficients with degree below 40 the differences were significantly larger than the formal error of SGM100h, which meant prominent mismatch between them. This point might hint the possibility of SELENE data in improving long wavelength part of GRAIL models.

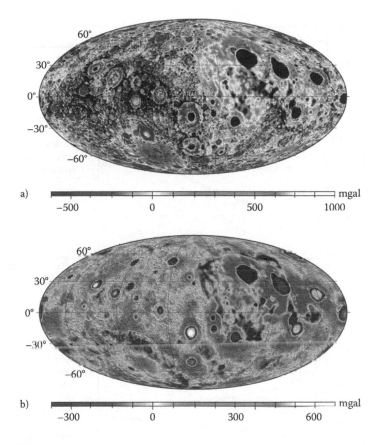

a)

b)

FIGURE 11.18
a) Free-air and b) Bouguer gravity anomaly maps from GRAIL lunar gravity model GL0420A, to spherical harmonic degree and order 420. (From Zuber, M.T., et al. Gravity field of the moon from the Gravity Recovery and Interior Laboratory (GRAIL) mission. *Science* 339(6120), 668–671, 2013b.)

11.5 Conclusions

The development of lunar gravity field model since the 1960s is introduced here, as well as the theory of lunar spacecraft precision orbit determination and gravity field recovery using dynamic method. The precision orbit determination of the Chinese first lunar exploration mission Chang'E-1 was studied, and the significant contribution of VLBI to the Chang'E-1 short-arc orbit accuracy was found. Based on the precision orbit determination of Chang'E-1 tracking data, the model CEGM-01 with degree and order of 50 was developed. This model only used the Chang'E-1 tracking data in the normal phase, and it provided more gravity field information compared with GLGM-2 model. By considering these data, we solved the model

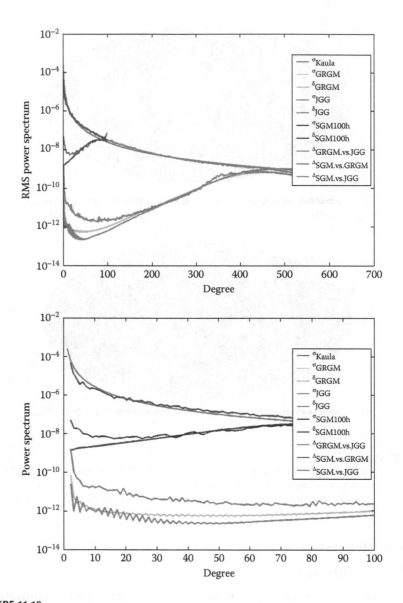

FIGURE 11.19

Models power spectrum and model difference power spectrum. GRGM is the abbreviation of GRGM660PRIM and JGG is abbreviation of GL0660B.

CEGM02 further by combining it with the tracking data of SELENE and other historical spacecraft. By comparing with SGM100h, we found there were obvious improvements in low degrees of the coefficients, and it was validated by gravity anomaly error distribution, short and long arc orbit determination of LP, as well as orbit determination of Rstar and Vstar of

SELENE. The model CEGM02 also showed better consistency of lunar moments of inertia, which indicated its better constraint on lunar internal structure inversion.

After the successful implementation of the GRAIL mission, the knowledge of lunar gravity field is greatly improved, as the accuracy has been improved more than three orders of magnitude using the tracking data of its normal phase. The GRAIL model will be expanded in higher degree and order and with even better accuracy after including the tracking data in the extended phase. It will help us have a more rigorous understanding of lunar internal structure and its thermal evolution history.

Acknowledgments

We greatly thank the VLBI group at Shanghai Astronomical Observatory for providing VLBI data and Beijing Command Center for providing USB data. We acknowledge NASA/GSFC, USA, for the permission to use the GEODYN II/SOLVE system. This research was supported by a grant from the National Natural Science Foundation of China (41374024, 41174019) and the Opening Project of Shanghai Key Laboratory of Space Navigation and Position Techniques.

References

Andrews-Hanna, J.C., Asmar, S.W., Head, J.W., et al. Ancient igneous intrusions and early expansion of the Moon revealed by GRAIL gravity gradiometry. *Science* 339(6120), 675–678, 2013.

Asmar, S.W., Konopliv, A.S., Watkins, M.M., et al. The scientific measurement system of the Gravity Recovery and Interior Laboratory (GRAIL) mission. *Space Sci. Rev.* 174(1), 2013.

Bills, B.G., Ferrari A.J. A harmonic analysis of lunar gravity. *J. Geophys. Res.* 85(B2), 1013–1025, 1980.

Bills, B.G. Discrepant etimates of moments of inertia of the Moon. *J. Geophys. Res.* 100(E12), 26297–26303, 1995.

Floberghagen, R., Visser, P., Weischede, F. Lunar albedo force modeling and its effect on low lunar orbit and gravity field determination. *Adv. Space Res.* 23(4), 733–738, 1999.

Frey, H.V. Previously unknown large impact basins on the Moon: Implications for lunar stratigraphy. *Geol. Soc. Am. Special Papers* 477(1), 53–75, 2011.

Goossens, S., Matsumoto, K., Liu, Q., et al. Lunar gravity field determination using SELENE same-beam differential VLBI tracking data. *J. Geod.* 85(4), 205–228, 2011.

Goossens, S., Matsumoto, K. Lunar degree 2 potential Love number determination from satellite tracking data. *Geophys. Res. Lett.* 3 (2), L02204, 2008.

Heiskanen, W.A., Moritz, H. *Phys. Geodesy.* W.H. Freeman (Eds), San Francisco, California, 1967.

Jin, S.G., Arivazhagan, S., Araki, H., New results and questions of lunar exploration from SELENE, Chang'E-1, Chandrayaan-1 and LRO/LCROSS. *Adv. Space Res.* 52(2), 285–305, 2013.

Kaula, W.M. *Theory of Satellite Geodesy.* Blaisdell Publishing Co., Waltham, MA, 1966.

Konopliv, A.S., Binder, A.B., Hood, L.L., et al. Improved gravity field of the Moon from Lunar Prospector. *Science* 281(5382), 1476–1480, 1998.

Konopliv, A.S., Park, R.S., Yuan, D.N., et al. The JPL lunar gravity field to spherical harmonic degree 660 from the GRAIL primary mission. *J. Geophys. Res. Planets* 118(7), 2013.

Konopliv, A.S., Asmar, S.W., Carranza, E., et al. Recent gravity models as a result of the Lunar Prospector mission. *Icarus* 150(1), 1–18, 2001.

Konopliv, A.S., Binder, A.B., Hood, L.L., et al. Improved gravity field of the Moon from Lunar Prospector. *Science* 281(5382), 1476–1480, 1998.

Lemoine, F.G., Smith, D.E., Zuber, M.T., et al. GLGM-2: A 70th degree and order lunar gravity model from Clementine and historical data. *J. Geophys. Res.* 102(E7), 16339–16359, 1997.

Lemoine, F.G, Goossens, G., Sabaka, T.J., et al. High-degree gravity models from GRAIL primary mission data. *J. Geophys. Res.* 118(8), 1676–1698, 2013.

Matsumoto, K., Goossens, S., Ishihara, Y., et al. An improved lunar gravity field model from SELENE and historical tracking data: Revealing the farside gravity features. *J. Geophys. Res.* 115(E06007), 1–22, 2010.

Mazarico, E., Lemoine, F.G., Han, S.C., et al. GLGM-3: A degree-150 lunar gravity model from the historical tracking data of NASA Moon orbiters. *J. Geophys. Res.* 115(E05001), 1–14, 2010.

Namiki, N., Iwata, T., Matsumoto, K., et al. Farside gravity field of the Moon from four-way Doppler measurements of *SELENE* (Kaguya). *Science* 323(5916), 900–905, 2009.

Rowlands, D.D., Marshall, J.A., Mccarthy, J., et al. *GEODYN II System Description.* Vols. 1–5, Contractor report, Hughes STX Corp., Greenbelt, MD, 2000.

Standish, E.M. JPL planetary and lunar ephemerides, DE405/LE405, JPL IOM 312, F-98-127, 1998.

Tapley, B., Bettadpur, S., Watkins, M.M., et al. The Gravity Recovery and Climate Experiment: mission overview and early results. *Geophys. Res. Lett.* 31(9), L09607, 2004a.

Tapley, B, Schutz, B.E., Born, G.H. *Statistical Orbit Determination*, Elesevier Academic Press, 2004b.

Ulman, R.E. SOLVE program: Mathematical formulation and guide to user input, Hughes/STX Report, NAS5-31760, Goddard Space Flight Center, Greenbelt, 1994.

Wei, E., Yan, W., Jin, S.G., Liu, J., Cai, J. Improvement of Earth orientation parameters estimate with Chang'E-1 ΔVLBI Observations. *J. Geodyn.* 2013.

Wieczorek, M.A., Neumann, G.A., Nimmo, F., et al. The crust of the Moon as seen by GRAIL. *Science* 339, 671–675, 2013.

Wilhelms, D.E., Howard, K.A., Wilshire, H.G. Geologic map of the south side of the Moon. U.S. Geological Survey, I-1162, 1979.

Williams, J.G., Boggs, D.H., Ratcliff, J.T. A larger lunar core? 40th Lunar and Planetary Science Conference, Abstract 1452, 2009.

Yan, J.G., Ping, J.S., Li, F., et al. Chang'E-1 precision orbit determination and lunar gravity field solution. *Adv. Space Res.* 46(1), 50–57, 2010.

Yan, J.G., Goossens, S., Matsumoto, K., et al. CEGM02: An improved lunar gravity model using Chang'E-1 orbital tracking data. *Planet. Space Sci.* 62(1), 1–9, 2012.

Yan, J.G., Zhong, Z., Li, F., et al. Comparison analyses on the 150×150 lunar gravity field models by gravity/topography admittance, correlation and precision orbit determination. *Adv. Space Res.* 52(3), 512–520, 2013a.

Yan, J.G., Baur, O., Li, F., Ping, J.S. Long-wavelength lunar gravity field recovery from simulated orbit and inter-satellite tracking data. *Adv. Space Res.* 52(11), 1919–1928, 2013b.

Zuber, M.T., Smith, D.E., Lehman, D.H., et al. Gravity Recovery and Interior Laboratory (GRAIL): Mapping the lunar interior from crust to core. *Space Sci. Rev.* 178(1), 3–24, 2013a.

Zuber, M.T., Smith, D.E., Watkins, M.M., et al. Gravity field of the moon from the Gravity Recovery and Interior Laboratory (GRAIL) mission. *Science* 339(6120), 668–671, 2013b.

12

Martian Crust Thickness and Structure from Gravity and Topography

Tengyu Zhang, Shuanggen Jin, and Robert Tenzer

CONTENTS

12.1 Introduction 293
12.2 Martian Gravity Field and Topography 295
    12.2.1 Topography 295
    12.2.2 Gravity Field 296
12.3 Estimation of Martian Crust Thickness 297
    12.3.1 Inversion Method 297
    12.3.2 Anomalies Correction 299
    12.3.3 Inversion with Constraint 300
12.4 Localized Admittance Analysis 300
    12.4.1 Method 300
12.5 Results and Discussion... 303
    12.5.1 Crustal Dichotomy 303
    12.5.2 Crustal Thickness Variations 303
    12.5.3 Localized Analysis 306
12.6 Summary and Conclusions... 306
References... 307

12.1 Introduction

Mars is the second smallest terrestrial planet in the solar system with distinct surface features, such as impact craters, volcanoes, and polar ice caps. In the last few decades, a number of missions have been launched to explore Mars, which greatly contributed to improving our knowledge on this planet. Currently, there are five working spacecrafts: three in orbit—Mars Odyssey (ODY), Mars Express, and Mars Reconnaissance Orbiter (MRO), and two on the surface—Mars Exploration Rover, Opportunity, and Mars science laboratory, Curiosity. Over a long geological time, Mars has experienced Noachian, Hesperian, and Amazonian periods, indicating complicated formation and evolution of Mars. Similar to the Earth, the structure of Mars consists of crust, mantle, and core, while Mars has a different geological evolution when compared to the Earth (Jin, Park, and Zhu 2007), which is probably due to its smaller size. Furthermore, Mars is a single plate planet with a thick and rigid outer shell while the Earth has a convective cooling process driving the movement of the surface plates. The knowledge of crust and mantle of a planetary body can be used to constrain the magmatic processes for its formation, and the bulk composition and origin of the planet. The orbital global geophysical measurements of topography and gravity field can help understand the evolution of the Martian crust and mantle. For the Earth, lots of direct measurements are used to constrain the crust thickness. However, for Mars, there were almost no in situ measurements. Moreover, the past estimate of the average crust thickness ranged from a few kilometers to over 250 km.

A number of observations can be used to directly constrain the mean Martian crustal thickness, including the moment of inertia, the viscous relaxation of topography and geochemical mass balance calculations based on the composition of the Martian meteorites and soils. Since it is more convenient to calculate and model with spherical harmonic functions, the gravity and topography have traditionally been expanded in terms of spherical harmonics. Earlier studies (Philips, Saunders, and Conel 1973; Bills and Ferrari 1978; Frey et al. 1996; Kiefer, Bills, and Nerem 1996) used the spherical harmonic gravity and topography models of a relatively low spectral resolution and accuracy, even with long wavelength errors of up to about 5 km in topography. Using Viking-era gravity and topography models, Bills and Nerem (1995) estimated the mean crustal thickness to be between 50 and 200 km based on the assumption of the Airy isostatic model. However, when the new gravity data were observed from Mars Global Surveyor (MGS) mission, Yuan et al. (2001) estimated that the global crustal thickness exceeds 100 km. Based on the data from Mars orbiter laser altimeter (MOLA) and radio science investigation of the MGS spacecraft, the first reliable model of the structure of the Martian crust and upper mantle was proposed by Zuber et al. (2000). A uniform crustal density model was assumed and the global variations in crustal thickness were solved using a gravity field derived from the preliminary MGS tracking data (Smith et al. 1999). Due to the presence of noise, the crust thickness model has a spatial resolution of ~180 km, or degree 60, though the spherical harmonic model of gravity was truncated at degree 80. By utilizing the viscous relaxation of dichotomy boundary combined with crustal thickness inversion, Zuber et al. (2001) presented a global crust thickness map with the spatial resolution better than 100 km. The map revealed two distinctive crustal zones that did not correlate globally with the geologic dichotomy. Apparently due to that the geologic expression of the dichotomy boundary was not a fundamental feature of the Martian internal structure. Therefore, it is not very proper to assume that the entire planet was compensated by the same isostatic mechanism and density contrast. Thus, with additional correction for the anomalous densities of the polar caps, major volcanoes and the hydrostatic flattening of the core, Neumann et al. (2004) investigated the crustal structure by applying a finite-amplitude terrain correction. With an assumption of mantle density contrast of 600 kg m^{-3}, the global mean crustal thickness was constrained to be larger than 45 km.

From gravity/topography admittances (Simons, Solomon, and Hager 1997) and correlation with specific assumption for some regions, more geophysical parameters can be derived by comparing them to those predicted from models of the lithospheric flexure (McGovern et al. 2002). For the Hellas area, the crustal thickness ranges from 38 to 62 km with an uncertainty of 12 km. With the constraint of geoid-to-topography ratios (GTRs) over the extensive ancient southern highlands of Mars, Wieczorek and Zuber (2004) interpreted these structures by means f applying the spectrally weighted admittance model (Wieczorek & Phillips 1997). Assuming that the highlands are compensated by the Airy mechanism, the mean Martian crustal thickness between 33 and 81 km was found. On the other hand, more localized admittance studies have been applied to Mars (McKenzie et al. 2002; Belleguic, Lognonne, and Wieczorek 2005; Wieczorek 2008). By utilizing the admittance estimates from line-of-sight acceleration profiles of the MGS, Nimmo (2002) modeled the mean crustal thickness, the elastic thickness, and the surface density for a region centered on the hemispheric dichotomy. Wieczorek (2008) applied the same method to polar caps of Mars, where it has been covered by unknown proportions of water ice, solid CO_2, and dust. For some regions that are homogeneous with respect to lateral density variations and compensation state, Pauer and Breuer (2008) constrained a maximum density of crust with a two-layer model by combining two different methods, namely computing the geoid-topography ratio and using the Bouguer gravity data inversion. In this study, we compile new models of the Martian crustal thickness and its structure. The topography and gravity field models are compared and analyzed in Section 12.2. The numerical methods are reviewed in Section 12.3. Results are presented and discussed in Section 12.4. Summary and conclusions are given in Section 12.5.

12.2 Martian Gravity Field and Topography

12.2.1 Topography

Prior to 1990s, the Martian topography models were constructed by combining with different measurements, like Earth-based radar, spacecraft radio occultations, and stereo and photoclinometric observations. However, all these models have limited use, because of their large uncertainties and limited spatial resolution. Until 1997, the MGS spacecraft with MOLA has been launched into orbit. The digital topography model has been greatly improved with a collection of more than 640 million ranges to the surface by MOLA

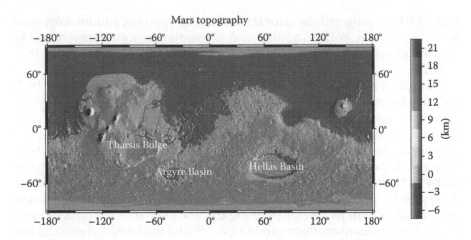

FIGURE 12.1
The global Martian topography.

over a mission period of four years (Figure 12.1) (Smith et al. 2001). The spot size of laser at the surface was approximately 168 m, and these were spaced every 300 m in the along-track direction of the spacecraft orbit. Elevations measured from MOLA have provided a high accuracy topographic model of Mars. The highest spatial resolution of the DEM is 1/128° covering from 88S to 88N, 180W to 180E. In the near future, a more accurate topography can be expected with the accumulated observations collected by the high-resolution stereo camera onboard Mars Express (Gwinner et al. 2010).

Long-wavelength topography reflects the processes that have shaped Mars on a global scale drawing lots of interest in the evolution of Mars. The short-wavelength topography is dominated by regionally remarkable features, like volcanoes and basins. The Tharsis Bulge includes some prominent superposed volcanoes, which have tremendous height. The highest elevation corresponding to the volcano Olympus Mons is almost 22 km above the surface. Some giant impact basins, Hellas and Argyre, cover a large area on the Martian surface. Another impressive feature on Mars is the dichotomy in elevations between the northern and southern hemispheres. All these factors contribute to a 3.3 km offset of the geometrical center from the center of mass.

12.2.2   Gravity Field

The gravity field of Mars reflects the internal and external process over billions of years, which will contribute to understanding the Mars interior. Many missions have been successively launched to obtain the global gravity field, which has greatly improved the accuracy from the accumulated tracking data by the Mariner 9, Viking 1 and 2, MGS, and ODY missions. The old Martian gravity fields have the limited spherical harmonic degrees with up to 50 and large uncertainties in high-degree terms, for example, GMM1

(Smith et al. 1993) and Mars50c (Konopliv & Sjogren 1995). With a much higher quality of the modern X-band tracking technique employed in the mission of MGS and ODY, it has become possible to obtain higher resolution gravity field models. Recently, the most precise and highest resolution gravity field model of Mars is the JPL Mars gravity field MRO110B, which used 2 years of tracking data from the MRO spacecraft, which greatly improved the high-frequency part of the Mars gravity field (Konopliv et al. 2011). Based on the MRO110B, Hirt et al. (2012) proposed a new Mars Gravity Model 2011 (MGM2011) that employed the Newtonian forward-modeling and the MOLA topography model to estimate the short-scale gravity field.

The gravity field is modeled by a spherical harmonic series, which is written at each point P as:

$$U - \frac{GM}{r} \sum_{l=2}^{l=L} (\frac{R}{r})^l \sum_{m=0}^{l} (C_{lm} \cos m\lambda + S_{lm} \sin m\lambda) P_{lm}(\sin \psi) \tag{12.1}$$

where G is the gravitational constant, M is the mass of Mars, R the equatorial reference radius, C_{lm} and S_{lm} are the dimensionless harmonic coefficients of degree l and order m, and r, φ, λ are the spherical coordinates of the point P in a reference system fixed with respect to the planet; and the P_{lm} is the Legendre function (polynomials when m = 0).

Figure 12.2 shows the image of MGM2011 free-air gravity field. The major features in the free-air gravity field include a broad but geometrically complex anomaly associated with the Tharsis province, positive gravity anomalies with large area of volcanoes, and localized negative anomalies associated with the Valles Marineris canyon system. Except for these structures, the gravitational field is elsewhere relatively smooth.

12.3   Estimation of Martian Crust Thickness

Mars has a relatively low-density crust that is chemically distinct from primitive solar system bodies. The early differentiation processes and cumulative effects of Mars' magmatic evolution result in the formation of Martian crust. Since the seismometers are not available on Mars, topography and gravity models are the primary data sets to constrain the global structure of Mars.

12.3.1   Inversion Method

Because the Bouguer anomalies reflect the crust thickness variations in the interior Mars, normally the first step is to correct the gravity in the global gravity field. The degree one zonal term is the largest single component of

the Bouguer anomaly, which represents a mass excess in the northern hemisphere and a corresponding mass deficiency in the southern hemisphere.
Aside from the degree 0, the potential can be separated into surface and
subsurface components:

$$U = U_{crust} + U_{local} + U_{core} + U_{mantle} \tag{12.2}$$

where U_{local} represents isolated crustal density anomalies in an otherwise
homogenous crust. U_{crust}, U_{core} and U_{mantle} are, respectively, the gravitational effects from crust, core and mantle. Before we use a uniform crustal
density model to estimate the Moho undulations, some density anomalies in the polar caps, some major volcanoes, and the hydrostatic flattening of the core should be removed firstly. Wieczorek and Phillips (1998)
developed a method for computing potential anomalies on a sphere due
to finite amplitude. The topography can be raised to the n-th power and
expanded into spherical harmonics, and then potential anomalies due to
topography on a spherical density interface can be computed to a certain
precision. The potential signal of an interface correlates with the Moho
undulations represented by spherical harmonic coefficients, which can be
written as follows:

$$C_{ilm} = \frac{4\pi\Delta\rho D^3}{M(2l+1)} \sum_{n=1}^{l+3} \frac{{}^n h_{ilm}}{D^n n!} \frac{\Pi_{j=1}^{n}(l+4-j)}{(l+3)} \tag{12.3}$$

where ${}^n h_{ilm}$ is the spherical harmonic coefficient of the nth power of the interface undulations, $\Delta\rho$ is the density contrast between crust and mantle, D
is the radius of referenced sphere, and M is the total mass of Mars. From
this equation, we can calculate the anomalous potential U_{crust} from surface
topography. The method to evaluate the gravity signal of the density interface with a fixed shape as well as of the iteratively adjusted crust–mantle
interface relief uses the higher-order approximation formalism, which can
be derived using the following relationship (Wieczorek & Phillips 1998):

$$h_{ilm} = \omega_l [\frac{C_{ilm}^{BA}M(2l+1)}{4\pi\Delta\rho D^2}(\frac{R}{D})^l - D \sum_{n=2}^{l+3} \frac{{}^n h_{ilm}}{D^n n!} \frac{\Pi_{j=1}^{n}(l+4-j)}{(l+3)}] \tag{12.4}$$

where ω_l is the stabilizing downward continuation filter applied to the
function

$$\omega_l = \left\{1 + A[\frac{M(2l+1)}{4\pi\Delta\rho D^2}(\frac{R}{D})^l]^2\right\}^{-1} \tag{12.5}$$

The Lagrange multiplier λ determines how much the relief will be minimized. If λ equals to zero, then the relief is not filtered at all, whereas the larger the value of λ is, the more the short wavelength topography will be filtered.

12.3.2 Anomalies Correction

Mars is normally believed to have a similar structure as Earth with an interior core. The Martian core has greater density than the mantle, which contributes to the degree 2 zonal harmonic potential. The effect from the mantle convection and dynamic core–mantle topography was considered very small to the Bouguer anomaly (Zhong 2002), and cannot be uniquely constrained from gravity and topography. The gravitational potential contains a large degree term, which is consisted of rotational flattening, topographic flattening, and hydrostatic flattening of the core mantle boundary. Based on the analysis of Neumann et al. (2004), approximately 1–3% of the degree 2 potential arises from the density contrast at the core–mantle boundary.

There are some volumetrically largest volcanoes on the Martian surface, which produce large positive free-air gravity anomalies. It can be clearly seen from Figure 12.2 that Olympus Mons and Tharsis Montes have positive Bouguer signatures at a scale of ~300 km. Alba Patera is similar in volume to Olympus Mons but differs markedly in profile with a negative Bouguer signature. If all these volcanoes were flexurally compensated, the expected Bouguer anomaly would be negative. For the localized analysis by McGovern et al. (2002), the density of the major domical edifices of Tharsis for the best-fit models is larger than the average 2900 kg m^{-3} crustal density. Kiefer (2004) proposed that local densities exceeded 3300 kg m^{-3} in extinct

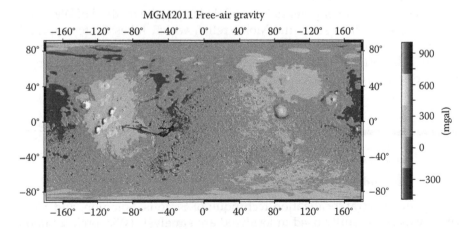

FIGURE 12.2
The Martian free-air gravity model (MGM2011).

magma chambers underlying shallower volcanoes, such as Nili Patera. Using a series of geometric models representing the volcanoes with anomalous density, Neumann et al. (2004) defined the extent of the additional mass concentrations without resolving the details of the volcanic structure.

On Mars, two polar caps are the largest reservoirs affecting the Martian climate. The polar layered terrains, averaged to 3 km thick, are less dense than the average density of the rocky crust. Assuming a lithospheric flexure model with localized spectral analysis for polar cap, the best-fit density of the volatile-rich south polar layered deposits was found to be 1271 kg m^{-3} (Wieczorek et al. 2008). Only the volume of the polar ice emplaced above the trend of the surrounding plains was considering into correction because of complex polar gravity signals. With these local corrections, the residual Bouguer potential variance at high degrees is reduced, because these localized loads have dominated effects in short-wavelength gravity field.

12.3.3 Inversion with Constraint

The residual Bouguer anomalies after correction are mostly accounted for the density contrast between crust–mantle interfaces, so the undulations of the mantle can be solved at the Moho with matching the residual anomaly. As it is downward continuation to an average crust–mantle interface depth, the mean crustal thickness is assumed firstly. Using geochemical arguments, Sohl and Spohn (1997) showed that the mean crust thickness ranges from 100 to 200 km, but Norman (1999) estimated this value to be less than 45 km. Nimmo (2002) inferred that the mean crustal thickness is likely 55 km with admittance estimate, while an average 50 km thickness was proposed by Wiezorek and Zuber (2004) on a combination of geophysical and geochemical studies. Neumann et al. (2004) suggested a mean thickness of 45 km following Zuber et al. (2000).

The residual Bouguer potential is modeled using the method of Wieczorek and Philips (1998) to solve the Moho relief, which takes finite amplitude effects into account. The finite amplitude effects should be included in the gravity calculations, otherwise which would result in misunderstanding in the modeling process (McGovern et al. 2002).

12.4 Localized Admittance Analysis

12.4.1 Method

The spatio-spectral localization technique of Simons, Solomon, and Hager (1997) was successfully used in localized area analysis (Wiezorek & Simons 2005, 2007). This approach is aimed at obtaining the cross-spectral estimates of the radial gravity and topography localized to a specific region on the surface of a spherical planet and to compare these with a similarly localized geophysical model. The localization window $\Phi(\theta,\phi)$ is applied to eliminate signals from outside the region of interest, where the resulting gravity and topography fields are given by

$$\Psi(\theta,\phi) = \Phi(\theta,\phi)h(\theta,\phi) \tag{12.6}$$

$$\Gamma(\theta,\phi) = \Phi(\theta,\phi)g(\theta,\phi) \tag{12.7}$$

where θ and ϕ are the longitude and latitude of the planet, respectively. For the localized analysis, the amplitude of the windows exterior to the region of interest should be close to zero in order to ensure that the local spectral estimates are sufficiently localized in space. Furthermore, to minimize the leakage of power from a given spherical harmonic degree into adjacent degrees, the spectral bandwidth of the window cap should be chosen as small as possible.

It is convenient to work in the spherical harmonic domain, and $\Gamma(\theta,\phi)$ and $\Psi(\theta,\phi)$ can be expanded in terms of standard normalized spherical harmonics. Here the l and $-l < m < l$ in the resulting expansion coefficients by Ψ_{lm} and Γ_{lm} are the spherical harmonic degree and order, respectively, and the localized cross-spectral power is given by

$$S_{\Gamma\Psi} = \sum_{m=-l}^{l} \Gamma_{lm}\Psi_{lm} \tag{12.8}$$

with the localized admittance

$$Z(l) = \frac{S_{\Gamma\Psi}(l)}{S_{\Psi\Psi}(l)} \tag{12.9}$$

and the correlation

$$\gamma(l) = \frac{S_{\Gamma\Psi}(l)}{\sqrt{S_{\Gamma\Gamma}(l)S_{\Psi\Psi}(l)}} \tag{12.10}$$

If a model describing a planetary gravity and topography is to be considered successfully, then it must satisfy both these functions. If the functions cannot be fit for a given degree, it clearly indicates that either the model assumption or input model data are not sufficiently accurate.

The lithosphere of a planet is normally approximated by a thin elastic spherical shell, loaded at the surface by an infinitesimally thin mass sheet of surface density. The topography is considered as loading on the lithosphere resulting in the lithospheric flexure. In this case, the flexure can be described using the thin-shell flexure equation (Kraus 1967). In the spectral harmonic domain, the lithospheric flexure ω_{lm} is approximately considered to be linearly related to the magnitude of the mass sheet σ_{lm} (Turcotte et al. 1981; Wilemann & Turcotte 1981; Wiezorek 2008)

$$\omega_{lm} = -C_l^s \frac{\sigma_{lm}}{\rho_m} \tag{12.11}$$

where C_l^s is a function depending on the elastic thickness T_e, E is Young's modulus, ν is Poisson's ratio, T_c is crust thickness, R is planetary radius, ρ_c is the density of crust, and ρ_m is the density of mantle. The radial free-air gravitational acceleration at the surface is directly related to the mass anomalies associated with the surface deflection, the deflection of the crust–mantle interface, and the load itself, which can be written as follows:

$$g_{lm} = \frac{4\pi G(l+1)}{(2l+1)} \times [1 - C_l^s \frac{\Delta\rho}{\rho_m} (\frac{R-T_c}{R})^{l+2}]\sigma_{lm} \tag{12.12}$$

where $\Delta\rho = \rho_m - \rho_c$ is the density contrast. Finally, the relation of the gravity and topography field coefficients can be expressed by the linear relation of the form:

$$g_{lm} = \frac{Q_l(l+1)}{R} h_{lm} \tag{12.13}$$

where the linear transfer function is

$$Q_l = \frac{4\pi G\rho_l(l+1)}{(2l+1)} [\frac{1 - C_l^s \frac{\rho_c}{\rho_m} - C_l^s \frac{\Delta\rho}{\rho_m} (\frac{R-T_c}{R})^{l+2}}{1 - C_l^s \frac{\rho_l}{\rho_m}}] \tag{12.14}$$

To determine the load density and elastic thickness, the observed admittance should be compared to that from a model, which is given by a set of model parameters. For the adopted loading model, the spectral localized correlation between gravity and topography should be close to unity at all degrees because of the assumed linear relation. The deviation can be interpreted as the unmodeled gravitational signals. In fact, it should be noted that if a given model of the lithospheric loading is an accurate description, it must fit both the admittance and correlation function at the same time. Unfortunately, most of the published results presented only the admittance or coherence function.

12.5   Results and Discussion

12.5.1   Crustal Dichotomy

The crustal thickness map correlates with the principal features of surface topography. One of the major features is crustal dichotomy, which shows a progressive thinning of the crust from high southern latitudes toward the north. The older heavily cratered southern highland was separated from the young smooth northern hemisphere lowlands by the crustal transition boundary. For a geologic record, a boundary zone exists with about 700 km wide (Zuber et al. 2000). Most of the dichotomy boundary did not correspond to the crustal thickness transition, which may indicate that the geologic expression of the dichotomy boundary was not a fundamental feature of Martian internal structure. There are several hypotheses explaining the formation of hemispheric dichotomy, with the most famous giant-impact hypothesis (Wilhelms & Squyres 1984; McGill 1989; Frey & Schultz 1988). Nimmo et al. (2008) proposed a numeric model to simulate the giant impact on the Mars, which predicted that the impact-generated melt forming the northern lowlands crust derived from a deep, depleted mantle source. However, Zuber et al. (2000) disagreed with this hypothesis because some specific features from gravity and topography cannot be explained by this theoretical model.

12.5.2   Crustal Thickness Variations

From the gravity anomalies after successive gravity correction, the global crustal thickness can be obtained assuming a 600 kg m^{-3} density contrast between the crust and the mantle (Figure 12.3). Due to the absence of seismic measurements, the crustal thickness cannot be determined the same as a depth of seismic velocity discontinuity. Therefore, the global mean crustal thickness is assumed by the calculation of the minimum value of the lower crustal viscosity as a constraint. The crustal thickness varies from 2.9 to 93.3 km, and the global mean crustal thickness chosen for inversion is 50 km, corresponding to 4.4% of the planetary volume. Because the amplitude of the crustal thickness correlates with the density contrast, different assumption of the density differences between the crust and mantle contributes to variations of the mean crustal thickness as much as several tens of percent.

An equal-area histogram of the Martian global thickness shown in Figure 12.4 revealed two major peaks in the crustal thickness structure,

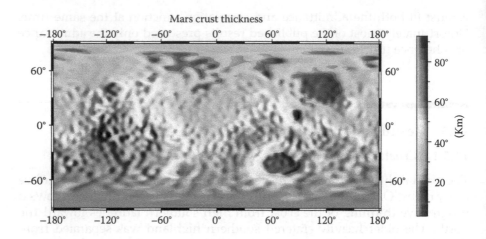

FIGURE 12.3
The global Martian crustal thickness.

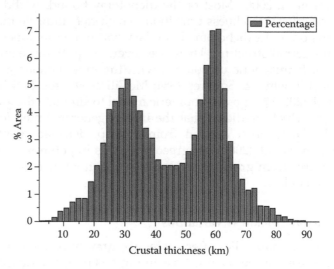

FIGURE 12.4
Histogram of the crustal thickness.

approximately at 30 and 60 km. As we can see in Figure 12.3, the largest peak of the crustal thickness mainly comes from the heavily cratered southern highlands, where most regions have crustal thickness exceeding 60 km. While the southern polar area has a relatively uniform crustal thickness, the basins and highlands have extremely various crustal thickness structures. Alba Patera, Olympus, and Ascraeus Mons are the areas in the north with the thickest crust, and the Hellas and Argyre Basins have the thinnest crust.

Crustal structure profiles along longitudes show that transitions in thickness between crustal provinces in eastern Mars are abrupt, corresponding

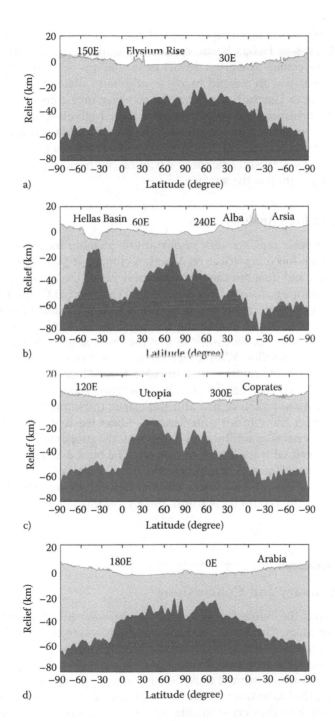

FIGURE 12.5
The profile of crustal structure.

to a relatively steep topographic signature (Figure 12.5). Where the profile at 60°E crosses Hellas Basin, crustal thinning is prominent within the 500 km of the center. The crustal thickness changes dramatically along the edge of the basin with a rapid increase at greater distance to the center. It is also similar to Utopia Basin. The Tharsis province consists of extremely thick crust, corresponding to its predominantly high elevation and volcanic mechanism formation. The crust under the broad Arabia is thickening more gradually. In the Arabia and Elysium regions, the dichotomy boundary is compensated by a crustal thickness variation, with more pronounced relief along the Moho than at the surface.

12.5.3 Localized Analysis

The Martian polar caps are very important to decipher the volatile evolution of Mars, which has a big effect on Martian climate. It is believed that water ice, solid CO_2, and dust are the main components of the polar caps, but their bulk compositions remain mostly unknown. Wiezorek (2007) used a localized spectral analysis combined with a lithosphere flexure model of ice cap loading for the south polar cap. Considering the relief of the south polar cap with a few kilometers, it is not necessary to use the finite amplitude method to calculate gravity anomalies. With a localization window of 8°, the gravity and topography harmonic coefficients truncated from 33 to 47 are used to calculate the admittance and correlation. The best fitting model was obtained with the load density $\rho_1 = 1271$ kg m^{-3} and the elastic thickness $T_e = 140$ km. With that assumption, three major constituents—water ice, dry ice, and dust—are the only components of south-layered deposits, the proportion of their volume can be estimated with their known density. The bulk density is very low for the deposits, which means rich in volatile. If the south polar cap is composed of a mixture of dry ice and water ice, the CO_2 by volume would be ~55%. In contrast, the dust could be 14% and 28% by volume if it is free of CO_2.

12.6 Summary and Conclusions

We have used the refined gravity and topography to retrieve the crustal structure of Mars. The results improve our understanding on the Martian crust and mantle. A single layer elastic crust model was applied into the interpretation of the terrain-corrected gravity anomalies. The global average crust thickness was assumed to be 50 km, which plays an important role in determining the Moho undulation. Some distinct provinces have been found with very thick or thin crust in this model, corresponding to their specific topographic features. The feature of dichotomy also remains pronounced in the crustal thickness model, and the impact redistribution of the crust appears to be the reason of its formation. However, many fundamental questions still remain unknown due to the lack of data and in situ observations. In the future, the seismometer on Mars will greatly contribute to answering these issues.

References

Belleguic, V., P. Lognonne, and M. Wieczorek (2005), Constraints on the Martian lithosphere from gravity and topography data, *J. Geophys. Res.*, 110, E11005.

Bills, B. G., and A. J. Ferrari (1978), Mars topography harmonics and geophysical implications, *J. Geophys. Res.*, 83, 3497–3508.

Bills, B. G., and R. S. Nerem (1995), A harmonic analysis of Martian topography, *J. Geophys. Res.*, 100, 26317–26326.

Frey, H. V., and R. A. Schultz (1988), Large impact basins and the mega-impact origin for the crustal dichotomy on Mars, *Geophys. Res. Lett.*, 15, 229–232.

Frey, H. V., B. G. Bills, R. S. Nerem, and J. H. Roark (1996), The isostatic state of Martian topography—Revisited, *Geophys. Res. Lett.*, 23, 721–724.

Gwinner, K., F. Scholten, F. Preusker, S. Elgner, T. Roatsch, M. Spiegel, R. Schmidt, J. Oberst, R. Jaumann, C. Heipke. 2010. Topography of Mars from global mapping by HRSC high-resolution digital terrain models and orthoimages: Characteristics and performance. *Earth Planet. Sci. Lett.*, 294, 506–519.

Hirt, C., S. J. Claessens, M. Kuhn, W. E. Featherstone (2012), Kilometer-resolution gravity field of Mars: MGM2011. *Planet. Space Sci.*, 67(1), 147–154.

Jin, S. G., P. H. Park, and W. Y. Zhu (2007), Micro-plate tectonics and kinematics in Northeast Asia inferred from a dense set of GPS observations, *Earth Planet. Sci. Lett.*, 257, 486–496.

Kiefer, W. S. (2004), Gravity evidence for an extinct magma chamber beneath Syrtis Major, Mars: A look at the magmatic plumbing system, *Earth Planet. Sci. Lett.*, 222, 349–361.

Kiefer, W. S., B. G. Bills, and R. S. Nerem (1996), An inversion of gravity and topography for mantle and crustal structure on Mars, *J. Geophys. Res.*, 101, 9239–9252.

Konopliv, A. S., S. W. Asmar, W. M. Folkner, Ö. Karatekin, D. C. Nunes, S. E. Smrekar, C. F. Yoder, M. T. Zuber, 2011. Mars high resolution gravity fields from MRO, Mars seasonal gravity, and other dynamical parameters. *Icarus*, 211, 401–428.

Konopliv, A. S., W. L. Sjogren, 1995, The JPL Mars gravity field, Mars50c, based upon Viking and Mariner 9 Doppler tracking data, JPL Publ. 95–5, Jet Propul. Lab., Pasadena, California.

Kraus, H., 1967, *Thin Elastic Shells: An Introduction to the Theoretical Foundations and the Analysis of Their Static and Dynamic Behavior*. Wiley, New York.

McGill, G. E. (1989), Buried topography of Utopia, Mars: Persistence of a giant impact depression, *J. Geophys. Res.*, 94, 2753–2759.

McGovern, P. J., S. C. Solomon, D. E. Smith, M. T. Zuber, M. Simons, M. A. Wieczorek, R. J. Phillips, G. A. Neumann, O. Aharonson, and J. W. Head (2002), Localized gravity/topography admittance and correlation spectra on Mars: Implications for regional and global evolution, *J. Geophys. Res.*, 107, 5136.

McKenzie, D., D. N. Barnett, and D.-N. Yuan (2002), The relationship between Martian gravity and topography, *Earth Planet*. Sci. Lett., 195, 1–16.

Neumann, G. A., M. T. Zuber, M. A. Wieczorek, P. J. McGovern, F. G. Lemoine, and D. E. Smith (2004), Crustal structure of Mars from gravity and topography, *J. Geophys. Res.*, 109, E08002.

Nimmo, F. (2002), Admittance estimates of mean crustal thickness and density at the Martian hemispheric dichotomy, *J. Geophys. Res.*, 107, 5117.

Nimmo, F., S. D. Hart, D. G. Korycansky, and C. B. Agnor (2008), Implications of an impact origin for the Martian hemispheric dichotomy, *Science*, 453, 1220–1224.

Norman, N. D. (1999), The composition and thickness of the crust of Mars estimated from rare earth elements and neodymium-isotopic composition of Martian meteorites, *Meteorit. Planet. Sci.*, 34, 439–449.

Pauer, M., D. Breuer (2008), Constraints on the maximum crustal density from gravity topography modeling: Applications to the southern highlands of Mars. *Earth Planet. Sci. Lett.*, 276 (3–4), 253–261.

Phillips, R. J., R. S. Saunders, and J. E. Conel (1973), Mars: Crustal structure inferred from Bouguer anomalies, *J. Geophys. Res.*, 78, 4815–4820.

Simons, M., S. C. Solomon, and B. H. Hager (1997), Localization of gravity and topography: Constraints on the tectonics and mantle dynamics of Venus, *Geophys. J. Int.*, 131, 24–44.

Smith, D. E., W. L. Sjogren, G. L. Tyler, G. Balmino, F. G. Lemoine, and A. S. Konopliv (1999), The gravity field of Mars: Results from Mars Global Surveyor, *Science*, 286, 94–96.

Smith, D. E., F. J. Lerch, R. S. Nerem, M. T. Zuber, G. B. Patel, S. K. Fricke, F. G. Lemoine (1993), An improved gravity model for Mars: Goddard Mars model 1. *J. Geophys. Res.*, 98, 20871–20889.

Smith, D. E., M. T. Zuber, M. T. Frey, J. B. Garvin, J. W. Head, D. O. Muhleman, G. H. Pettengill, R. J. Phillips, S. C. Solomon, H. J. Zwally, W. B. Banerdt, W. B. Duxbury, M. P. Golombek, F. G. Lemoine, G. A. Neumann, D. D. Rowlands, O. Aharonson, P. G. Ford, A. B. Ivanov, C. L. Johnson, P. J. McGovern, J. B. Abshire, R. S. Afzal, X. Sun (2001), Mars orbiter laser altimeter: Experiment summary after the first year of global mapping of Mars, *J. Geophys. Res.* 106, 23689–23722.

Sohl, F., and T. Spohn (1997), The interior structure of Mars: Implications from SNC meteorites, *J. Geophys. Res.*, 102, 1613–1635.

Turcotte, D. L., R. J. Willemann, W. F. Haxby, J. Norberry (1981), Role of membrane stresses in the support of planetary topography, *J. Geophys. Res.*, 86, 3951–3959.

Wieczorek, M. A., and M. T. Zuber (2004), Thickness of the Martian crust: Improved constraints from geoid-to-topography ratios, *J. Geophys. Res.*, 109, E01009.

Wieczorek, M. A., and R. J. Phillips (1997), The structure and compensation of the lunar highland crust, *J. Geophys. Res.*, 102, 10933–10943.

Wieczorek, M. A. (2007), Gravity and topography of the terrestrial planets. In: Spohn, T., Schubert, G. (Eds.), *Treat. Geophys.*, vol. 10. Elsevier-Pergamon, Oxford, pp. 165–206.

Wieczorek, M. A. (2008), Constraints on the composition of the martian south polar cap from gravity and topography. *Icarus* 196, 506–517.

Wieczorek, M.A., R. J. Phillips (1998), Potential anomalies on a sphere: Applications to the thickness of the lunar crust, *J. Geophys. Res.*, 103, 1715–1724.

Wieczorek, M. A., F. J. Simons (2005), Localized spectral analysis on the sphere. *Geophys. J. Int.*, 162, 655–675.

Wieczorek, M. A., F. J. Simons (2007), Minimum-variance multitaper spectral estimation on the sphere. *J. Fourier Anal. Appl.*, 13, 665–692.

Wieczorek, M. A., M. T. Zuber (2004), Thickness of the martian crust: Improved constraints from geoid-to-topography ratios. *J. Geophys. Res.*, 109, E01009.

Wilhelms, D. E., S. W. Squyres (1984), The Martian hemispheric dichotomy may be due to a giant impact, *Nature*, 309, 138–140.

Willemann, R. J., D. L. Turcotte (1981), Support of topographic and other loads on the moon and on the terrestrial planets. *Proc. Lunar Sci. Conf. B* 12, 837–851.

Yuan, D. N., W. L. Sjogren, A. S. Konopliv, and A. B. Kucinskas (2001), Gravity field of Mars: A 75th degree and order model, *J. Geophys. Res.*, 106, 23377–23401.

Zhong, S. (2002), Effects of lithosphere on the long-wavelength gravity anomalies and their implications for the formation of the Tharsis rise on Mars, *J. Geophys. Res.*, 107, 5054.

Zuber, M. T. (2001). The crust and mantle of Mars. *Nature*, 412, 237–244

Zuber, M. T., S. C. Solomon, R. J. Phillips, D. E. Smith, G. L. Tyler, O. Aharonson, G. Balmino, W. B. Banerdt, J. W. Head, F. G. Lemoine, P. J. McGovern, G. A. Neumann, D. D. Rowlands, and S. Zhong (2000), Internal structure and early thermal evolution of Mars from Mars Global Surveyor topography and gravity. *Science*, 287, 1788–1793.

13

Theory of the Physical Libration of the Moon with a Liquid Core

Yuri Barkin, Hideo Hanada, José M. Ferrándiz, Koji Matsumoto, Shuanggen Jin, and Mikhail Barkin

CONTENTS

13.1 Main Geodynamic Parameters of the Lunar System 312
    13.1.1 Model of the Liquid Core of the Moon 314
    13.1.2 Masses and Moments of Inertia of the Shell of the Moon 314
    13.1.3 Dynamic Parameter of Influencing Liquid Core on the
        Moon Rotation 316
    13.1.4 Dynamic Oblateness of the Core.. 316
    13.1.5 Geometric Compressions and the Difference between
        the Polar and Equatorial Axes of the Liquid Core 317
    13.1.6 Estimates of the Dynamic Compressions of the Core 318
13.2 Main Problems of the Physical Libration of the Moon 319
    13.2.1 Main Coordinate System and Variables................................. 319
    13.2.2 Euler Angles and Andoyer's Variables 320
    13.2.3 Kinetic Energy and Canonical Equations of
        Rotational Motion of the Moon with a Liquid Core in
        Eulerian Variables 321
    13.2.4 Canonical Equations of Rotational Motion of the Moon
        with Liquid Core in Andoyer Variables 323
    13.2.5 Canonical Equations of Rotational Motion of the Moon
        with an Ellipsoidal Liquid Core in Andoyer's Variables 326
    13.2.6 Canonical Equations of the Physical Libration of the
        Moon in Poincaré's Variables 327
    13.2.7 Noninertial Part of the Hamiltonian 328
13.3 Development of the Force Function and the Hamiltonian in
    Poincaré's Variables 329
    13.3.1 Expansions of Spherical Functions of the Coordinates in
        the Expression of the Force Function 329
    13.3.2 Force Functions in the Andoyer Variables............................ 331
    13.3.3 Hamiltonian of the Problem in the Poincaré's Variables 332
    13.3.4 Structure of the General Solution of the Lunar Physical
        Libration with the Ellipsoidal Liquid Core 336

13.4 Motion of the Moon with Liquid Core under Cassini's Laws...........338
    13.4.1 Unperturbed Rotational Motion of the Moon with
        Liquid Core 338
    13.4.2 Cassini's Laws as Conditions of Existence of the
        Conditionally Periodic Intermediate Solution of the
        Problem...339
    13.4.3 Determination of Cassini's Angle 339
    13.4.4 Cassini's Laws of the Moon Rotation.......................................341
13.5 Forced Physical Libration of the Moon in Andoyer's and
    Poincaré's Variables 342
    13.5.1 Perturbations of First Order of the Andoyer–Poincaré
        Variables 342
    13.5.2 Perturbations of the First Order of the Motion of the
        Pole of the Axis of Lunar Rotation..345
    13.5.3 Perturbations of First Order of the Duration of the
        Lunar Day...348
    13.5.4 Perturbations of First Order of the Motion of the Pole of
        the Lunar Core 349
    13.5.5 Perturbations of First Order of the Axial Rotation of the
        Lunar Core..349
    13.5.6 Perturbations of the Classical Variables in Terms of the
        Perturbations of Andoyer–Poincaré's Variables 350
13.6 Free Oscillations of the Pole of the Moon...352
    13.6.1 Main Resonance Effects in the Rotational Motion of the
        Moon with a Molten Core in Andoyer Variables...................353
    13.6.2 Free Libration in the Classical Variables of the Theory of
        the Moon's Rotation...355
    13.6.3 Determination of the Amplitudes and Phases of the Free
        Libration of the Moon...356
13.7 Conclusions...359
Acknowledgments 360
References..360
13.A1 Tables Forced Variations of the Projections of the Angular
    Velocity of the Moon at Its Principal Central Axis of Inertia..........363

13.1 Main Geodynamic Parameters of the Lunar System

The Moon is thought to consist of a solid mantle and a liquid core ellipsoid filled by an ideal fluid, which makes a simple motion as described by Poincaré (Poincaré 1910; Barkin and Ferrándiz 2000). In this study, the solid core is not considered. To study the rotational motion of a two-layer Moon, we must first introduce the inertia moments A, B, C, of the Moon and A_c, B_c, C_c, of its core, which can be used to evaluate the dynamic oblatenesses of the core $\varepsilon_D = 1 - A_c/C_c$ and $\mu_D = 1 - B_c/C_c$ ($C_c > B_c > A_c; \varepsilon_D > \mu_D$). The above parameters are very important in this problem, since they define as the forced and free libration of the Moon and its core in the gravitational field of the Earth and other celestial bodies.

The orbital motion of the Moon is described by a high-precision theory called DE/LE-406 (Kudryavtsev 2007). Based on this theory, we have developed the spherical functions of the lunar spherical orbital coordinates in form of Poisson series. These series are presented in trigonometric form with respect to main arguments of the lunar orbital theory l_M, l_s, F, and D (Kudryavtsev 2007; Barkin, Kudryavtsev, and Barkin 2009). The latest long-term numerical ephemerides of the Moon and planets DE/LE-406 are used as the source of disturbing bodies coordinates (Kudryavtsev 2007; Barkin et al. 2009). In this chapter, we do not take into account their influence of the third and higher order harmonics of the force function of gravitational attraction of the Moon and Earth and the gravitational influence of the physical libration of Sun on the Moon.

The recent lunar missions provided good constraints on the lunar structure, for example, Apollo, SELENE, and GRAIL (Jin, Arivazhagan, and Araki 2013; Wei et al. 2013). Very important results on the internal structure of the Moon have been obtained from seismometers (see Figure 13.1; Weber et al. 2011) and an accurate empirical theory of the physical libration of the Moon were built on the basis of data obtained from laser observations for a long period of about 40 years (Rambaux and Williams 2011). These works formed the basis for our study on the Moon rotation. To determine the ellipsoidalities of the liquid core, we have used some determination of the ratios of the inertia moments of the core and the whole Moon (Williams, Boggs, and Ratcliff 2010, 2011, 2012).

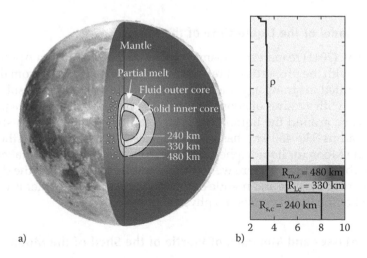

a)   b)

FIGURE 13.1
a) The modern model of the internal structure of the Moon. (From Weber, R.C., et al. (2011) Seismic detection of the lunar core. *Science* 331, 309.) and b) density distribution with depth.

TABLE 13.1

Main Parameters of the Ellipsoidal Core of Earth, Mars, and Moon

Parameters	Earth	Mars	Moon
$L_c = C_c/C$	0.1139	0.0297	0.000426
$\varepsilon_D + \mu_D$	3.53×10^{-3}	4.88×10^{-3}	7.01×10^{-4}
ε_D	2.53×10^{-3}	4.28×10^{-3}	3.74×10^{-4}

By comparing our analytical theory of the lunar physical libration, in particular the free librations, with the available empirical theory of the Moon's rotation (Rambaux and Williams 2011), we have identified the period, amplitude, and the initial phase of the fourth mode of free libration of the Moon, caused by its liquid core. For the first time the period of free core nutation was defined by $T_\Pi = 77757.032d$ and $\Pi_0 = -132^0$ as amplitude and phase (Barkin et al. 2012). Based on this estimated total value of the period T_Π for the sum of two meridian oblatenesses of the core, the following value has been obtained $\varepsilon_D + \mu_D = 7.00 \times 10^{-4}$. On the basis of assumption about similarity and identity of ratios of oblatenesses for the Moon and its core, we have determined their individual values (Barkin et al. 2012) (see Table 13.1).

Table 13.1 shows the results of the determination of the ellipsoidalities, the core of the Moon, based on the current available data on the compression of the core (like laser observations) and data observed by LLR librations of the Moon. New updated estimations of the same parameters for the Earth and Mars are given also in the table.

13.1.1 Model of the Liquid Core of the Moon

Weber et al. (2011) reanalyzed seismic signals on the Moon from Apollo lunar missions, with the properties of reflected and converted signals from the core, which resulted in strong arguments in favor of the existence of a solid and a liquid core with a radius of about 240 and 330 km. The existence of a partially molten zone around the liquid core with a radius of 480 km, or a spherical shell of radius 330–480 km, has been confirmed. It is shown that the liquid core of the Moon (or liquid spherical shell of radius 240–330 km) takes about ~60% of the whole core. Here we use these results to determine the dynamic parameters of the core and mantle of the Moon in order to further investigate the effect of the liquid core on its physical libration.

13.1.2 Masses and Moments of Inertia of the Shell of the Moon

Weber et al. (2011) constructed a three-layer model of the Moon, which was that built using seismographic data from the Apollo expeditions, consisting of a solid core, liquid core (spherical shell), and mantle. For the mean radius of the solid core $R_{s,c}$ and the liquid core $R_{l,c}$, we have the following estimates (Weber et al. 2011):

$$R_{l,c} = 330 \pm 20 \text{ km}, \ R_{s,c} = 240 \pm 10 \text{ km}. \tag{13.1}$$

From the distribution curve of the density of the Moon with the depth (Weber et al. 2011), average densities of the lunar liquid core and the solid core are estimated, respectively, as

$$\delta_{l,c} = 5.11 \ \Gamma/\text{cm}^3, \ \delta_{s,c} = 8.04 \ \Gamma/\text{cm}^3 \tag{13.2}$$

Thus, the density excess of the solid core compared to the density of the liquid core is $\Delta\delta_{s,l} = 2.93$ g/cm³. Furthermore, we neglect the errors in the determination of the density and masses of the polar and axial moments of inertia of the considered lunar shells.

We emphasize that the spherical model of a homogeneous solid core and a liquid shell is in good agreement with seismic observations (Figure 13.1). Therefore, a number of dynamic characteristics of the solid and liquid core (polar moment of inertia, dynamic compressions, etc.) derived by us are based on the corresponding homogeneous models considering homogeneous spheres or ellipsoids and using additional observational data, such as lunar laser ranging (Williams, Boggs, and Ratcliff 2011, 2012).

We present the results of the axial moments of inertia of the solid core, considered as a homogeneous sphere, and the liquid core as a homogeneous spherical shell. For the rigid (or unchangeable) core of the Moon, considered as a homogeneous body with a spherical radius $R_{s,c} = 240 \pm 10$ km, the mass and polar moment of inertia are calculated using simple formulas and are equal to:

$$m_{s,c} = \frac{4}{3}\pi\delta_{s,c} R_{s,c}^3 = (4.653 \pm 0.582) \cdot 10^{23} \ \Gamma,$$

$$C_{s,c} = \frac{8}{15}\pi\delta_{s,c} R_{s,c}^5 = (1.072 \pm 0.225) \cdot 10^{38} \ \Gamma/\text{cm}^2. \tag{13.3}$$

Similarly, we determine the mass and moment of inertia of the liquid core, considered as a homogeneous spherical layer of liquid, by the formulas:

$$m_{l,c} = \frac{4}{3}\pi\delta_{l,c}\left(R_{l,c}^3 - R_{s,c}^3\right) = (4.729 \pm 1.780) \cdot 10^{23} \ \Gamma,$$

$$C_{l,c} = \frac{8}{15}\pi\delta_{l,c}\left(R_{l,c}^5 - R_{s,c}^5\right) = (2.667 \pm 0.813) \cdot 10^{38} \ \Gamma \cdot \text{cm}^2. \tag{13.4}$$

The total mass and polar moment of inertia of the entire core, including a liquid core and a solid core, will be:

$$m_c = m_{s,c} + m_{l,c} = (9.382 \pm 1.610) \cdot 10^{23} \text{ kg,}$$

$$C_c = C_{s,c} + C_{l,c} = (3.74 \pm 1.04) \cdot 10^{38} \text{ g} \cdot \text{cm}^2. \tag{13.5}$$

The mass of the considered model of the Moon's core is determined with a relative error of 25.2%. Importantly, the inertia moment of the liquid core is about 71.3% of the full core (including the solid one), which means that the liquid core plays a significant role in the rotational of the Moon.

This justifies our choice of a two-layer model of the Moon for the first studies of the effect of the liquid core. However, in subsequent studies we intend to consider the characteristics of the dynamics of a three-layer model of the Moon, consisting of a mantle (rigid or elastic), a liquid core, and a solid core. For the considered model of the core, its mass is determined with a relative error of 17.2%, and the polar moment of inertia is determined with a relative error of 27.8%. The mean density of the homogeneous spherical liquid core is 6.233 g/cm³. The mass and polar moment of inertia of the core agree with the values obtained from the data of seismometers.

13.1.3 Dynamic Parameter of Influencing Liquid Core on the Moon Rotation

To study the effects caused by a molten core in the lunar libration, let the parameter L_c be equal to the ratio of axial polar inertia moments of the core C_c and the Moon C. We take the known value of the polar moment of inertia of the Moon, $C = 0.873486 \times 10^{42}$ Γ × cm² (Araki et al. 2009). Based on the found value of inertia moment of the core, we can obtain the parameter magnitude:

$$L_c = C_c / C = (4.28 \pm 1.19) \cdot 10^{-4} \text{ (with error 27.8\%).} \tag{13.6}$$

In fact, this estimate is a fundamental parameter of the Moon derived from seismic data (Barkin et al. 2012). Parameter L_c is defined here most accurately than in previous estimates. However, the uncertainty in the determination of the dynamic parameter L_c is quite significant. Therefore, the first studies of the effects on the rotation of the Moon, due to the molten core, are restricted to the two-layers model.

13.1.4 Dynamic Oblateness of the Core

Dynamic effects of the liquid core on the rotational motion of the Moon will be studied assuming a Poincaré model of the body with an ellipsoidal cavity filled by an ideal fluid, based on the special forms of the equations of motion in the Andoyer's and Poincaré's variables introduced in previous studies (Ferrándiz and Barkin 2000, 2001). An ellipsoidality of the liquid core is the dominant factor in studying the effects on the free nutation of the Moon due to the hydrodynamic effect of the liquid core. It can be characterized by the semiaxe of the ellipsoidal cavity, which contains the liquid core, $a_c > b_c > c_c$. The principal moments of inertia A_c, B_c, C_c and Poincaré's parameters D_c, E_c, F_c are determined by simple formulae (Poincaré 1910, Lamb 1947):

$$A_c = \frac{1}{5} m_c \left(b^2 + c^2 \right), \; B_c = \frac{1}{5} m_c \left(a^2 + c^2 \right), \; C_c = \frac{1}{5} m_c \left(a^2 + b^2 \right),$$

$$D_c = \frac{2}{5} m_c ab, \; E_c = \frac{2}{5} m_c ac, \; F_c = \frac{2}{5} m_c bc, \tag{13.7}$$

where m_c is the mass of the liquid core. The relations $C_c > B_c > A_c$ and $a > b > c$ are satisfied and thus the following dynamic oblatenesses of ellipsoidal core are positive:

$$\varepsilon_D = 1 - A_c/C_c, \text{ and } \mu_D = 1 - B_c/C_c. \tag{13.8}$$

Along with these assumptions, similar dynamic characteristics can be introduced as

$$\varepsilon_c = 1 - c_c^2/b_c^2 < 1, \; \mu_c = 1 - c_c^2/a_c^2 < 1, \tag{13.9}$$

which are associated with dynamic oblatenesses obtained by Misha Barkin (2012) as:

$$\varepsilon_c = \frac{2\varepsilon_D}{1 + \varepsilon_D - \mu_D}, \; \mu_c = \frac{2\mu_D}{1 + \mu_D - \varepsilon_D}. \tag{13.10}$$

13.1.5 Geometric Compressions and the Difference between the Polar and Equatorial Axes of the Liquid Core

The reference model of the core is a homogeneous ellipsoid with semiaxes $a_c > b_c > c_c$, where the c_c is the polar axis, b_c is the middle equatorial axis, and a_c the largest semiaxis of co-ellipsoids. Geometric and dynamic compressions of the core have approximately same values

$$\varepsilon_D = \frac{C_c - A_c}{C_c} = \frac{a_c^2 - c_c^2}{2a_c^2} \approx \frac{a_c - c_c}{a_c} \approx \frac{a_c - c_c}{R_c}$$

$$\mu_D = \frac{B_c - A_c}{C_c} = \frac{a_c^2 - b_c^2}{2a_c^2} \approx \frac{a_c - b_c}{a_c} \approx \frac{a_c - b_c}{R_c}, \tag{13.11}$$

where $R_c = R_{l,c} = 330 \pm 20$ km and hence the contraction parameters of the Moon are:

$$a_c - c_c = 140.9 \text{ m}, a_c - b_c = 90.0 \text{ m}. \tag{13.12}$$

Seismographic methods currently do not allow to identify and evaluate the oblateness of the core $\varepsilon_D = 1 - A_c/C_c$ and $\mu_D = 1 - B_c/C_c$. However, studies of the physical libration on the Moon can be performed from long-term series of data of laser observations.

13.1.6 Estimates of the Dynamic Compressions of the Core

The dynamic effects of the liquid core on the physical libration of the Moon depend on the relative moment of the liquid core $L_c = C_c/C$ and its oblateness $\varepsilon_D + \mu_D$. One of the first estimates of the parameter $L_c = 6 \times 10^{-4}$ was obtained for a simple model of a homogeneous iron core with a radius of about 340 km, and the oblateness of the core (for axially symmetric model with flattenings $\varepsilon_D = \mu_D$) has been estimated as 3×10^{-4}. It was noted that the long period of free nutation of the liquid core should be about two centuries (Williams et al. 2001). If the dynamic compression of the core (or the amount of meridional compression) is larger in reality, the period will be short. Conversely, if such flattening is lower, then the period will have a greater value. In this chapter, we obtain consistent estimation of the dynamical flattening of the core, including the developed analytical theory and empirical theory based on laser observations (Rambaux and Williams 2011).

Recent studies (Williams, Boggs, and Ratcliff 2010, 2011, 2012) attempt to determine firstly the dynamical compression of the core from the dynamical analysis of the rotation of the Moon using high-precision laser observations. Williams et al. (2010) obtained the core parameters in a similar estimate for the following combination of input to our work:

$$\frac{1}{2}L_c\left(\varepsilon_D + \mu_D\right) = (2.3 \pm 0.8) \cdot 10^{-7} \text{ (error of 35\%),} \tag{13.13}$$

which is an estimate of the amount of the dynamical oblateness of the lunar core $\varepsilon_D + \mu_D$ resulting from comparing the analytical theory of the lunar physical libration with liquid core and the empirical theory of physical libration built on the basis of long-term series of laser observations (Rambaux and Williams 2011). From the former value (see Table 13.1), we estimate the parameter $L_c = (6.57 \pm 2.29) \times 10^{-4}$ by analyzing the ratio of lunar libration. In addition, Williams, Boggs, and Ratcliff (2012) obtained another parameter by setting

$$\frac{1}{2}L_c\left(\varepsilon_D+\mu_D\right) = \left(1.6\pm0.7\right)\cdot10^{-7} \text{ (error 44\%).} \tag{13.14}$$

Using the value $\varepsilon_D + \mu_D = 7.00 \times 10^{-4}$ obtained from the analysis of the lunar libration derived from the relationship between contractions and the period of the Poincaré oscillation, we estimate the parameter solving Equation 13.14

$$L_c = \left(4.57\pm2.00\right)\cdot10^{-4} \text{ (error is 44\%).} \tag{13.15}$$

This value is consistent with the above value for seismographic data, but with a little lower error L_c = (4.28 ± 1.19) × 10^{-4} (about 27.8%). In the calculations we have neglected the uncertainty in the determination of the core oblatenesses. Table 13.1 presents the known values of the core parameters for the three celestial bodies, Earth, Mars, and the Moon.

13.2 Main Problems of the Physical Libration of the Moon

We develop an analytical theory of the physical libration of the Moon as a system of interacting core and mantle in the gravitational field of Earth and other celestial bodies. The core is modeled by an ellipsoid with an ideal homogeneous fluid. The mantle is considered as a nonspherical solid body, but taking into account its elastic properties. It is assumed that the centers of mass of the core and mantle are coincident and besides the principal central axes of inertia are aligned. The theory is developed using the canonical equations in Andoyer's variables and in Poincaré's variables and perturbation theory for the construction of quasi-periodic solutions and the investigation of their neighborhood (based on the corresponding equations in variations).

The values of the amplitudes and periods of perturbations of the first order for Andoyer's variables describing the libration of the Moon and its core, for the variations of the angular velocity components of rotation of the Moon, are provided. In the construction of the libration of the Moon we adopted high-precision orbit of the Moon DE/LE-406 and took into account the second order harmonic of the gravitational potential of the nonspherical the Moon according to the modern model of its gravitational field and dynamic structure (Matsumoto et al. 2010). The dynamic characteristics of the body are introduced and discussed in the next section.

13.2.1 Main Coordinate System and Variables

The basis of this study is the Poincaré study on the rotation of a body with an ellipsoidal cavity filled with ideal fluid. We use the canonical form of the equations of motion in Andoyer variables (Sevilla and Romero 1987, Ferrándiz and Barkin 2000). Let us consider the rotational motion of a celestial body with an ellipsoidal cavity filled by an incompressible homogeneous fluid around its center of mass in the gravitational field of another celestial body. The outer shell is called the mantle, and liquid body—the core. In general, the mantle is elastic, but in this chapter we consider the mantle as an absolutely rigid and nonspherical shell. The core behaves like an ideal fluid, fully occupying the ellipsoidal cavity.

$CXYZ$ is the Cartesian coordinate system with its origin located at the center of mass C of the body and with the axes referred to the fixed space. $C\xi\eta\zeta$ is a system of reference with axes directed along the principal central axes of inertia of the model of the Moon. Equatorial axes of inertia $O\xi$, $O\eta$ and the polar axis of inertia $O\zeta$ of the Moon correspond to the principal central moments of inertia A, B, and C ($C > B > A$) of the Moon.

The axes of the coordinate system $O\xi_c\eta_c\zeta_c$ are directed along the principal central axes of inertia of the core and the moments of inertia A_c, B_c and C_c, as well as other characteristics D_c, E_c and F_c, we introduced above. We assume that the axes of the ellipsoidal cavity $O\xi_c\eta_c\zeta_c$ coincide with the principal central axes of inertia of the Moon $O\xi\eta\zeta$ and the liquid core. Axes $O\xi_c$, $O\eta_c$ and $O\zeta_c$ correspond to the moments of inertia A_c, B_c and C_c of the core and to semi-axes of the core ellipsoidal cavity a, b and c.

13.2.2 Euler Angles and Andoyer's Variables

Here we describe the main sets of variables we use for the two-layer model of the Moon—the Euler angles, Andoyer's variables, and their modifications. These variables and the corresponding equations of motion of a liquid core–mantle system (solid or elastic) are used to study the rotational motion of the Moon. Since the orientation and rotation of the lunar mantle P_m (or the coordinate system of the main central axes of inertia $C\xi_m\eta_m\zeta_m$) in the main selenocentric ecliptic coordinate system $CXYZ$ are defined by the Euler angles Ψ_m, Θ_m, Φ_m and the components p_m, q_m, r_m of the angular velocity vector of lunar rotation ω_m:

$$\Psi_m, \Theta_m, \Phi_m; \ p_m, q_m, r_m. \tag{13.16}$$

To describe the motion of the core, we use the classical model of a simple motion of an ideal fluid in an ellipsoidal cavity (in the sense of Poincaré) and introduce a new coordinate system $Ox_cy_cz_c$ connected with the core, the relation to which is determined by a simple motion of the liquid. We call it as the Poincaré's coordinate system. In order to preserve the symmetry of the variables in the notation, we will consider the moving coordinate system $Ox_cy_cz_c$ as the base for the liquid core, and the system's main central axes of inertia of the core $O\xi_c\eta_c\zeta_c$ will be considered as rotating with a certain angular velocity

ω_c with respect to system $Ox_cy_cz_c$. The corresponding Euler angles and the projections of the vector ω_c on the axes of the ellipsoidal cavity $O\xi_c\eta_c\zeta_c$ are denoted as:

$$\Psi_c, \Theta_c, \Phi_c; \; p_c, q_c, r_c. \tag{13.17}$$

In our notation, the projections p_c, q_c, r_c of the angular velocity ω_c of rotation of the ellipsoidal cavity (with respect to the Poincaré's coordinate system $Ox_cy_cz_c$) are different from the classical notation by their opposite sign (Lamb 1947).

The Euler angles in Equations 13.16 and 13.17 have standard notations: Ψ, the precession angle, Θ, - the angle of nutation, Φ, the angle of proper rotation, and the components introduced in consideration of angular velocities in Equations 13.16 and 13.17 are determined by the kinematic Euler equations:

$$p_s = \sin\Phi_s\sin\Theta_s\dot{\Psi}_s + \cos\Phi_s\dot{\Theta}_s$$

$$q_s = \cos\Phi_s\sin\Theta_s\dot{\Psi}_s - \sin\Phi_s\dot{\Theta}_s$$

$$r_s = \cos\Theta_s\dot{\Psi}_s + \dot{\Phi}_s \; (s = m, c). \tag{13.18}$$

To describe the relative simple motion of a liquid (in the sense of Poincaré) in the cavity P_c, we introduce a reference frame $C_cx_cy_cz_c$ associated with the core. Its orientation and rotation are defined by the Euler angles. To satisfy the symmetry in notations, we assume that the coordinate system $C_cx_cy_cz_c$ is the basic coordinate system and the coordinate system $Cxyz$ rotates with a certain angular velocity ω_c relative to the first. The corresponding Euler angles and projections of the vector ω_c on the axes of coordinate frame $Cxyz$ are denoted in Equations 13.16 and 13.17. From a geometrical point of view the variables are similar. The components p_c, q_c, r_c that define the vector ω_c are different from the classical notation of Poincaré by opposite sign (Poincaré 1910).

13.2.3 Kinetic Energy and Canonical Equations of Rotational Motion of the Moon with a Liquid Core in Eulerian Variables

The kinetic energy of a planet with a molten core in Equations 13.16–13.18 is defined by the well-known expression (Poincaré 1910):

$$2T = Ap_m^2 + Bq_m^2 + Cr_m^2 + A_cp_c^2 + B_cq_c^2 + C_cr_c^2 - 2F_cp_mp_c - 2E_cq_mq_c - 2D_cr_mr_c. \tag{13.19}$$

Assuming that the body moves under the action of potential forces, the force function U is

$$U = U(\Psi_m, \Theta_m, \Phi_m, \Psi_c, \Theta_c, \Phi_c t). \tag{13.20}$$

The canonical momenta conjugate to U is generalized coordinates (Euler angles) are defined by:

$$p_{\Psi_s} = \frac{\partial T}{\partial \dot{\Psi}_s} = \lambda_s \sin \Theta_s \sin \Phi_s + \mu_s \sin \Theta_s \cos \Phi_s + \nu_s \cos \Theta_s$$

$$p_{\Theta_s} = \frac{\partial T}{\partial \dot{\Theta}_s} = \lambda_s \cos \Phi_s - \mu_s \sin \Phi_s$$

$$p_{\Phi_s} = \frac{\partial T}{\partial \dot{\Phi}_s} = \nu_s \ (s = m, c) \tag{13.21}$$

where

$$\lambda = A p_m - F_c p_c, \ \mu = B q_m - E_c q_c, \ \nu = C r_m - D_c r_c \tag{13.22}$$

and

$$\lambda_c = A_c p_c - F_c p_m, \mu_c = B_c q_c - E_c q_m, \nu_c = C_c r_c - D_c r_m \tag{13.23}$$

are the projections of the total angular momentum of the rotational motion of the celestial body with a liquid core (Equation 13.22) $\mathbf{G}_m$ (with respect to its center of mass), and the components of the angular momentum of the liquid core (Equation 13.23) $\mathbf{G}_c$ (with respect to the center of gravity of the planet) on the axis of a Cartesian coordinate system $Cxyz$ and $C_c x_c y_c z_c$, accordingly.

In the canonical variables of Equations 13.21–13.23, the equations of motion of the Poincaré's problem are as follows:

$$\frac{d(\Psi_s, \Theta_s, \Phi_s)}{dt} = \frac{\partial K}{\partial (p_{\Psi_s}, p_{\Theta_s}, p_{\Phi_s})},$$

$$\frac{d(p_{\Psi_s}, p_{\Theta_s}, p_{\Phi_s})}{dt} = -\frac{\partial K}{\partial (\Psi_s, \Theta_s, \Phi_s)}. \tag{13.24}$$

Based on Equations 13.19, 13.22, 13.23, the Hamiltonian problem we has the following expression:

$$K = T - U = \frac{1}{2}\left(\Lambda \lambda^2 + M\mu^2 + N\nu^2 + \Lambda_c \lambda_c^2 + M_c \mu_c^2 + N_c \nu_c^2\right)$$

$$+ P_c \lambda \lambda_c + Q_c \mu \mu_c + R_c \nu \nu_c - U(\Psi_m, \Theta_m, \Phi_m, \Psi_c, \Theta_c, \Phi_c, t), \tag{13.25}$$

where

$$\lambda_s = p_{\Theta_s}\cos\Phi_s + \frac{\sin\Phi_s}{\sin\Theta_s}\left(p_{\Psi_s} - p_{\Phi_s}\cos\Theta_s\right),$$

$$\mu_s = -p_{\Theta_s}\sin\Phi_s + \frac{\cos\Phi_s}{\sin\Theta_s}\left(p_{\Psi_s} - p_{\Phi_s}\cos\Theta_s\right),$$

$$\nu_s = p_{\Phi_s},$$

$$G_s = \sqrt{p_{\Theta_s}^2 + p_{\Phi_s}^2 + \csc^2\Theta_s(p_{\Psi_s} - p_{\Phi_s}\cos\Theta_s)^2}. \tag{13.26}$$

The constant coefficients in the Hamiltonian function (Equation 13.25) are given by:

$$\Lambda = \frac{A_c}{\Delta_1}, \ M = \frac{B_c}{\Delta_2}, \ N = \frac{C_c}{\Delta_3}$$

$$\Lambda_c = \frac{A}{\Delta_1}, \ M_c = \frac{B}{\Delta_2}, \ N_c = \frac{C}{\Delta_3}$$

$$P_c = \frac{F_c}{\Delta_1}, \ Q_c = \frac{F_c}{\Delta_2}, \ R_c = \frac{D_c}{\Delta_3} \tag{13.27}$$

$$\Delta_1 = A_cA - F_c^2, \ \Delta_2 = B_cB - E_c^2, \ \Delta_3 = C_cC - D_c^2. \tag{13.28}$$

In the classical formulation of the problem, the corresponding axes of the reference systems $O\xi_c\eta_c\zeta_c$ and $C\xi_m\eta_m\zeta_m$ are the same and rotate with the same angular velocity $\omega = \omega_m$ ($p = p_m, q = q_m, r = r_m$) with respect to the ecliptic selenocentric coordinate system $CXYZ$. On the basis of Euler variables we can introduce the canonical variables of the Euler angles and their conjugate canonical momenta as:

$$\Psi_s, \Theta_s, \Phi_s; p_{\Psi_s}, p_{\Theta_s}, p_{\Phi_s} (s = m, c). \tag{13.29}$$

13.2.4 Canonical Equations of Rotational Motion of the Moon with Liquid Core in Andoyer Variables

Now let us consider the two groups of Andoyer's variables (Barkin et al. 2012):

$$G, \theta, \rho, l, g, h; G_c, \theta_c, \rho_c, l_c, g_c, h_c \tag{13.30}$$

where the first is connected with the angular momentum vector of the whole system—the Moon $\mathbf{G}$ (with taking into account the dynamics of the liquid core), and the second is determined by an angular momentum vector $\mathbf{G}_c$ of the simple relative motion of the fluid. In the list of variables (Equation 13.30) G and $\mathbf{G}_c$ are the modules of the angular momentum vectors $\mathbf{G}$ and $\mathbf{G}_c$, and ρ and ρ_c are the angles of inclinations of vectors $\mathbf{G}$ and $\mathbf{G}_c$ with respect to the normal of the main coordinate plane of ecliptic of date (the angles between the axis CZ and the angular momentum vectors). We introduce two intermediate coordinate systems $CG_1G_2G_3$ and $CG_{c1}G_{c2}G_{C3}$, also associated with the vectors $\mathbf{G}$ and $\mathbf{G}_c$. The axes CG_3 and CGc_3 are directed along the corresponding vectors $\mathbf{G}$ and $\mathbf{G}_c$. The other two axes CG_1 and CG_{c1} are located in the plane CXY of the base ecliptic, selenocentric coordinate system. Axis CG_1 is along the lines of intersection of the coordinate planes CG_1G_2, CXY is in the direction of the ascending node of plane CG_1G_2, and the axis CGc_1 is directed along the line of intersection of the planes $CG_{c1}G_{c2}$ and CXY (Figure 13.2).

Andoyer variables associated with the angular momentum vector $\mathbf{G}$ of rotational motion of the Moon, are denoted as:

$$G, \theta, \rho, l, g, h \tag{13.31}$$

$CG_1G_2G_3$ is an intermediate coordinate system associated with the vector $\mathbf{G}$. The axis CG_3 is directed along the vector $\mathbf{G}$ and axis CG_1 is situated in the plane CXY of the base coordinate system and directed along the line of intersection of the planes CG_1G_2 and Cxy in the side of the ascending

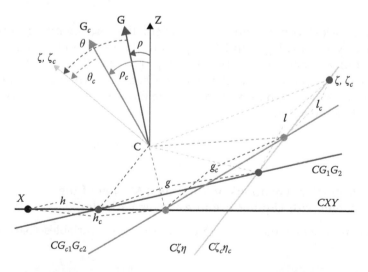

FIGURE 13.2
Andoyer's variables for the Poincaré problem.

node of the plane CG_1G_2 (Figure 13.2). $G = |\mathbf{G}|$ is the modulus of the angular momentum vector, p and h are the angles which determine the orientation of the axes $CG_1G_2G_3$ with respect to the coordinate system $Cxyz$, p is the angle between the axis Cz and the angular momentum vector G and h is the angle between the positive directions of the coordinate axes Cx and CG_1 (h is a longitude of the ascending node of the intermediate plane CG_1G_2).

Andoyer's variables l, g, θ and l_c, g_c, θ_c determine the orientation of the principal axes of inertia of the Moon $C\xi\eta\zeta$ (and, accordingly, the principal axes of inertia of the liquid core $C\xi_c\eta_c\zeta_c$, because they are the same) relative to the former two intermediate Andoyer's coordinate systems $CG_1G_2G_3$ and $CG_{c1}G_{c2}G_{c3}$ (Figure 13.2). These variables are the Euler angles that define the orientation of the principal axes of inertia is central relative to the two intermediate coordinate systems: θ and θ_c are the nutation angle (angle between the axis CG_3 and CG_{c3} and the polar axis of inertia of the Moon $C\zeta$), g and g_c are the precession angles (the angles measured from the axes CG_1 and CG_{c1} of the corresponding intermediate systems of coordinates to the lines of intersection of the coordinate planes CG_1G_2 and $CG_{c1}G_{c2}$, and with the equatorial plane of the Moon $C\xi\eta$), and l and l_c are the angles of the proper rotation of the principal axes of inertia of the Moon (and its core) with respect to the intersection lines of the coordinate planes CG_1G_2 and $CG_{c1}G_{c2}$, with the equatorial plane $C\xi\eta$ (the angles between pointed lines of intersections and the equatorial axis of inertia of the Moon $C\xi$).

Thus, if the angles h, p and h_c, p_c determine the orientation of the intermediate coordinate systems $CG_1G_2G_3$ and $CG_{c1}G_{c2}G_{c3}$ with respect to the main selenocentric ecliptic coordinate system, the Euler angles g, θ, l and g_c, θ_c, l_c and specifying the orientation of the coordinate axes related to the Moon $C\xi\eta\zeta$ relative to the two intermediate coordinate systems $CG_1G_2G_3$ and $CG_{c1}G_{c2}G_{c3}$. The geometric meaning of the variables is also well illustrated in Figure 13.1. Additionally, another Andoyer canonical moments previously considered are defined as:

$$L_s = G_s \cos\theta_s, G_s = |\mathbf{G}_s|\cos\rho_s, \; H_s = G_s \cos\rho_s. \tag{13.32}$$

Now we perform a canonical transformation of variables to the canonical Andoyer set:

$$L_s, G_s, H_s, l_s, g_s, h_s \; (s = m, c). \tag{13.33}$$

The canonical transformation from Euler to Andoyer's variables was discussed in detail (e.g., Arkhangelskii 1975; Barkin 2000; and so on). Here the dynamic and clear geometric interpretation of Andoyer variables is given. If necessary, we will use the relation between these variables and the expression of different dynamical characteristics in Andoyer variables for the model of the Moon with a liquid core. So for the projections of the angular momentum vectors $\mathbf{G}$ and $\mathbf{G}_c$ and the principal central axis of inertia of the the Moon, we have the following expressions:

$$\lambda_s = G_s \sin\theta_s \sin l_s, \mu_s = G_s \sin\theta_s \cos l_s, \nu_s = G_s \cos\theta_s \, (s=\nearrow, c) \,, \quad (13.34)$$

where $\cos\theta_s = L_s/G_s$, $\sin\theta_s = \sqrt{G_s^2 - L_s^2}/G_s$.

Euler angles and direction cosines of the inertia axes of the Moon in Andoyer's variables are expressed by well-known formulas (Barkin 1987, 2000). The first Andoyer variables in the model of the rotation of Earth with a liquid core were introduced by Romero and Sevilla (987) based on the approach by Sasao et al. (1980). The generalized Poincaré problem of the motion equation in Andoyer variables were obtained by Ferrándiz and Barkin (2000).

13.2.5 Canonical Equations of Rotational Motion of the Moon with an Ellipsoidal Liquid Core in Andoyer's Variables

In the development of the analytical theory of rotational motion of the Moon, we use the Hamiltonian formalism and perturbation theory based on the equations of motion in the canonical Andoyer's and Poincaré's variables (Ferrándiz and Barkin 2000, 2001). In the canonical Andoyer variables

$$\mathbf{z} = \left(l,g,h,l_c,g_c,h_c\right), \ \mathbf{Z} = \left(L,G,H,L_c,G_c,H_c\right). \quad (13.35)$$

The equations of rotational motion of the considered of the Moon model (solid mantle, liquid ellipsoidal core) are:

$$\frac{d\mathbf{z}}{dt} = \frac{\partial K}{\partial \mathbf{Z}}, \frac{d\mathbf{Z}}{dt} = -\frac{\partial K}{\partial \mathbf{z}}, \quad (13.36)$$

where

$$K = T - U \cdot \quad (13.37)$$

This is the Hamiltonian of the problem, T is the kinetic energy of rotational motion of the Moon with liquid core, and U is the force function of gravitational interaction of the nonspherical the Moon with the Earth and other external celestial bodies. In this chapter, we restrict the analysis of the second harmonic of the force function of the Moon. All functions pointed here must be submitted as explicit functions of the Andoyer variables in Equation 13.35. Transformation of variables formula 13.32–13.34 allows us to write Equation 13.36 and the Hamiltonian Equation 13.37 as follows:

$$\frac{d(l_s, g_s, h_s)}{dt} = \frac{\partial K}{\partial(L_s, G_s, H_s)}$$

$$\frac{d(L_s, G_s, H_s)}{dt} = -\frac{\partial K}{\partial(l_s, g_s, h_s)} (s=m,c) \tag{13.38}$$

where

$$K = \frac{1}{2}G^2\left[\left(\frac{A_c}{\Delta_1}\sin^2 l + \frac{B_c}{\Delta_2}\cos^2 l\right)\sin^2\theta + \frac{C_c}{\Delta_3}\cos^2\theta\right]$$

$$+ \frac{1}{2}G_c^2\left[\left(\frac{A}{\Delta_1}\sin^2 l_c + \frac{B}{\Delta_2}\cos^2 l_c\right)\sin^2\theta_c + \frac{C}{\Delta_3}\cos^2\theta_c\right]$$

$$+ GG_c\left[\left(\frac{F_r}{\Delta_1}\sin l\sin l_c + \frac{F_e}{\Delta_2}\cos l\cos l_c\right)\sin\theta\sin\theta_c + \frac{D_c}{\Delta_3}\cos\theta\cos\theta_c\right]$$

$$- U(l,g,h,L,G,H;l_c,g_c,h_c,L_c,G_c,H_c;t). \tag{13.39}$$

Here θ and θ_c are the angles between vectors of angular kinetic momentums of the Moon and its liquid core (G and G_c), and the corresponding coordinate axes $C\zeta$ and Cz_c: $\cos\theta = L/G$, $\cos\theta = L_c/G_c$, $\sin\theta = \sqrt{G^2 - L^2}/G$, $\sin\theta = \sqrt{G_c^2 - L_c^2}/G_c$.

The force function in this chapter is identified with the second harmonic of the gravitational potential of the Moon. The corresponding expansion of this function in trigonometric series in multiples of Andoyer angular variables and arguments of the theory of orbital motion of the Moon is given below.

13.2.6 Canonical Equations of the Physical Libration of the Moon in Poincaré's Variables

The direct use of equations of motion in Andoyer variables has certain difficulties in their singularity at zero values of the variables θ and θ_c. However, it can be circumvented by using a special redesignation of the principal axes of inertia of the Moon. A similar device, apparently, first proposed by Beletskii (1975), was effectively used in the construction of the analytical theory of the physical libration of the rigid Moon directly to the canonical Andoyer's variables (Barkin 1987, 1989). In this chapter, we propose a different method to construct a theory of the physical libration of the Moon on the basis of the canonical equations in Poincaré's variables. These variables are expressed in terms of canonical Andoyer's variables using simple relations:

$$\varLambda = G,H,h, \lambda = l+g,$$

$$\xi = \sqrt{2(G-L)}\cos l, \eta = -\sqrt{2(G-L)}\sin l, \tag{13.40}$$

$$\Lambda_c = G_c, H_c, h_c \, \lambda_c = l_c + g_c$$

$$\xi_c = \sqrt{2(G_c + L_c)} \cos l_c, \; \eta_c = \sqrt{2(G_c + L_c)} \sin l_c. \tag{13.41}$$

Thus, in variables

$$\mathbf{z} = (\lambda, \lambda_c, h, h_c, \xi, \xi_c), \; \mathbf{Z} = (\Lambda, \Lambda_c, H, H_c, \eta, \eta_c) \tag{13.42}$$

equations of motion again retain the canonical form (Equations 13.36 and 13.42) in which the Hamiltonian of the problem (the kinetic energy and the force function) should be represented by a function of Poincaré's variables (Equation 13.42).

13.2.7 Noninertial Part of the Hamiltonian

With the development of the analytical theory of the rotation of the solid the Moon, the Andoyer variables (Equation 13.35) can be referred to the fixed ecliptic plane (the ecliptic of epoch) or to the moving plane of the ecliptic of date (Kinoshita 1977; Barkin 1989). In both cases, the Hamiltonian is written in the Andoyer variables, preserving the structure of the expressions of the kinetic energy and the force function, but in the second case noninertial terms must be added:

$$F_{ni} = G \frac{d\Pi_1}{dt} \left[\sin \pi_1 \sin \rho \cos(h + \Omega_0 - \Pi_1) - \cos \pi_1 \cos \rho \right]$$

$$- G \sin \rho \frac{d\pi_1}{dt} \sin(h + \Omega_0 - \Pi_1) + G \cos \rho \left(\frac{d\Pi_1}{dt} - \frac{d\Omega_0}{dt} \right) \tag{13.43}$$

where angles π_1 and Π_1 define the orientation of the plane of the ecliptic of date with respect to the plane of the ecliptic of epoch, and Ω_0 is the mean longitude of the ascending node on the ecliptic plane of the ecliptic date of the epoch Barkin 1989. The variable h is measured from the mean position of the node of the lunar orbit to the ecliptic of date. Here we consider only this effect and the inertial component of the Hamiltonian (Equation 13.43) is approximated by

$$F_{ni} = -n_\Omega G \cos \rho, \tag{13.44}$$

where n_Ω is a constant angular velocity of the change of longitude of lunar orbit Ω_0.

13.3 Development of the Force Function and the Hamiltonian in Poincaré's Variables

13.3.1 Expansions of Spherical Functions of the Coordinates in the Expression of the Force Function

The data represent the new expansions of the functions (the spherical coordinates of the orbital motion of the Moon r, φ and λ_m) from the paper (Kinoshita 1977). The developments are made in the time interval 1000 AD–3000 AD based on long-term numerical lunar ephemeris LE-406 (the coordinate system ICRF) and presented in the form of the following Poisson series (Kudryavtsev 2007; Barkin, Kudryavtsev, and Barkin 2009). These expansions determine a more accurate formula when compared with series of Kinoshita and take into account new effects in the orbital motion of the Moon:

$$\frac{1}{2}\left(\frac{a}{r}\right)^3 \left(1-3\sin^2\varphi\right) = \sum_{\nu}\left(A_{\nu;0}^{(0)} + A_{\nu;1}^{(0)}\cdot t + A_{\nu;2}^{(0)}\cdot t^2\right)\cos\Theta_\nu$$

$$+ \left(a_{\nu;0}^{(0)} + a_{\nu;1}^{(0)}\cdot t + a_{\nu;2}^{(0)}\cdot t^2\right)\sin\Theta_\nu \tag{13.45}$$

$$\left(\frac{a}{r}\right)^3 \sin\varphi\cos\varphi\sin\lambda_m = \sum_{\nu}\left(A_{\nu;0}^{(1)} + A_{\nu;1}^{(1)}\cdot t + A_{\nu;2}^{(1)}\cdot t^2\right)\cos\Theta_\nu$$

$$+ \left(a_{\nu;0}^{(1)} + a_{\nu;1}^{(1)}\cdot t + a_{\nu;2}^{(1)}\cdot t^2\right)\sin\Theta_\nu \tag{13.46}$$

$$\left(\frac{a}{r}\right)^3 \sin\varphi\cos\varphi\cos\lambda_m = \sum_{\nu}\left(b_{\nu;0}^{(1)} + b_{\nu;1}^{(1)}\cdot t + b_{\nu;2}^{(1)}\cdot t^2\right)\cos\Theta_\nu$$

$$+ \left(B_{\nu;0}^{(1)} + B_{\nu;1}^{(1)}\cdot t + B_{\nu;2}^{(1)}\cdot t^2\right)\sin\Theta_\nu \tag{13.47}$$

$$\left(\frac{a}{r}\right)^3 \cos^2\varphi\sin 2\lambda_m = \sum_{\nu}\left(b_{\nu;0}^{(2)} + b_{\nu;1}^{(2)}\cdot t + b_{\nu;2}^{(2)}\cdot t^2\right)\cos\Theta_\nu$$

$$+ \left(B_{\nu;0}^{(2)} + B_{\nu;1}^{(2)}\cdot t + B_{\nu;2}^{(2)}\cdot t^2\right)\sin\Theta_\nu \tag{13.48}$$

$$\left(\frac{a}{r}\right)^3 \cos^2\varphi\cos 2\lambda_m = \sum_{\nu}\left(A_{\nu;0}^{(2)} + A_{\nu;1}^{(2)}\cdot t + A_{\nu;2}^{(2)}\cdot t^2\right)\cos\Theta_\nu$$

$$+ \left(a_{\nu;0}^{(2)} + a_{\nu;1}^{(2)}\cdot t + a_{\nu;2}^{(2)}\cdot t^2\right)\sin\Theta_\nu \tag{13.49}$$

where a is a unperturbed value of major semiaxis of lunar orbit with $a = 383397772.5$ m, $\lambda_m = \lambda - h$, and t is the time measured in Julian centuries from epoch of J2000.0 (JED2451545.0). Here Θ_v is any linear combination of the classical arguments of the theory of orbital motion of the Moon

$$\Theta_v = \nu_1 l + \nu_2 l' + \nu_3 F + \nu_4 D + \nu_5 \Omega, \, \mathbf{v} = (\nu_1, \nu_2, ..., \nu_5) \qquad (13.50)$$

where $v = (v_1, v_2, ..., v_5)$ are integer coefficients, l, l', F, D, Ω are Delaunay's arguments and the mean longitude of the ascending node of the orbit of the Moon, respectively (Simon et al. 1994; Kudryavtsev 2007). In the Poisson's series (Equations 13.45–13.49), the coefficients

$$A_{v;n}^{(j)}, \, a_{v;n}^{(j)} \left(j = 0,1,2 \right), \, B_{v;n}^{(j)}, \, b_{v;n}^{(j)} \left(j = 1,2 \right) \qquad (13.51)$$

are numerical coefficients (amplitude) ($j = 0,1,2; n = 0,1,2$). The files obtained by Kudryavtsev (2007) contain these coefficients for each of the five spherical functions (Equations 13.45–13.49):

$$A_{v;0}^{(j)}, \, a_{v;0}^{(j)} \left(j = 0,1,2 \right), \, B_{v;0}^{(j)}, \, b_{v;0}^{(j)} \left(j = 1,2 \right)$$

$$A_{v;1}^{(j)}, \, a_{v;1}^{(j)} \left(j = 0,1,2 \right), \, B_{v;1}^{(j)}, \, b_{v;1}^{(j)} \left(j = 1,2 \right)$$

$$A_{v;2}^{(j)}, \, a_{v;2}^{(j)} \left(j = 0,1,2 \right), \, B_{v;2}^{(j)}, \, b_{v;2}^{(j)} \left(j = 1,2 \right) \qquad (13.52)$$

The first column indicates the number of sets of coefficients. In the next five columns by rows are 5 sets of integer indices $v = (v_1, v_2, v_3, v_4, v_5)$—the coefficients in the linear combinations of the classical arguments of the theory of orbital motion of the Moon (Equation 13.50). The seventh column contains the values of periods (in days) of the corresponding perturbation or trigonometric expansions terms (provided that the age-old neglected quadratic variation of the arguments with time). It does not provide a table of values of these expansions coefficients in Equations 13.45–13.52.

In his theory of Earth's rotation, Kinoshita provided additional expansions of certain functions of spherical geocentric coordinates of the Moon. In the quadratic terms with respect to time with the coefficients $A_{v;2}^{(j)}$, $a_{v;2}^{(j)} \left(j = 0,1,2 \right)$, $B_{v;2}^{(j)}$, $b_{v;2}^{(j)}, (j = 1,2)$, Kinoshita is not considered. In our work, we also reject the simplified assumptions:

$$A_{v;0}^{(1)} = -B_{v;0}^{(1)}, \, A_{v;0}^{(2)} = -B_{v;0}^{(2)}, \qquad (13.53)$$

which were previously used by Kinoshita (1977) and Barkin (1989). For Equation 13.53 though executing with a high degree of accuracy as confirmed by our construction of the expansions (Equations 13.45–13.52), however, the theory developed in the rotation of the Moon is rejected. The Equation 13.44, though executing with a high degree of accuracy as confirmed by our construction of the expansions (Equations 13.45–13.52), however, in our developed theory of the rotation of the Moon is rejected.

13.3.2 Force Functions in the Andoyer Variables

As noted earlier in this chapter we restrict it to the second harmonic of the force function of the problem. The development of this harmonic in trigonometric series (and Poisson series) at multiple angular Andoyer variables in the theory of rotation of the solid the Moon was built by Barkin (1987, 1989), and even earlier in the theory of the rotation of Earth (Kinoshita 1977; Getino and Ferrándiz 1991), where the simplified assumption (Equation 13.53) was used. Here, we present a similar expansion obtained without such limitations. In this chapter we will not explore the role of secular time variations of the coefficients in the representations (Equations 13.45–13.49), which lead to perturbations of the mixed type in the rotation of the Moon. Here is the expansion of the force function in the problem of the libration of the Moon (its second harmonic) in the Andoyer variables (Barkin 1987, 1989):

$$
U_2 = -\frac{3}{2}Cn_0^2 \frac{J_2}{I}\left[(3\cos^2\theta - 1)\sum_{|v|} B_v \cos\Theta_v - \sin 2\theta \sum_{\varepsilon=\pm 1}\sum_{|v|} C_v^{(\varepsilon)}\cos(g - \varepsilon\Theta_v)\right.
$$

$$
+\frac{1}{2}\sin^2\theta \sum_{\varepsilon=\pm 1}\sum_{|v|} D_v^{(\varepsilon)}\cos(2g - \varepsilon\Theta_v)\right] - \frac{3}{2}Cn_0^2 \frac{C_{22}}{I}\left[3\sin^2\theta \sum_{\varepsilon=\pm 1}\sum_{|v|} B_v\right.
$$

$$
\cos(2l - \varepsilon\Theta_v) + 2\sum_{\sigma=\pm 1}\sin\theta(\sigma + \cos\theta)\sum_{\varepsilon=\pm 1}\sum_{|v|} C_v^{(\varepsilon)}\cos(g + 2\sigma l - \varepsilon\Theta_v)
$$

$$
\left. +\frac{1}{2}\sum_{\sigma=\pm 1}(1 + \sigma\cos\theta)^2 \sum_{\varepsilon=\pm 1}\sum_{|v|} D_v^{(\varepsilon)}\cos(2g + 2\sigma l - \varepsilon\Theta_v)\right].
$$

(13.54)

In Equation 13.54 we use the notation $n_0^2 = f\,\dfrac{m_E}{a^3}$, where f is the gravitational constant, m_E is the mass of Earth, and $a = 383397772.5$ m is the value of the semimajor axis of the lunar orbit, which was adopted in the construction of the expansions in Equations 13.45–13.49. In the expression (Equation 13.54) we have introduced the inclination functions of the angle ρ:

$$B_v = -\frac{1}{6}\left(3\cos^2\rho - 1\right)A_v^{(0)} - \frac{1}{2}\sin 2\rho A_v^{(1)} - \frac{1}{4}\sin^2\rho A_v^{(2)},$$

$$C_v^{(\varepsilon)} = -\frac{1}{4}\sin 2\rho A_v^{(0)} + \frac{1}{2}\left(1+\varepsilon\cos\rho\right)\left(-1+2\varepsilon\cos\rho\right)A_v^{(1)} + \frac{1}{4}\varepsilon\sin\rho\left(1+\varepsilon\cos\rho\right)A_v^{(2)}$$

$$D_v^{(\varepsilon)} = -\frac{1}{2}\sin^2\rho A_v^{(0)} + \varepsilon\sin\rho\left(1+\varepsilon\cos\rho\right)A_v^{(1)} - \frac{1}{4}\left(1+\varepsilon\cos\rho\right)^2 A_v^{(2)}.$$

$$(13.55)$$

where $v = (v_1, v_2, v_3, v_4, v_5)$, $\varepsilon = \pm 1$. The functions of inclination and the expansion of the force function were applied in the construction of the theory of rotation of the rigid the Moon (Barkin 1987).

The expansion coefficients of the factors include the second harmonic of the gravitational potential of the Moon $J_2 = -C_{20}$ and C_{22}. For these factors, the mean radius r_0 and dimensionless moment of inertia of the Moon $I = C / m_0 r_0^2$ as reference, we use values in accordance with the Moon's gravitational field model constructed by Matsumoto et al. (2010)

$$C_{20} = -203.4495 \cdot 10^{-6}, \ C_{22} = 22.3722 \cdot 10^{-6}, \ r_0 = 1738.0 \text{ km}, \ I = 0.39348, \quad (13.56)$$

The highly accurate specified values $J_2 = -C_{20}$ and C_{22} correspond to the principal central axes of inertia of the the Moon.

13.3.3 Hamiltonian of the Problem in the Poincaré's Variables

These equations have the canonical form (Equation 13.36 and Equations 13.40–13.42):

$$\frac{d\left(\lambda, \lambda_c, h, h_c, \xi, \xi_c\right)}{dt} = \frac{\partial K}{\partial\left(\Lambda, \Lambda_c, H, H_c, \eta, \eta_c\right)}$$

$$\frac{d\left(\Lambda, \Lambda_c, H, H_c, \eta, \eta_c\right)}{dt} = -\frac{\partial K}{\partial\left(\lambda, \lambda_c, h, h_c, \xi, \xi_c\right)}. \quad (13.57)$$

The Hamiltonian of the problem $K = T - U$ should be written in the canonical variables. After we make the change of variables in the expression of the kinetic energy, we get

$$T = \frac{1}{2}\Lambda^2\left\{\frac{A_c}{\Delta_1}\frac{\eta^2}{\Lambda}\left(1-\frac{\xi^2+\eta^2}{4\Lambda}\right)+\frac{B_c}{\Delta_2}\frac{\xi^2}{\Lambda}\left(1-\frac{\xi^2+\eta^2}{4\Lambda}\right)+\frac{C_c}{\Delta_3}\left[1-2\frac{\xi^2+\eta^2}{2\Lambda}+\frac{(\xi^2+\eta^2)^2}{4\Lambda^2}\right]\right.$$

$$+\frac{1}{2}\Lambda_c^2\left[\frac{A}{\Delta_1}\frac{\eta_c^2}{\Lambda_c}\left(1-\frac{\xi_c^2+\eta_c^2}{4\Lambda_c}\right)+\frac{B}{\Delta_2}\frac{\xi_c^2}{\Lambda_c}\left(1-\frac{\xi_c^2+\eta_c^2}{4\Lambda_c}\right)+\frac{C}{\Delta_3}\left[1-2\frac{\xi_c^2+\eta_c^2}{2\Lambda_c}+\frac{(\xi_c^2+\eta_c^2)^2}{4\Lambda_c^2}\right]\right]$$

$$+\sqrt{\Lambda\Lambda_c}\sqrt{1-\frac{\xi^2+\eta^2}{4\Lambda}}\sqrt{1-\frac{\xi_c^2+\eta_c^2}{4\Lambda_c}}\left(-\frac{F_c}{\Delta_1}\eta_c+\frac{E_c}{\Delta_2}\xi\xi_c\right)+\Lambda\Lambda_c\frac{D_c}{\Delta_3}\left(1-\frac{\xi^2+\eta^2}{2\Lambda}\right)\left(-1+\frac{\xi_c^2+\eta_c^2}{2\Lambda_c}\right)\right\}$$

(13.58)

A similar expression can be written for the force function of the problem based on the above expression for the second harmonic of the force function of the non-spherical the Moon and the change of variables:

$$U_2 = -mr_0^2 n_0^2 J_2 \left\{\frac{3}{2}\left[2-3\frac{\xi^2+\eta^2}{\Lambda}\left(1-\frac{\xi^2+\eta^2}{4\Lambda}\right)\right]\sum_{|\nu|}B_\nu\cos\Theta_\nu\right.$$ (13.59)

$$-3\frac{\xi}{\sqrt{\Lambda}}\left(1-\frac{\xi^2+\eta^2}{2\Lambda}\right)\sqrt{1-\frac{\xi^2+\eta^2}{4\Lambda}}\sum_{\varepsilon=\pm1}\sum_{|\nu|}C_\nu^{(\varepsilon)}\cos(\lambda-\varepsilon\Theta_\nu)$$

$$+3\frac{\eta}{\sqrt{\Lambda}}\left(1-\frac{\xi^2+\eta^2}{2\Lambda}\right)\sqrt{1-\frac{\xi^2+\eta^2}{4\Lambda}}\sum_{\varepsilon=\pm1}\sum_{|\nu|}C_\nu^{(\varepsilon)}\sin(\lambda-\varepsilon\Theta_\nu)$$

$$+\frac{3}{4\Lambda}\left(1-\frac{\xi^2+\eta^2}{4\Lambda}\right)(\xi^2-\eta^2)\sum_{\varepsilon=\pm1}\sum_{|\nu|}D_\nu^{(\varepsilon)}\cos(2\lambda-\varepsilon\Theta_\nu)$$

$$\left.-\frac{3}{2}\frac{\xi\eta}{\Lambda}\left(1-\frac{\xi^2+\eta^2}{4\Lambda}\right)\sum_{\varepsilon=\pm1}\sum_{|\nu|}D_\nu^{(\varepsilon)}\sin(2\lambda-\varepsilon\Theta_\nu)\right\}$$

$$+mr_0^2 n_0^2 C_{22}\left\{-\frac{9}{\Lambda}\left(1-\frac{\xi^2+\eta^2}{4\Lambda}\right)(\xi^2-\eta^2)\sum_{|\nu|}B_\nu\cos\Theta_\nu\right.$$

$$-3\frac{\xi}{\sqrt{\Lambda}}\left(2-\frac{\xi^2+\eta^2}{2\Lambda}\right)\sqrt{1-\frac{\xi^2+\eta^2}{4\Lambda}}\sum_{\varepsilon=\pm1}\sum_{\|v\|}C_v^{(\varepsilon)}\cos\left(\lambda-\varepsilon\,\Theta_v\right)$$

$$-3\frac{\eta}{\sqrt{\Lambda}}\left(2-\frac{\xi^2+\eta^2}{2\Lambda}\right)\sqrt{1-\frac{\xi^2+\eta^2}{4\Lambda}}\sum_{\varepsilon=\pm1}\sum_{\|v\|}C_v^{(\varepsilon)}\sin\left(\lambda-\varepsilon\,\Theta_v\right)$$

$$+\frac{3\xi}{2\Lambda\sqrt{\Lambda}}\sqrt{1-\frac{\xi^2+\eta^2}{4\Lambda}}\left(\xi^2-3\eta^2\right)\sum_{\varepsilon=\pm1}\sum_{\|v\|}C_v^{(\varepsilon)}\cos\left(\lambda-\varepsilon\,\Theta_v\right)$$

$$+\frac{3\eta}{2\Lambda\sqrt{\Lambda}}\sqrt{1-\frac{\xi^2+\eta^2}{4\Lambda}}\left(\eta^2-3\xi^2\right)\sum_{\varepsilon=\pm1}\sum_{\|v\|}C_v^{(\varepsilon)}\sin\left(\lambda-\varepsilon\,\Theta_v\right)$$

$$-\frac{3}{4}\left(2-\frac{\xi^2+\eta^2}{2\Lambda}\right)^2\sum_{\varepsilon=\pm1}\sum_{\|v\|}D_v^{(\varepsilon)}\cos\left(2\lambda-\varepsilon\,\Theta_v\right)$$

$$-\frac{3}{16\Lambda^2}\left(\xi^4-6\xi^2\eta^2+\eta^4\right)\sum_{\varepsilon=\pm1}\sum_{\|v\|}D_v^{(\varepsilon)}\cos\left(2\lambda-\varepsilon\,\Theta_v\right)$$

$$\left.+\frac{3}{4\Lambda^2}\xi\eta\left(\xi^2-\eta^2\right)\sum_{\varepsilon=\pm1}\sum_{\|v\|}D_v^{(\varepsilon)}\sin\left(2\lambda-\varepsilon\,\Theta_v\right)\right\}.$$

We make now an important assumption of the canonical variables ξ, η and ξ_c, η_c. For these motions of the Moon, the angles formed by vectors of the angular momentums $\mathbf{G}$ and $\mathbf{G}_c$ with the polar axis of lunar inertia $C\zeta$ are small. After simple transformations, we get the basic terms in the expansion of Equation 13.58 with respect to variables ξ, η, and ξ_c, η_c. The zero-order and second-order terms with respect to these variables in terms of the kinetic energy can be written as:

$$T_0=\frac{1}{2\Delta_3}\left(C_c\Lambda^2+C\Lambda_c^2-2D_c\Lambda\Lambda_c\right),\tag{13.60}$$

$$T_2=\frac{1}{2}P\xi^2+\frac{1}{2}Q\eta^2+\frac{1}{2}P_c\xi_c^2+\frac{1}{2}Q_c\eta_c^2+R\xi\xi_c-T\eta\eta_c,\tag{13.61}$$

where we have used the new notations:

$$P = \Lambda\left(\frac{B_c}{\Delta_2} - \frac{C_c}{\Delta_3}\right) + \Lambda_c \frac{D_c}{\Delta_3}, \; Q = \Lambda\left(\frac{A_c}{\Delta_1} - \frac{C_c}{\Delta_3}\right) + \Lambda_c \frac{D_c}{\Delta_3}, \; R = \sqrt{\Lambda\Lambda_c}\frac{E_c}{\Delta_2}$$

$$P_c = \Lambda_c\left(\frac{B}{\Delta_2} - \frac{C}{\Delta_3}\right) + \Lambda\frac{D_c}{\Delta_3}, \; Q_c = \Lambda_c\left(\frac{A}{\Delta_1} - \frac{C}{\Delta_3}\right) + \Lambda\frac{D_c}{\Delta_3}, \; T = \sqrt{\Lambda\Lambda_c}\frac{F_c}{\Delta_1}$$

$$\Delta_1 = A_c A - F_c^2, \; \Delta_2 = B_c B - E_c^2, \; \Delta_3 = C_c C - D_c^2. \tag{13.62}$$

In terms of the force function (Equation 13.59) we choose the zero-, first- and second-order (relatively to the variables ξ,η):

$$U_2 = V_0 + V_{10}\xi + V_{01}\eta + \frac{1}{2}\left(V_{20}\xi^2 + V_{02}\eta^2 + 2V_{11}\xi\eta\right), \tag{13.63}$$

where

$$V_0 = -3mr_0^2 n_0^2 \left[J_2 \sum_{\mathbf{v};v_5} B_{\mathbf{v};v_5} \cos\Theta_{\mathbf{v};v_5} + C_{22} \sum_{\varepsilon=\pm1} \sum_{\mathbf{v};v_6} D_{\mathbf{v};v_5}^{(\varepsilon)} \cos\left(2\lambda - \varepsilon\Theta_{\mathbf{v};v_5}\right)\right], \tag{13.64}$$

$$V_{10} = \frac{3}{\sqrt{\Lambda}}mr_0^2 n_0^2 \left(J_2 - 2C_{22}\right)\sum_{\mathbf{v}}\sum_{\varepsilon=\pm1}\sum_{v_5} C_{\mathbf{v};v_5}^{(\varepsilon)} \cos\left(\Theta_{\mathbf{v};v_5} - \varepsilon\lambda\right),$$

$$V_{01} = \frac{3}{\sqrt{\Lambda}}mr_0^2 n_0^2 \left(J_2 + 2C_{22}\right)\sum_{\mathbf{v}}\sum_{\varepsilon=\pm1}\sum_{v_5} \varepsilon C_{\mathbf{v};v_5}^{(\varepsilon)} \sin\left(\Theta_{\mathbf{v};v_5} - \varepsilon\lambda\right). \tag{13.65}$$

$$V_{20} = \frac{3}{2\Lambda}mr_0^2 n_0^2 \left(J_2 - 2C_{22}\right)\sum_{\mathbf{v};v_5}\left[6B_{\mathbf{v};v_5} \cos\Theta_{\mathbf{v};v_5} - \sum_{\varepsilon=\pm1} D_{\mathbf{v};v_5}^{(\varepsilon)} \cos\left(2\lambda - \varepsilon\Theta_{\mathbf{v};v_5}\right)\right],$$

$$V_{02} = \frac{3}{2\Lambda}mr_0^2 n_0^2 \left(J_2 + 2C_{22}\right)\sum_{\mathbf{v};v_5}\left[6B_{\mathbf{v};v_5} \cos\Theta_{\mathbf{v};v_5} + \sum_{\varepsilon=\pm1} D_{\mathbf{v};v_5}^{(\varepsilon)} \cos\left(2\lambda - \varepsilon\Theta_{\mathbf{v};v_5}\right)\right],$$

$$V_{11} = -\frac{3}{2\Lambda}mr_0^2 n_0^2 J_2 \sum_{\varepsilon=\pm1}\sum_{\|\mathbf{v}\|} \varepsilon D_{\mathbf{v};v_5}^{(\varepsilon)} \sin\left(\Theta_{\mathbf{v};v_5} - 2\varepsilon\lambda\right). \tag{13.66}$$

All six functions depend on four Andoyer's variables Λ, H and λ, h. The arguments of the trigonometric functions are defined as

$$\Theta_\nu = \nu_1 l_M + \nu_2 l_S + \nu_3 F + \nu_4 D - \nu_5 h, \ \mathbf{v} = (\nu_1, \nu_2, \nu_3, \nu_4, \nu_5)^T, \ \nu_1 \geq 0, \lambda = F + \lambda_0,$$

$$(13.67)$$

and the summation in Equations 13.64–13.66 is carried out over all five indices $\nu1, \nu2, \dots, \nu5$,

$$\sum_{\mathbf{v};\nu_5} = \sum_{\nu_1} \sum_{\nu_2} \sum_{\nu_3} \sum_{\nu_4} \sum_{\nu_5} \text{ and the values of } \varepsilon = \pm 1. \qquad (13.68)$$

The functions of the inclination angle $\rho.B_{\mathbf{v};\nu_5} \ C_{\mathbf{v};\nu_5}^{(\varepsilon)}$ and $D_{\mathbf{v};\nu_5}^{(\varepsilon)}$ are given by Equation 13.55.

13.3.4 Structure of the General Solution of the Lunar Physical Libration with the Ellipsoidal Liquid Core

Differential equations of lunar rotational motion with the ellipsoidal liquid core (assuming a simple fluid motion on the Poincaré) are reduced to a system of differential equations of order 8 (corresponding to four degrees of freedom). The procedure for constructing an approximate solution of this problem was carried out by an analytical method of constructing a theory of rotational motion of the Moon as a solid body (Barkin 1987), but with some additions and modifications associated with higher-order equations and the additional terms in the unperturbed Hamiltonian. Due to limited space, we do not give full details on the construction of the analytical theory of the rotational motion of the Moon with an ellipsoidal liquid core. Therefore, here we will give some short review of the main stages of the theory. We will focus on the analysis of free (resonant) librations of the Moon in the first place due to the hydrodynamic effect of the liquid core.

Pre-point to the overall structure of the solution of the physical libration of the Moon:

$$\mathbf{Z}(t) = \mathbf{Z}_0 + \langle \mathbf{Z} \rangle + \tilde{\mathbf{Z}}(l_M, l_S, F, D, \dots) + \mathbf{Z}_{res}(P, Q, R, S, U_p, U_q, U_r, U_s, l_M, l_S, F, D, \dots, t)$$

$$(13.69)$$

Here we have used vector notation for the variables used in the analytical theory of the Moon's rotation:

$$\mathbf{Z}(t) = (\mathbf{Z}_A, \mathbf{Z}_P, \mathbf{Z}_\omega, \mathbf{Z}_{\omega c}, \mathbf{Z}_{cl})$$

We use a wide list of variables for studying the Moon rotation: $\mathbf{Z}_A = (L, G, H, l, g, h; p, \theta)$ are Andoyer's variables, $\mathbf{Z}_p = (\Lambda, \Lambda_c, H, H_c, \eta, \eta_c; \lambda, \lambda_c, h, h_c, \xi, \xi_c)$ are Andoyer's and Poincaré's variables; $\mathbf{Z}_\omega = (p, q, r, \omega, LOD)$ are the components of the angular velocity of the Moon (along its principal axes of inertia) and LOD is the duration of the lunar day; $\mathbf{Z}_\omega = (p_c, q_c, r_c, \omega_c, LOD_c)$ are components of the relative angular velocity of rotation of the lunar core LOD_c (of the Poincaré's reference frame relative to the principal axes of the Moon), LOD_c is the duration of the lunar core axial rotation; $\mathbf{Z}_{cl} = (P_1, P_2, \tau, \rho, I\sigma)$ are the classical variables of the theory of physical libration of the Moon and some others dynamical characteristics. l_M, l_S, F, D are classical variables (arguments) of the orbital motion of the Moon (Sagitov 1979, Kudryavtsev 2007). The geometric and dynamic meaning of all the above variables is well known (Barkin et al. 2012).

The basic equations of the Moon motion are written in canonical form, but in general and the nonlinear neighborhood is developed. These methods were originally developed for the rigid model of the Moon (Barkin 1987), and in this chapter we develop the more general model of a celestial body with an ellipsoidal liquid core. It uses a small parameter method, which is introduced by the assumption of the small dynamic compressions of the Moon and by the assumption about nearly concentric distribution of its density.

The first three terms of the solution in Equation 13.69 describe the intermediate quasi-periodic solutions and do not actually contain the initial conditions of the problem. Here the solution of the problem is based primarily on Andoyer's variables on integer powers of a small parameter. In Equation 13.69, the $\mathbf{Z}_0$ is a base and main solution, describing the rotational motion on the Cassini's laws. This solution is determined as a result of analyzing the terms of Hamiltonian of the problem of the first order relative to the oblatenesses of the Moon—terms of the second harmonic of the force function of the gravitational interaction of the nonspherical the Moon and Earth. Here the main regularities of the Moon's motion (by Cassini's laws) have been fully explained. The main result here is a determination of the Cassini angle ρ, which is the angle between the normal to the plane of the ecliptic and the axis of rotation of the Moon. The value of this angle depends on the basic parameters of gravitational potential of the the Moon.

The more difficult problem is the construction of analytical expressions for the constant components of variables $\langle \mathbf{Z} \rangle$. These constant additions to the generating values of the Andoyer (and others variables) $\mathbf{Z}_0$ are determined by the analysis of second and higher-order approximations, which are sequentially addressed in the construction of conditionally periodic solution in which $\check{\mathbf{Z}}$ (l_M, l_S, F, D,...) is a purely quasiperiodic component. It is the sum of trigonometric terms with arguments that are linear combinations of the classical arguments of the theory of orbital motion of the Moon.

The fourth term of the solution in Equation 13.69 is

$$\mathbf{Z}_{res}(P,Q,R,S,U_p,U_q,U_r,U_s,l_M,l_S,F,D,....,t) \tag{13.70}$$

which describes free and resonant libration of the Moon with an ellipsoidal liquid core. Here P,Q,R,S are constant amplitudes and U_p, U_q, U_r, U_s are arguments of free libration of the Moon, which are linear functions of time with frequencies n_p, n_q, n_r and n_s:

$$U_p = n_p \cdot t + U_p^{(0)}, \, U_q = n_q \cdot t + U_q^{(0)}, \, U_r = n_r \cdot t + U_r^{(0)}, \, U_s = n_s \cdot t + U_s^{(0)} \qquad (13.71)$$

$U_p^{(0)}$, $U_r^{(0)}$, $U_r^{(0)}$ and $U_s^{(0)}$ are initial values of corresponding arguments (phases of these arguments for given epoch 2000.0). Thus, the solution of the problem of the physical libration of the Moon (69) (for the given two-layer Poincaré model) contains eight initial conditions:

$$P, Q, R, S; U_p^{(0)}, U_q^{(0)}, U_r^{(0)}, U_s^{(0)}. \qquad (13.72)$$

They represent the amplitude and phase of the free libration in longitude $(P, U_p^{(0)})$, inclination $(R, U_r^{(0)})$, pole motion $(Q, U_q^{(0)})$, and free core nutation $(S, U_s^{(0)})$. The first six initial conditions of the problem of physical libration have been firstly determined from lunar laser ranging observations by Calame (1976). More precisely, these six initial conditions were identified in the current study (Rambaux and Williams 2011). Despite the fact that a number of new librations were observed with certain periods but unknown nature, they were not able to determine the parameters $(S, U_s^{(0)})$. These unidentified librations of the empirical theory (Rambaux and Williams 2011) contain about 50 terms in expressions of classical variables $Z = (P_1, P_2, \tau, \rho, I\sigma)$. In this chapter, we first identify a fourth mode of the free libration of the Moon in the solution and then identify the period of free libration and determine its amplitude and initial phase (Barkin et al. 2012).

Because the resonant character of the Moon has a solution of the equations in variations of the problem Z_{res} $(P, Q, R, S, U_p, U_q, U_r, U_s, l_M, l_s, F, D, \ldots, t)$ is constructed as a series of integer and fractional powers of a small parameter (in $\sqrt{\mu}$). A detailed description of the analytical solution in Equation 13.69 will be given in subsequent papers. Here we focus on the description of the effects of the free motion on the pole axis of lunar rotation, caused by its ellipsoidal liquid core.

13.4 Motion of the Moon with Liquid Core under Cassini's Laws

13.4.1 Unperturbed Rotational Motion of the Moon with Liquid Core

The analytic theory of the Moon's rotation in this chapter is based on the specific methods of construction of quasi-periodic solutions of Hamiltonian systems with resonances developed earlier in analytical theory of the Moon's rotation for a rigid model by Barkin (1987, 1989). These methods were proposed in the event of a natural development of a more general model of the Moon with the liquid core. A generalization of the method is related primarily to the study of the perturbed motion of the poles of the Moon and the axes of rotation of the liquid core. But main properties of the lunar rotation (main regularities) for the solid model and the model of the Moon with the same liquid core coincide for both theories. Later, the new provisions on the nature of the unperturbed motion of the core in accordance with Cassini's laws of motion here were added (Barkin and Ferrándiz 2004).

13.4.2 Cassini's Laws as Conditions of Existence of the Conditionally Periodic Intermediate Solution of the Problem

The motion of the Moon with a molten core under Cassini's laws is determined from conditions of existence of conditionally periodic solutions of the lunar rotational motion equations (from the first approximation for the variables Λ, λ and ρ, h). The motion of the liquid core in the Moon's motion by Cassini's laws, in turn, is determined by the analysis of the first approximation to the Poincaré's variables Λ_c, λ_c, ξ, η and ξ_c, η_c The above equations give the necessary conditions for the existence of the resonant motion of the Moon by Cassini's laws:

$$\Lambda_0 = Cn_F, \frac{\partial \langle V_0 \rangle}{\partial h_0} = 0, \frac{\partial \langle V_0 \rangle}{\partial \lambda_0} = 0, -n_\Omega + \frac{1}{A \sin \rho} \frac{\partial \langle V_0 \rangle}{\partial \rho} = 0, \quad (13.73)$$

where $\langle V_0 \rangle$ is an average resonance value of the perturbing function V_0 (13.64), defined by the formula

$$\langle V_0 \rangle = -3mR^2n_0^2 \left[J_2 \sum_{v_5} B_{0,0,0,0,v_5} \cos(v_5 h_0) + C_{22} \sum_{\varepsilon=\pm 1} \sum_{v_5} D_{0,0,2\varepsilon,0,v_5}^{(\varepsilon)} \cos(v_5 h_0 + 2\varepsilon\lambda_0) \right]$$

$$(13.74)$$

where h_0, λ_0, $\Lambda = \Lambda_0$ and $\rho = \rho_0$ stand for the generating (initial) values of the corresponding variables. The first three equations of system 13.73–13.74 are easily solved, and allow us to determine the values of:

$$\Lambda_0 = Cn_F, h_0 = \pi, \lambda_0 = 0. \quad (13.75)$$

13.4.3 Determination of Cassini's Angle

The fourth equation in Equations 13.73 and 13.61 is one of the necessary conditions for the existence of conditionally periodic rotational motion of the

Moon in the neighborhood of the unperturbed motion. It can be converted to the form:

$$B\cos 2\rho + A\sin 2\rho + Y\cos\rho + U\sin\rho = 0 \qquad (13.76)$$

where

$$B = -J_2 \sum_{v_5} (-1)^{v_5} A^{(1)}_{0,0,0,0,v_5} + C_{22} \sum_{\varepsilon=\pm} \sum_{v_5} (-1)^{v_5} A^{(1)}_{0,0,2\varepsilon,0,v_5},$$

$$A = \frac{1}{4} J_2 \sum_{v_5} (-1)^{v_5} \left(2A^{(0)}_{0,0,0,0,v_5} - A^{(2)}_{0,0,0,0,v_5}\right) - \frac{1}{4} C_{22} \sum_{\varepsilon=\pm} \sum_{v_5} (-1)^{v_5} \left(2A^{(0)}_{0,0,2\varepsilon,0,v_5} - A^{(2)}_{0,0,2\varepsilon,0,v_5}\right),$$

$$Y = -C_{22} \sum_{\varepsilon=\pm} \sum_{v_5} (-1)^{v_5} \varepsilon B^{(1)}_{0,0,2\varepsilon,0,v_5}, \quad U = N + \frac{1}{2} C_{22} \sum_{\varepsilon=\pm} \sum_{v_5} (-1)^{v_5} \varepsilon B^{(2)}_{0,0,2\varepsilon,0,v_5}, \qquad (13.77)$$

and

$$N = \frac{n_\Omega \Lambda}{3mR^2 n_0^2} = I \frac{n_\Omega n_F}{3n_0^2} < 0, n_0^2 = \frac{fm_E}{a^3} \qquad (13.78)$$

We take the value of the fundamental constant $fm_E = 398\,600.5 \times 10^9$ m^3/s^2. A value of semimajor axis $a = 383397772.5$ m was used by us in the construction of Poisson series (Equations 13.45–13.49). For the calculation of the frequency n_0 from Equation 13.78 and frequencies of the orbital motion of the Moon n_Ω, n_F, we take

$$n_0 = 17311058''6854464 \ 1/\text{yr}$$

$$n_\Omega = -69679''193631 \ 1/\text{yr}, \qquad (13.79)$$

$$n_F = 17395272''628478 \ 1/\text{yr},$$

with the value of the coefficient $N = -53.050380896 \times 10^{-5}$.

In Equations 13.76–13.78 there are 22 orbital coefficient of expansions (Equations 13.45–13.49) shown in Table 13.2, depending on the coefficients that are determined and a Cassini's angle that is $\rho = \rho_0$.

For these values of the coefficients in Table 13.2, we find the numerical values of the fundamental equation in the theory of physical libration of the Moon:

$$B = 1.00459140 \cdot 10^5, A = 5.53048377 \cdot 10^5,$$

$$Y = 0.0987623006 \cdot 10^5, U = -51.9537976 \cdot 10^5. \qquad (13.80)$$

TABLE 13.2

Coefficients of the Orbital Motion of the Moon in the Equation for Determining of the Angle of Cassini $\rho = \rho_0$

$A^{(0)}_{0,0,0,0,0} = 0.4924353$	$A^{(0)}_{0,0,2,0,0} = 0.0059115$	$A^{(0)}_{0,0,0,0,-1} = 0.0000055$	$A^{(0)}_{0,0,2,0,-1} = -0.0000002$
			$A^{(0)}_{0,0,2,0,1} = -0.0000053$
$A^{(1)}_{0,0,0,0,0} = -0.0000205$	$A^{(1)}_{0,0,2,0,0} = 0.0000014$	$A^{(1)}_{0,0,2,0,2} = 0.0000188$	$A^{(1)}_{0,0,2,0,1} = -0.0440367$
			$A^{(1)}_{0,0,2,0,-1} = 0.0000879$
$A^{(1)}_{0,0,0,0,-1} = 0.0445224$	$B^{(1)}_{0,0,2,0,0} = -0.0000016$	$B^{(1)}_{0,0,2,0,2} = -0.0000188$	$B^{(1)}_{0,0,2,0,1} = 0.0440368$
			$B^{(1)}_{0,0,2,0,-1} = 0.0000879$
$A^{(2)}_{0,0,0,0,-2} = 0.0040117$	$A^{(2)}_{0,0,0,0,-1} = -0.0000037$		
$B^{(2)}_{0,0,2,0,2} = 0.9803163$	$B^{(2)}_{0,0,2,0,-2} = -0.0000039$	$B^{(2)}_{0,0,2,0,1} = -0.0000320$	$B^{(2)}_{0,0,2,0,3} = 0.0000354$
			$B^{(2)}_{0,0,2,0,4} = -0.0000001$

By solving the Equations 13.76–13.79, we obtain a generating value of Cassini's angle of inclination in the plane of the ecliptic of date

$$\rho = 1^0543871 \,(S) = 1^032'37''9356 = 5557''9356 \tag{13.81}$$

This value is in good agreement with the observed average value of the inclination angle of the rotation axis of the Moon relative to the normal to the plane of the ecliptic $I = 5553''60965$ (Rambaux and Williams 2011).

13.4.4 Cassini's Laws of the Moon Rotation

The generating periodic solution for conditionally periodic solution corresponding to the Moon's perturbed rotation is determined by Equations 13.75 and 13.81 and describes the following fundamental regularities in the Moon motion:

1. The Moon and its liquid core rotate as one rigid body with constant angular velocity synchronously with orbital motion.

2. Within a mean period of time (equal to the draconic month $T_{\text{draconic}} = 27.22$ days) between two consecutive passages of the center of mass of the Moon through a mean ascending node of its orbit on the ecliptic of date, the Moon completes one revolution about its polar axis of inertia $C_M \bar{\eta}$ in the coordinate reference system $C_M XYZ$ connected with the mean moveable node of the lunar orbit.

3. The axis $C_M \bar{\eta}$ corresponds to the maximal moment of inertia B, and when the center of mass crosses the mean orbit node of the Moon,

another axis of inertia $C_M \overline{\xi}$, corresponding to the minimal moment of inertia A, is directed along the line of nodes.

4. In Cassini's motion the angular momentum vector of the Moon rotation **G** and its angular velocity vector **ω** coincide with the polar inertia axis of the Moon $C_M \overline{\eta}$;

5. Vector **G** forms a constant angle ρ_0 with the normal vector to the ecliptic plane of date, which describes in the space a cone with half-angle $\rho_0 = 1^05439$ with the time period $T_\Omega = 2\pi / |n_\Omega| = 18.5995$ years.

6. The orbit plane, the intermediate plane Q_G normal to the angular momentum vector **G**, and the equatorial plane of the Moon (plane $C_M \overline{\xi}\zeta$ in our notations) have the same line of nodes on the ecliptic of date NN', with the longitude of the general descending nodes of the planes Q_G and $C_M \overline{\xi}\zeta$ being equal to the mean longitude of the ascending node of the orbit plane of the center of mass of the Moon.

7. The polar inertia axis of the Moon $C_M \overline{\eta}$, the vector of normal to the ecliptic plane of date n_E, the angular momentum vector of the Moon rotation **G** and the angular velocity vector of the Moon rotation **ω** lie in the same plane, orthogonal to the line of nodes NN' on the ecliptic of date.

8. The value of the angle between the normal to ecliptic plane of date n_E and vectors **G**, **ω** coinciding with the Moon axis $C_M \overline{\eta}$ is obtained from Equation 13.14 and 13.15 depending on the dynamic parameters J_2 and C_{22} (oblatenesses of the Moon), from the parameter of mobility of the Moon's orbital plane F (its precession), and on corresponding parameters $A_\nu^{(j)}$ characterizing the perturbed lunar orbital motion (numerical value of the parameter $\rho_0 = 1^032'37''94$).

The regularities mentioned, 1–7 in particular, thoroughly explain all statements of classical empirical Cassini's laws, and contain some additions and specifications related to the orientation of the vectors **G** and **ω**, of the lunar inertia ellipsoid orientation and stability of Cassini's motions. The last aspect of the problem and general dynamics of the the Moon were not considered by Cassini.

13.5 Forced Physical Libration of the Moon in Andoyer's and Poincaré's Variables

13.5.1 Perturbations of First Order of the Andoyer–Poincaré Variables

We take the unperturbed motion—the motion of the Moon by Cassini's laws, as described above. We ignore the intermediate steps and give the final formulas for the first-order perturbations of the canonical variables

Equation 13.42 and variable ρ. It is about pure quasiperiodic components of these variables, which is denoted as $\tilde{z}_1$. Perturbations of the first order (forced libration of the Moon) for the six variables Λ, λ, Λ_c, λ_c, ρ, h are determined by the simple quadratures:

$$\tilde{\Lambda}^{(1)} = \int \frac{\partial \tilde{V}_0}{\partial \lambda_0} dt, \tilde{\lambda}^{(1)} = \frac{C_c}{\Delta_3} \int \tilde{\Lambda}^{(1)} dt - \frac{\cos\rho}{\Lambda \sin\rho} \int \frac{\partial V_0}{\partial \rho} dt,$$

$$\tilde{\Lambda}_c^{(1)} = 0, \tilde{\lambda}_c^{(1)} = -\frac{D_c}{\Delta_3} \int \tilde{\Lambda}^{(1)} dt,$$

$$\tilde{\rho}^{(1)} = -\frac{1}{\Lambda \sin\rho} \int \frac{\partial \tilde{V}_0}{\partial h} dt + \frac{\cos\rho}{\Lambda \sin\rho} \int \frac{\partial \tilde{V}_0}{\partial \lambda} dt, \tilde{h}^{(1)} = \frac{1}{\Lambda \sin\rho} \int \frac{\partial \tilde{V}_0}{\partial \rho} dt. \quad (13.82)$$

In the right-hand side of Equation 13.82 and in their integrands of all Andoyer–Poincaré variables, they take generating values in accordance with the motion by Cassinis's laws. The integrals in Equation 13.82 can be easily calculated, and the perturbations of the first order are represented by trigonometric functions with respect to a certain symmetric structure of sine and cosines of a wide range of arguments. Note that the perturbations are determined by the values of the partial derivatives of a single (primary) function V_0 of the overall presentation of the strength function (Equation 13.64). Here we just present the final expression for the first-order perturbation of the considering six variables:

$$\tilde{\Lambda}^{(1)} = \Lambda_0 \sum_{|v|>0} \tilde{\Lambda}_v^{(1)} \cos\theta_v, \tilde{\Lambda}_v^{(1)} = 6\frac{C_{22}}{I} F_v \sum_{\varepsilon=\pm1} \sum_{v5} \varepsilon(-1)^{v5} D_{v1,v2,v3+2\varepsilon,v4,v5}^{(\varepsilon)},$$

$$(13.83)$$

$$\tilde{\lambda}^{(1)} = \sum_{|v|>0} \lambda_v^{(1)} \sin\theta_v,$$

$$\lambda_v^{(1)} = \frac{CC_c}{\Delta_3} F_v \tilde{\Lambda}_v^{(1)} + 3\frac{\cos\rho}{I \sin\rho} F_v \left[J_2 \sum_{v5} (-1)^{v5} \frac{\partial B_{v1,v2,v3+2\varepsilon,v4,v5}}{\partial \rho} \right.$$

$$\left. + C_{22} \sum_{\varepsilon=\pm1} \sum_{v5} (-1)^{v5} \frac{\partial D_{v1,v2,v3+2\varepsilon,v4,v5}^{(\varepsilon)}}{\partial \rho} \right]$$

$$\tilde{\Lambda}_c^{(1)} = 0,$$

$$\tilde{\lambda}_c^{(1)} = \sum_{|v|>0} \lambda_{cv}^{(1)} \sin\theta_v, \lambda_{cv}^{(1)} = -\frac{CD_c}{\Delta_3} \cdot \frac{n_F}{\omega_v} \tilde{\Lambda}_v^{(1)},$$

$$\tilde{\rho}^{(1)} = \sum_{|\mathbf{v}|>0} \rho_{\mathbf{v}}^{(1)} \cos\theta_{\mathbf{v}},$$

$$\rho_{\mathbf{v}}^{(1)} = \frac{3}{I\sin\rho} F_{v_1,v_2,v_3,v_4} \left[-\sum_{v_5} v_5 (-1)^{v_5} J_2 B_{v_1,v_2,v_3,v_4;v_5} \right.$$

$$\left. + \sum_{v_5}\sum_{\varepsilon=\pm1} C_{22}\varepsilon \left(2\cos\rho - v_5\varepsilon\right)(-1)^{v_5} D_{v_1,v_2,v_3+2\varepsilon,v_4;v_5}^{(\varepsilon)} \right],$$

$$\tilde{h}^{(1)} = \sum_{|\mathbf{v}|>0} h_{\mathbf{v}}^{(1)} \sin\theta_{\mathbf{v}},$$

$$\tilde{h}_{\mathbf{v}}^{(1)} = -3\frac{1}{I\sin\rho} F_{v_1,v_2,v_3,v_4} \left(J_2 \sum_{v_5} (-1)^{v_5} \frac{\partial B_{v_1,v_2,v_3,v_4,v_5}}{\partial\rho} \right.$$

$$\left. + C_{22} \sum_{v_5}\sum_{\varepsilon=\pm1} (-1)^{v_5} \frac{\partial D_{v_1,v_2,v_3+2\varepsilon,v_4,v_5}^{(\varepsilon)}}{\partial\rho} \right).$$

For convenience, the formulas contain the dimensionless coefficients

$$F_{\mathbf{v}} = F_{v_1,v_2,v_3,v_4} = \frac{n_0^2}{n_F \left(v_1 n_M + v_2 n_S + v_3 n_F + v_4 n_D\right)}. \tag{13.84}$$

Unperturbed values of variables ξ, η and ξ_c, η_c, are assumed to be zero, corresponding to the axial synchronous rotation of the Moon and its core with a constant angular velocity n_F. Calculation of first-order perturbations for the canonical variables ξ, η and ξ_c, η_c is a more difficult task. It implies solving the following system of linear inhomogeneous differential equations:

$$\frac{d\xi^{(1)}}{dt} = -Q\eta^{(1)} + T\eta_c^{(1)} + \bar{V}_{01} + \tilde{V}_{01},$$

$$\frac{d\eta^{(1)}}{dt} = P\xi^{(1)} + R\xi_c^{(1)} - \bar{V}_{10} - \tilde{V}_{10},$$

$$\frac{d\xi_c^{(1)}}{dt} = -Q_c\eta_c^{(1)} + T\eta^{(1)},$$

$$\frac{d\eta_c^{(1)}}{dt} = P_c\xi_c^{(1)} + R\xi^{(1)}, \tag{13.85}$$

where $\bar{V}_{01}$ and $\bar{V}_{10}$ are purely conditionally periodic functions defined by Equation 13.65 for generating the values of the variables (Equation 13.75). In the derivation of the first approximation (Equation 13.85) we have used the property of the functions (Equation 13.65) that has no constant terms in their expressions $\bar{V}_{01} = \bar{V}_{10} = 0$ because of the nature of the lunar orbit perturbations.

Ignoring the procedure for solving a system of linear Equations 13.85, we present the final formula for the first-order perturbation in the variables ξ, η, and ξ_c, η_c:

$$\tilde{\xi}^{(1)} = \sum_{|v|>0} \xi_v^{(1)}(\rho)\cos\theta_v, \quad \tilde{\eta}^{(1)} = \sum_{|v|>0} \eta_v^{(1)}(\rho)\sin\theta_v,$$

$$\tilde{\xi}_c^{(1)} = \sum_{|v|>0} \xi_{cv}^{(1)}(\rho)\cos\theta_v, \quad \tilde{\eta}_c^{(1)} = \sum_{|v|>0} \eta_{cv}^{(1)}(\rho)\sin\theta_v,$$

$$(13.86)$$

where

$$\xi_v = -\frac{\sqrt{\Lambda}K}{D_v}\Big[(J_2+2C_{22})\omega_v(\Delta_v+RT)P_v+(J_2-2C_{22})(Q\Delta_v+T^2P_c)Q_v\Big],$$

$$\eta_v = -\frac{\sqrt{\Lambda}K}{D_v}\Big[(J_2+2C_{22})(P\Delta_v+R^2Q_c)P_v+(J_2-2C_{22})\omega_v(\Delta_v+RT)Q_v\Big],$$

$$\xi_{cv} = \frac{\sqrt{\Lambda}K}{D_v}\Big\{-(J_2+2C_{22})\omega_v(RQ_c-TP)P_v+(J_2-2C_{22})\Big[\omega_v^2T+R(T^2-Q_cQ)\Big]Q_v\Big\},$$

$$\eta_{cv} = -\frac{\sqrt{\Lambda}K}{D_v}\Big\{(J_2+2C_{22})\Big[\omega_v^2R+T(R^2-PP_c)\Big]P_v+(J_2-2C_{22})\omega_v(QR-TP_c)Q_v\Big\},$$

$$(13.87)$$

$$P_v = P_{v_1,v_2,v_3,v_4} = \sum_{v_5}\sum_{\varepsilon=\pm1}(-1)^{v_5}\varepsilon C_{v_1,v_2,v_3+\varepsilon,v_4,v_5}^{(\varepsilon)},$$

$$Q_v = Q_{v_1,v_2,v_3,v_4} = \sum_{v_5}\sum_{\varepsilon=\pm1}(-1)^{v_5} C_{v_1,v_2,v_3+\varepsilon,v_4,v_5}^{(\varepsilon)}. \quad (13.88)$$

Here $K = \dfrac{3n_0^2}{In_F}$ is some factor that has the dimension of a frequency.

13.5.2 Perturbations of the First Order of the Motion of the Pole of the Axis of Lunar Rotation

Let us derive the formulas for the calculation of first-order perturbations for the projections of the angular velocity of rotation of the Moon and its liquid core (for Poincaré coordinate system) to the principal central axes of the Moon inertia. The exact formulas for the projections of the lunar angular velocity p, q, and r (Equation 13.75) with the transformation of variables (Equations 13.40–13.41) (in Poincaré's variables) can take the following form:

$$p = \frac{1}{\Delta_1}\left(-\sqrt{\Lambda}A_c\eta\sqrt{1-\frac{\xi^2+\eta^2}{4\Lambda}} + \sqrt{\Lambda_c}F_c\eta_c\sqrt{1-\frac{\xi_c^2+\eta_c^2}{4\Lambda_c}}\right),$$

$$q = \frac{1}{\Delta_2}\left(\sqrt{\Lambda}B_c\xi\sqrt{1-\frac{\xi^2+\eta^2}{4\Lambda}} + \sqrt{\Lambda_c}E_c\xi_c\sqrt{1-\frac{\xi_c^2+\eta_c^2}{4\Lambda_c}}\right),$$

$$r = \frac{1}{\Delta_3}\left[C_c\Lambda\left(1-\frac{\xi^2+\eta^2}{2\Lambda}\right) + \Lambda_c D_c\left(-1+\frac{\xi_c^2+\eta_c^2}{2\Lambda_c}\right)\right]. \tag{13.89}$$

Similarly, for the projections p_c, q_c, and r_c (Equation 13.76) of the angular velocity of the rotation of the coordinate system $Cxyz$, the following expressions are written:

$$p_c = \frac{1}{\Delta_1}\left(-\sqrt{\Lambda}F_c\eta\sqrt{1-\frac{\xi^2+\eta^2}{4\Lambda}} + \sqrt{\Lambda_c}A\eta_c\sqrt{1-\frac{\xi_c^2+\eta_c^2}{4\Lambda_c}}\right)$$

$$q_c = \frac{1}{\Delta_2}\left(\sqrt{\Lambda}E_c\xi\sqrt{1-\frac{\xi^2+\eta^2}{4\Lambda}} + \sqrt{\Lambda_c}B\xi_c\sqrt{1-\frac{\xi_c^2+\eta_c^2}{4\Lambda_c}}\right)$$

$$r_c = \frac{1}{\Delta_3}\left[D_c\Lambda\left(1-\frac{\xi^2+\eta^2}{2\Lambda}\right) + C\Lambda_c\left(-1+\frac{\xi_c^2+\eta_c^2}{2\Lambda_c}\right)\right]. \tag{13.90}$$

In the first approximation the formulas (13.89) and (13.90) take the simple form:

$$p^{(1)} = \frac{1}{\Delta_1}\left(-\sqrt{\Lambda}A_c\eta^{(1)} + \sqrt{\Lambda_c}F_c\eta_c^{(1)}\right), \quad q^{(1)} = \frac{1}{\Delta_2}\left(\sqrt{\Lambda}B_c\xi^{(1)} + \sqrt{\Lambda_c}E_c\xi_c^{(1)}\right),$$

$$p_c^{(1)} = \frac{1}{\Delta_1}\left(-\sqrt{\Lambda}F_c\eta^{(1)} + \sqrt{\Lambda_c}A\eta_c^{(1)}\right), \quad q_c^{(1)} = \frac{1}{\Delta_2}\left(\sqrt{\Lambda}E_c\xi^{(1)} + \sqrt{\Lambda_c}B\xi_c^{(1)}\right),$$

$$r^{(1)} = \frac{C_c}{\Delta_3}\Lambda^{(1)}, \quad r_c^{(1)} = \frac{1}{\Delta_3}\left(D_c\Lambda^{(1)} - C\Lambda_c^{(1)}\right) = \frac{D_c}{\Delta_3}\Lambda^{(1)}. \tag{13.91}$$

For conditional periodic first-order perturbations of the equatorial components of the angular velocity of the Moon by the formulas (13.86) and (13.87) we obtain the following final formula:

$$\tilde{p}^{(1)} = \sum_{\|v\|>0} p_v^{(1)}(\rho)\sin\theta_v, \tilde{q}^{(1)} = \sum_{\|v\|>0} q_v^{(1)}(\rho)\cos\theta_v,$$

$$(13.92)$$

where the variations of the amplitudes of the equatorial components of the angular velocity are determined by the formulas:

$$\frac{p_v^{(1)}}{n_F} = (J_2 + 2C_{22})\Pi_v P_v + (J_2 - 2C_{22})T_v Q_v$$

$$\frac{q_v^{(1)}}{n_F} = (J_2 + 2C_{22})P_v P_v + (J_2 - 2C_{22})T_v Q_v, \qquad (13.93)$$

All these quantities are expressed in terms of certain combinations of functions of inclination (Equation 13.55) and are dimensionless. For constant coefficients in Equation 13.93 we have the following expressions:

$$\Pi_v = \frac{3}{I\Delta_1}\cdot\frac{n_0^2}{n_F D_v}\left\{CA_c\left(P\Delta_v + R^2 Q_c\right) - \sqrt{CD_c}\,F_c\left[\omega_v^2 R + T\left(R^2 - PP_c\right)\right]\right\}$$

$$T_v = \frac{3}{I\Delta_1}\cdot\frac{n_0^2}{n_F D_v}\left\{-CB_c\left(Q\Delta_v + T^2 P_c\right) + \sqrt{CD_c}\,F_c\left[\omega_v^2 T + R\left(T^2 - Q_c Q\right)\right]\right\}$$

$$P_v = -\frac{3}{I\Delta_1}\cdot\frac{n_0^2}{n_F D_v}\left[CB_c\left(\Delta_v + RT\right) + \sqrt{CD_c}\,F_c\left(RQ_c - TP\right)\right]\omega_v. \qquad (13.94)$$

The coefficients P, P_c, Q, Q_c and R, T given by Equation 13.62, are constant and here we write them in the form of:

$$P = \Omega\left(\frac{CB_c}{\Delta_2} - 1\right), Q = \Omega\left(\frac{CA_c}{\Delta_1} - 1\right), P_c = \Omega\frac{D_c B}{\Delta_2}, Q_c = \Omega\frac{D_c A}{\Delta_1}$$

$$R = \Omega\sqrt{CD_c}\,\frac{E_c}{\Delta_2}, T = \Omega\sqrt{CD_c}\,\frac{F_c}{\Delta_1}, \qquad (13.95)$$

where $\Omega = n_F$ is an unperturbed value of the angular velocity of the Moon. Some new notations are introduced:

$$\Delta_v = \omega_v^2 - P_c Q_c, D_v = \Delta_v\left(\omega_v^2 - PQ\right) + 2\omega_v^2 RT + T^2\left(R^2 - PP_c\right) - R^2 QQ_c \qquad (13.96)$$

Variations in the amplitude of the equatorial components of the angular velocity (Equations 13.91–13.96) in seconds of arc and their periods (in days) are given in Tables 13.A1 and 13.A2.

13.5.3 Perturbations of First Order of the Duration of the Lunar Day

Variations of the angular velocity of the lunar axial rotation are given by:

$$\tilde{r}^{(1)} = \sum_{|v|>0} r_v^{(1)}(\rho)\cos\theta_v \tag{13.97}$$

$$\frac{r_v^{(1)}}{n_F} = 6C_{22}\frac{CC_c}{I(CC_c - D_c^2)}F_v \cdot D_v \tag{13.98}$$

Taking into account the formula (13.10) for the above dynamic characteristics:

$$C_c > B_c > A_c,\ \varepsilon_D = 1-\frac{A_c}{C_c},\mu_D = 1-\frac{B_c}{C_c},\ \varepsilon_D > \mu_D, \Delta_3 = C_c(C-C_c)+(B_c-A_c)^2,$$

the factor in Equation 13.98 can be written as:

$$\frac{CC_c}{CC_c - D_c^2} = \frac{CC_c}{CC_c(1-C_c/C)+(B_c-A_c)^2} = \frac{1}{1-L_c+L_c(\varepsilon_D-\mu_D)^2} \tag{13.99}$$

and, accordingly,

$$\frac{r_v^{(1)}}{n_F} = \frac{6}{I}C_{22}\frac{1}{1-L_c+L_c(\mu_D-\varepsilon_D)^2}\cdot\frac{n_0^2}{n_F\omega_v}D_v \tag{13.100}$$

The formula (100) is accurate. It follows that the flattening of the very small core of the the Moon weakly influences the amplitude of the physical libration (on duration of the lunar day). Variation of the length of the lunar day T_w is determined by the simple relation with the variation of the polar component of the angular velocity:

$$T_\omega = \frac{2\pi}{n_F+\delta\omega} = \frac{2\pi}{n_F}\left(1-\frac{\delta\omega}{n_F}\right) \cong T_F - T_F\frac{\tilde{r}^{(1)}}{n_F} \tag{13.101}$$

Directly from Equation 13.91 with the importance of the period of $T_F = 27.21221$ days, we get a variation of the Moon LOD time in seconds:

$$\delta T_\omega = -T_F\frac{\tilde{r}^{(1)}}{n_F}. \tag{13.102}$$

Variations in the amplitude of the length of the lunar day and their periods are shown in Table 13.A3 (amplitude values are given in milliseconds of time).

13.5.4 Perturbations of First Order of the Motion of the Pole of the Lunar Core

Similar formulas are obtained for conditional periodic perturbations of the first order for the equatorial components of the angular velocity of the lunar core rotation:

$$\tilde{p}_c^{(1)} = \sum_{|v|>0} p_{cv}^{(1)}(\rho)\sin\theta_v, \quad \tilde{q}_c^{(1)} = \sum_{|v|>0} q_{cv}^{(1)}(\rho)\cos\theta_v \tag{13.103}$$

where the coefficients of the series are calculated by the nontrivial formulas:

$$p_{cv}^{(1)} = K\frac{\sqrt{\Omega}\sqrt{G}}{\Delta_1 D_v}\left\{\left\{\sqrt{C}F_c\left(P\Delta_v + R^2 Q_c\right) - \sqrt{D_c}A\left[\omega_v^2 R + T\left(R^2 - PP_c\right)\right]\right\}(J_2 + 2C_{22})P_v\right.$$

$$\left. + \omega_v\left[\sqrt{C}F_c(\Delta_v + RT) - \sqrt{D_c}A(QR - TP_c)\right](J_2 - 2C_{22})Q_v\right\},$$

$$q_{cv}^{(1)} = K\frac{\sqrt{\Omega}\sqrt{G}}{\Delta_2 D_v}\left\{-\omega_v\left[\sqrt{C}E_c(\Delta_v + RT) + \sqrt{D_c}B(RQ_c - TP)\right](J_2 + 2C_{22})P_v\right.$$

$$\left. + \left\{-\sqrt{C}E_c\left(Q\Delta_v + T^2 P_c\right)N_v + \sqrt{D_c}B\left[\omega_v^2 T + R\left(T^2 - Q_c Q\right)\right]\right\}(J_2 - 2C_{22})Q_v\right\}. \tag{13.104}$$

13.5.5 Perturbations of First Order of the Axial Rotation of the Lunar Core

Similar formulas are obtained for the projection of the angular velocity variations of the coordinate system of the core $Cxyz$ on the polar axis of inertia of the Moon:

$$\tilde{r}_c^{(1)} = \sum_{|v|>0} r_{cv}^{(1)}(\rho)\cos\theta_v, \tag{13.105}$$

$$r_{c;v}^{(1)}(\rho) = 6\frac{CD_c}{I\Delta_3 \omega_v} n_0^2 C_{22}\sum_{\varepsilon=\pm1}\sum_{v5}\varepsilon(-1)^{v5}D_{v_1,v_2,v_3+2\varepsilon,v_4,v_5}^{(\varepsilon)}(\rho). \tag{13.106}$$

It is easy to see that the expressions for the coefficients in the perturbations of the polar projections of the angular velocity of the Moon $r_v^{(1)}(\rho)$ and its core $r_{c;v}^{(1)}(\rho)$ are different only by the factors (moments of inertia C_c and D_c), and there is a simple relationship $C_c r_{c;v}^{(1)}(\rho) = D_c r_v^{(1)}(\rho)$ between them. If we neglect the small difference in the values of these moments of inertia and put $C_c = D_c$, we have $r_{c;v}^{(1)}(\rho) = r_v^{(1)}(\rho)$, which means that in a first approximation the liquid core rotates relative to the polar axis of inertia of itself as the Moon does, but in opposite direction. Or approximately the Poincaré coordinate system $Cxyz$, referred to the liquid core, to a first approximation does not rotate relative to the polar axis of inertia with respect to the space.

The above formulas correspond to the libration of the Moon with ellipsoidal liquid core and solid mantle. As a special case, on the basis of these general formulas, similar formulas for the variations of the components of the angular velocity of the solid moon are obtained. This allowed comparison of variations of the angular velocity for the two Moon models under consideration (with and without a molten core) and separately assess the effects of the influence of the liquid core on a variation of the angular velocity of the Moon. These results are briefly discussed in the Appendix (see Tables 13.5, 13.A1, and 13.A2).

13.5.6 Perturbations of the Classical Variables in Terms of the Perturbations of Andoyer–Poincaré's Variables

The expressions for the variations of the classic variables of the theory of lunar physical libration through a variation of the above Andoyer–Poincaré variables are obtained from the analysis of the direction cosines of the principal axes of the Moon in the ecliptic coordinate system $CXYZ$. Neglecting simple transformations and calculations, we give expressions for the first-order perturbation of the Euler angles through the variation of Andoyer–Poincaré variables (Barkin et al. 2012):

$$\Theta^{(1)} = \rho^{K(1)} = \rho_K^{(1)} + \frac{1}{\sqrt{\Lambda}}\left(\xi^{(1)}\cos F - \eta^{(1)}\sin F\right),$$

$$\Psi^{(1)} = \sigma_K^{(1)} = \frac{1}{\sqrt{\Lambda}\sin\rho_0}\left(\xi^{(1)}\sin F + \eta^{(1)}\cos F\right) - h^{(1)},$$

$$\Phi^{(1)} = \lambda^{(1)} + \frac{\cos\rho_0}{\sin\rho_0\sqrt{\Lambda}}\left(\xi^{(1)}\sin F - \eta^{(1)}\cos F\right). \qquad (13.107)$$

We now use the classical definition of the variables of the theory of lunar physical libration ρ^k, τ^k, σ^k, and for the variations of the Euler angles we obtain the simple relations:

$$\delta\Theta = \delta\rho_K, \quad \delta\Psi = \delta\sigma_K, \quad \delta\tau_K = \delta\Psi + \delta\Phi \tag{13.108}$$

and extend these relations to get expressions for the direction cosines of the polar axis of lunar inertia in the ecliptic coordinate system (rotating together with the center line of nodes of the lunar orbit)

$$P_1^{(K)} = \sin\Theta\sin\Phi, \quad P_2^{(K)} = \sin\Theta\cos\Phi. \tag{13.109}$$

There are simple relations between the variations of variables P_1, P_2 and Θ,Φ

$$\delta P_1 = \cos\Theta\sin\Phi\delta\Theta + \sin\Theta\cos\Phi\delta\Phi$$

$$\delta P_2 = \cos\Theta\cos\Phi\delta\Theta - \sin\Theta\sin\Phi\delta\Phi. \tag{13.110}$$

where $\Theta = \rho_0$ and $\Phi = F$. Substituting variations of Euler angles (Equation 13.107) in the formula (13.110), we obtain the expressions:

$$\delta P_1^{(K)} = \rho^{(1)}\cos\rho\sin F + \frac{1}{\sqrt{\Lambda}}\left(\xi^{(1)}\cos\rho\sin\Gamma\cos F - \eta^{(1)}\cos\rho\sin F\sin F\right)$$

$$+ \lambda^{(1)}\sin\rho\cos F + \frac{\cos\rho_0}{\sin\rho_0\sqrt{\Lambda}}\left(\xi^{(1)}\sin\rho\cos F\sin F - \eta^{(1)}\sin\rho\cos F\cos F\right)$$

$$\tag{13.111}$$

and finally we obtain the required expressions for the perturbations of all classical variables in terms of variations of the relevant Andoyer–Poincaré variables:

$$\delta\rho_K = \delta\rho + \frac{1}{\sqrt{\Lambda}}\left(\delta\xi\cos F - \delta\eta\sin F\right), \tag{13.112}$$

$$\sin\rho_0\delta\sigma_K = \frac{1}{\sqrt{\Lambda}}\left(\delta\xi\sin F + \delta\eta\cos F\right) - \delta h\sin\rho_0,$$

$$\delta\tau_K = \delta\lambda - \delta h + \frac{1+\cos\rho_0}{\sqrt{\Lambda}\sin\rho_0}\delta\xi\sin F + \frac{1-\cos\rho_0}{\sqrt{\Lambda}\sin\rho_0}\delta\eta\cos F$$

$$\delta P_1^{(K)} = -\sin\rho\cos F\delta g - \cos\rho\delta l - \cos\rho\sin F\delta\rho,$$

$$\delta P_2^{(K)} = \sin\rho\sin F\delta g - \cos\rho\delta\theta - \cos\rho\cos F\delta\rho,$$

Trigonometric series for the first-order perturbations for the classical variables are determined by the formulas:

$$\rho_K^{(1)} = \sum_{|\mathbf{v}|>0}\left[\rho_{\mathbf{v}}^{(1)} + \frac{1}{2\sqrt{\Lambda}}\sum_{\varepsilon=\pm1}\left(\xi_{\nu_1,\nu_2,\nu_3-\varepsilon,\nu_4}^{(1)} + \varepsilon\eta_{\nu_1,\nu_2,\nu_3-\varepsilon,\nu_4}^{(1)}\right)\right]\cos\theta_{\mathbf{v}},$$

$$\sigma_K^{(1)} = \sum_{|\mathbf{v}|>0}\left[-h_{\mathbf{v}}^{(1)} + \frac{1}{2\sin\rho_0\sqrt{\Lambda}}\sum_{\varepsilon=\pm1}\left(\varepsilon\xi_{\nu_1,\nu_2,\nu_3+\varepsilon,\nu_4}^{(1)} + \eta_{\nu_1,\nu_2,\nu_3+\varepsilon,\nu_4}^{(1)}\right)\right]\sin\theta_{\mathbf{v}},$$

(13.113)

$$\Phi^{(1)} = \sum_{|\mathbf{v}|>0}\left[\lambda_{\mathbf{v}}^{(1)} + \frac{\cos\rho_0}{2\sin\rho_0\sqrt{\Lambda}}\sum_{\varepsilon=\pm1}\left(\varepsilon\xi_{\nu_1,\nu_2,\nu_3+\varepsilon,\nu_4}^{(1)} - \eta_{\nu_1,\nu_2,\nu_3+\varepsilon,\nu_4}^{(1)}\right)\right]\sin\theta_{\mathbf{v}},$$

$$\theta_{\mathbf{v}} = \theta_{\nu_1,\nu_2,\nu_3,\nu_4} = \nu_1 l_M + \nu_1 l_S + \nu_1 F + \nu_1 D.$$

Similar formulas are obtained for the classical variables of the theory of physical libration of the Moon P_1, P_2 and τ (libration in longitude). Specified frequencies and corresponding periods are characterized by accurate analytical description and the observational data on the Moon libration are not required for their estimations, but the exact values of the dynamic parameters of the structure, its gravitational field, and the characteristics of the ellipsoidal core are needed. Conversely, the amplitudes of the free librations and their initial phases can be determined only from of observations of the libration of the Moon.

13.6 Free Oscillations of the Pole of the Moon

Free libration of Moon appearing in all variables z is briefly described above, particularly in Andoyer's variables. According to our approach, they are the solution of the equations in variations describing the rotational motion of the Moon in the vicinity of its intermediate conditionally periodic solutions. We point to the structure of this solution in the canonical Andoyer variables (Barkin and Ferrándiz 2007) as:

$$\delta z = P\left(z^{(p_0)}\cos U_p - z^{(p_0)*}\sin U_p\right) + Q\left(z^{(q_0)}\cos U_q - z^{(q_0)*}\sin U_q\right)$$
$$+ R\left(z^{(r_0)}\cos U_r - z^{(r_0)*}\sin U_r\right) + S\left(z^{(s_0)}\cos U_s - z^{(s_0)*}\sin U_s\right)$$

(13.114)

where $z = (l, g, h, l_c; L, G, H, L_c)$ is a vector of canonical Andoyer's variables, describing the rotational motion of Moon. The base of analytical studies consists of the canonical equations of rotational motion of a rigid body with an ellipsoidal cavity filled with an ideal fluid undergoing a simple motion of the Poincaré kind (Ferrándiz and Barkin 2000).

In solution (13.114) P, Q, R, S are amplitudes of free librations and U_p, U_q, U_r, U_s are arguments of free libration of the Moon, which are linear functions of time with frequencies n_p, n_q, n_r and n_s (Equation 13.71). Formula (13.114) describes resonant libration of the Moon (its free libration in longitude, in the oscillations of the pole, in the space nutations of the Moon and its core). Here $U_p^{(0)}$, $U_q^{(0)}$, $U_r^{(0)}$ and $U_s^{(0)}$ (72) are initial phases of corresponding resonant libration for a given period, for which we take the date of January 1, 2000. p, q, r, s are the frequencies of free librations and

$$T_p = \frac{2\pi}{p}, T_q = \frac{2\pi}{q}, T_r = \frac{2\pi}{r}, T_s = \frac{2\pi}{s} \qquad (13.115)$$

are the periods of the free librations. The specified frequencies and corresponding periods are characterized by an accurate analytical description and for their estimations the observational data on the Moon libration are not required, but they require the exact values of the dynamical parameters of the structure, its gravitational field, and the characteristics of the ellipsoidal core. In contrast, the amplitudes of the free librations and their initial phases can be determined only from observations of the Moon libration.

The pioneer determination of the amplitudes and phases of the free librations of the Moon using data of the laser ranging of reflectors installed on the lunar surface was performed by Calame et al. (1976). Highly accuracte determinations of the amplitudes and phases of the first three modes of free libration of the Moon were obtained by Rambaux and Williams (2011).

The determination of the first three modes in the solution (Equation 13.105) based on the modern theory of physical librations of the Moon was done by Barkin (1987, 1989) assuming a nonspherical rigid Moon. A good agreement has been demonstrated between our analytical studies and long-term data from laser ranging observations. Here we will present these results about free oscillations of the Moon for the first, second, and third modes and discuss the most urgent question of the definition of the fourth mode of free libration of the Moon, caused by a molten core.

13.6.1 Main Resonance Effects in the Rotational Motion of the Moon with a Molten Core in Andoyer Variables

In this part, we will study resonance effects in the rotational motion of the Moon in the neighborhood of an intermediate conditionally periodic motion. The Moon has a small core (see Section 13.1). Because of this feature, the first stage is taking advantage of the form of the general characteristic equation of the problem, which is split into two equations. The first equation defines three modes of free libration in longitude, tilt, and free oscillations of the pole with the frequencies p_0, q_0, r_0, which correspond to the three main periods of the resonant libration of the Moon (Barkin 1987; Barkin et al. 2012). The second equation determines the mode of free oscillation of the fourth pole of the Moon with frequency s_0 due to its ellipsoidal liquid core. The solution for the free libration of the Moon (Equation 13.114) is based on the variation equations in the vicinity of the intermediate conditionally periodic solutions $\mathbf{Z}(t) = \mathbf{Z}_0 + \langle \mathbf{Z} \rangle + \tilde{\mathbf{Z}}(l_M, l_S, F, D, ...)$ from (Equation 13.69). As the basic equations of motion are the canonical equations in the Andoyer variables, they are used in the method of constructing conditionally periodic solutions of the problem of the rotation of the Moon with the liquid core and the method of constructing solutions in its vicinity (Barkin 1987).

Quite intensive calculations for these constructions are not shown in this chapter, and we limit ourselves to present the final results of the study of free libration of the Moon, including the determination of the initial conditions of the problem by comparison with the empirical theory (Rambaux and Williams 2011) derived from the analysis of laser observations.

At the beginning, the solutions of the variational equations were obtained in Andoyer's variables $z = (l, g, h, l_c; L, G, H, L_c)$ and then were converted to the known classical variables $\mathbf{Z}_{cl} = (P_1, P_2, \tau, \rho, I\sigma)$. Also we used the angular Andoyer variables θ and ρ with $L = G\cos\theta$ and $H = G\cos\rho$, and ω is a module of angular velocity and similar variables for the motion of the liquid core: θ_c and ρ_c with $L_c = G_c \cos \theta_c$. So for variations $\delta z = (\delta l, \delta g, \delta h, \delta \omega, \delta \theta, \delta \rho, \delta l_c, \delta \theta_c)$ we have obtained the following compact representations:

$$\delta\omega/n_F = \Lambda\cos U, \; \delta\theta = T\cos W - T_s \sin U_s, \; \delta\rho = -M\cos U + N\cos V,$$

$$\delta l = -L\sin W + Q_s \cos U_s, \delta h = H\sin V, \delta g = E\sin U + \Gamma\sin V,$$

$$\delta l_c = Q_s^{(c)} \cos U_s, \; \delta\theta_c = -T_s^{(c)} \sin U_s \qquad (13.116)$$

where we have used the new notations for the amplitudes of the free libration described in Andoyer's variables:

$$\Lambda, T, T_s, M, N; \; L, Q_s, H, E, \Gamma; \; Q_s^{(c)} \; T_s^{(c)}. \qquad (13.117)$$

Arguments U, V, W and U_s are resonant arguments of free libration of the Moon. Here, we take them to cover those notations as in the empirical theory (Rambaux and Williams 2011). The amplitudes were obtained from analytical expressions, depending on the Moon model parameters and its perturbed orbital motion. These amplitudes are proportional to the free libration amplitudes of the Moon: P, Q, R, and S of all four modes considered libration. To compare our theory with the empirical theory (Rambaux and Williams 2011) in order to determine the initial conditions of the problem, we need to transform our solution in Andoyer variables (13.116) to the classical variables of the theory of the physical libration of the Moon.

13.6.2   Free Libration in the Classical Variables of the Theory of the Moon's Rotation

If the solution is built in Andoyer's variables, similar expressions can be obtained for the classical theory of physical libration variables and the angular velocity. Indeed, between the classical variables of theory of physical libration and Andoyer's variables there are simple geometric relationships that follow directly from the expressions of the direction cosines of the principal axes of inertia of the Moon in the main ecliptic coordinate system (Barkin 1987).

In the first stage of our theory, Andoyer's variables were used and then the solution of the variational equations have been built in the classical variables. In the linear approximation, these geometric relations in particular allow us to express the variation of the classical variables $Z = (P_1, P_2, \tau, \rho, I\sigma)$ in terms of the corresponding variations of the variables δl, $\delta \theta$, δg, δh and $\delta \rho$ (in the vicinity of Cassini's motion) (Barkin 1987):

$$\delta P_1^K = \sin \rho \cos F \delta g - \cos \rho \delta l - \cos \rho \sin F \delta \rho,$$

$$\delta P_2^K = \sin \rho \sin F \delta g - \cos \rho \delta \theta - \cos \rho \cos F \delta \rho,$$

$$\delta \tau_K = \delta g + \delta h + tg \frac{\rho}{2} (\sin F \delta \theta - \cos F \delta l),$$

$$\rho_K = \delta \rho + \sin F \delta l + \cos F \delta \theta,$$

$$I \sigma_K = - \cos F \delta l + \sin F \delta \theta + \sin \rho \delta h. \tag{13.118}$$

Substituting the solution of the free libration in Andoyer's variables into (13.118) we obtain the variation of the classic five variables:

$$\delta P_1^K = Q \cos \rho \sin W + Q_s \cos \rho \cos U_s$$

$$+ \frac{1}{2} (M \cos \rho - P \sin \rho) \sin (U + F) - \frac{1}{2} (M \cos \rho + P \sin \rho) \sin (U - F)$$

$$+ \frac{1}{2} (\Gamma \sin \rho - N \cos \rho) \sin (V + F) + \frac{1}{2} (\Gamma \sin \rho + N \cos \rho) \sin (F - V),$$

$$\tag{13.119}$$

$$\delta P_2^K = -T\cos\rho\cos W - T_s\cos\rho\sin U_s$$

$$+\frac{1}{2}(M\cos\rho - P\sin\rho)\cos(U+F)+\frac{1}{2}(M\cos\rho + P\sin\rho)\cos(U-F)$$

$$+\frac{1}{2}(\Gamma\sin\rho - N\cos\rho)\cos(V+F)-\frac{1}{2}(\Gamma\sin\rho + N\cos\rho)\cos(V-F),$$

$$\delta\tau_K = P\sin U + (R-\Gamma)\sin V$$

$$+\frac{1}{2}tg\frac{\rho}{2}\big[(T+Q)\sin(F+W)+(Q-T)\sin(W-F)\big]$$

$$+\frac{1}{2}tg\frac{\rho}{2}\big[(T_s-Q_s)\cos(U_s+F)-(T_s+Q_s)\cos(U_s-F)\big],$$

$$\delta\rho_K = -M\cos U + N\cos V$$

$$+\frac{1}{2}(T-Q)\cos(W-F)+\frac{1}{2}(T+Q)\cos(W+F)$$

$$-\frac{1}{2}(T_s-Q_s)\sin(U_s+F)-\frac{1}{2}(T_s+Q_s)\sin(U_s-F),$$

$$I\delta\sigma_K = \sin\rho R\sin V$$

$$+\frac{1}{2}(T+Q)\sin(W+F)+\frac{1}{2}(-T+Q)\sin(W-F)$$

$$+\frac{1}{2}(T_s-Q_s)\cos(U_s+F)-\frac{1}{2}(T_s+Q_s)\cos(U_s-F).$$

Equations 13.118 and 13.119 can be complemented by similar expressions for the projections of the angular velocity of the Moon at its principal central axis of inertia. However, in this chapter we will not consider them.

13.6.3 Determination of the Amplitudes and Phases of the Free Libration of the Moon

The constant amplitudes and initial phases of the arguments are

$$P,Q,R,S \text{ and } U_0,W_0,V_0,U_0^{(s)}$$

(13.120)

We can determine them as a result of comparing our previous analytical solution of the problem of free libration of the Moon (119) and the tables of libration (free and for the same variables) in the empirical theory (Rambaux and Williams 2011). As a result of the procedure, we determine the amplitudes and initial phases of the four modes of free libration of the Moon (in longitude, at an inclination, in the oscillations of the pole) (Table 13.3).

The calculated theoretical values of the free libration of the Moon, including those caused by the effect of the ellipsoidal liquid core, are shown in Tables 13.3 and 13.4. Thus, at this stage, we have identified the main free librations of the Moon, which are featured in the variations of the classical variables. The adopted initial time is the Julian date $t_0 = 2451545.0$. From Tables 13.1 and 13.2, we obtain the theoretical values of the amplitudes, and periods of free libration phases and are in good agreement with the corresponding parameters of the foresaid empirical theory (Rambaux and Williams 2011).

In this chapter we developed the first theory of the forced libration of the Moon with a molten core directly for components of the angular velocity vector of rotation of the Moon and its variations in the duration of the day. As a special case of the theory, similar perturbations have been calculated for a model of the solid moon without a liquid core. These results were confirmed by comparison with the calculations of the perturbation theory of the Moon rotation as a rigid body (Barkin 1987, 1989).

The fourth line shows the characteristics of the fourth vibration mode: $Us = -F-\Pi, U_s^{(0)} = -F_0 - \Pi_0$; from the estimated initial value $\Pi_0 - -128^0 + -139^0$ and the initial value of the argument of the theory of orbital motion $F_0 = 93^027$, we obtain the initial value of the argument $U_s^{(0)} = 39°$. For frequencies of pointed arguments we have a similar relation: $n_s = -n_F - n_\Pi$.

TABLE 13.3

Four Modes of Free Libration of the Moon with Liquid Core

Amplitude (")	Arguments	Period (days)	Phase (⁰)	Modes
$P = 1''735$	$U_p(t) = pt + U_p^{(0)}$	10517.13	$U_p^{(0)} = 207.01^0$	The free libration in longitude with a period of 2.99 years
$Q = 3''3072$	$U_q(t) = qt + U_q^{(0)}$	27257.27	$U_q^{(0)} = 161.60^0$	Free oscillations of the pole with a period of 74.3 years
$R = 1''1881$	$U_r(t) = rt + U_r^{(0)}$	8822.88	$U_r^{(0)} = 160.81^0$	Free oscillations of the angular momentum vector in space with a period of 24.3 years
$S = 0''0160$	$U_s(t) = st + U_s^{(0)}$	27.218	$U_s^{(0)} = 39^0$	Quasi daily variation of the moon with a period of 8 days

Table 13.4 provides a comparison of the theoretical results for the free libration with the determination of physical libration in the work by Rambaux and Williams (2011) and the analytical model of the Moon rotation (Barkin 1987; Barkin et al. 2012).

Table 13.4 used auxiliary arguments

$$\Xi = (n_\Omega + n_\Pi)t, \Sigma = (n_F + n_\Omega + n_\Pi)t$$

$$(13.121)$$

TABLE 13.4

Free and Resonant Librations of the Moon: Amplitude, Periods, and the Respective Arguments of Trigonometric Functions

No.	Variables	Theory: Analytical	Theory: Empirical	Trigonometic Terms	Periods (days)
1	δP_1	$-3''3060$	$-3''306$	$\sin W$	27257.273
2	δP_1	$-0''0320$	$-0''032$	$\sin(V-F)$	27.296
3	δP_1	$0''0250$	$0''025$	$\sin(U-F)$	27.932
4	δP_1	$0''0222$	—	$\sin(U+F)$	26.529
5	δP_1	$-0''00004$	—	$\sin(V+F)$	27.129
6	δP_1	$-0''001$	$0''000$	$\cos\Theta$	27.312
7	δP_1	$-0''016$	$-0''016$	$\sin\Theta$	27.312
8	δP_2	$8''1830$	$8''183$	$\cos W$	27257.273
9	δP_2	$0''0320$	$0''032$	$\cos(V-F)$	27.296
10	δP_2	$-0''0250$	$-0''025$	$\cos(U-F)$	27.932
11	δP_2	$0''0222$	$0''022$	$\cos(U+F)$	26.529
12	δP_2	$-0''00004$	—	$\cos(V+F)$	27.129
13	δP_2	$-0''016$	$-0''016$	$\cos\Theta$	27.312
14	δP_2	$0''001$	$0''002$	$\sin\Theta$	27.312
15	$\delta\tau_k$	$1''7570$	$1''735$	$\sin U$	1056.210
16	$\delta\tau_k$	$0''0774$	$0''077$	$\sin(W+F)$	27.185
17	$\delta\tau_k$	$-0''0328$	$-0''032$	$\sin(W-F)$	27.239
18	$\delta\tau_k$	$-0''0014$	—	$\sin V$	8822.883
19	$\delta\tau_k$	$-0''001$	—	$\cos\Xi$	7449.890
20	$\delta\rho_k$	$5''7402$	$5''753$	$\cos(W+F)$	27.185
21	$\delta\rho_k$	$2''4330$	$2''437$	$\cos(W-F)$	27.239
22	$\delta\rho_k$	$0''0320$	$0''029$	$\cos V$	8822.883
23	$\delta\rho_k$	$-0''0026$	$-0''003$	$\cos U$	1056.210
24	$\delta\rho_k$	$-0''007$	$-0''013$	$\cos\Xi$	7449.890
25	$\delta\rho_k$	$0''049$	$0''052$	$\sin\Xi$	7449.89
26	$I\delta\sigma_k$	$5''7402$	$5''758$	$\sin(W+F)$	27.185
27	$I\delta\sigma_k$	$-2''4330$	$-2''443$	$\sin(W-F)$	27.239
28	$I\delta\sigma_k$	$-0''0320$	$0''033$	$\sin V$	8822.883
29	$I\delta\sigma_k$	$-0''049$	$-0''045$	$\cos\Xi$	7449.890
30	$I\delta\sigma_k$	$-0''007$	$-0''002$	$\sin\Xi$	7449.890

They are used to describe the free libration of the Moon, due to the influence of its ellipsoidal liquid core (in Table 13.4, these libration are numbered by 6, 7; 13, 14; 19, 24, 25, 29, and 30). In a discussion of some of the authors with Dr. Rambaux during the conferences held in Mizusawa and Tokyo in 2013, we knew that the trigonometric term corresponding to free libration in the fourth line of Table 13.4 is related to a term found in the empirical theory (Rambaux & Willams 2011) with the same amplitude but with different argument.

13.7 Conclusions

In this chapter, an analytical theory of forced and free librations of a two-layer model of the Moon (with a liquid ellipsoidal core) has been constructed. The main resonant libration of the Moon in Andoyer's variables are defined and their amplitudes in arcseconds and their periods of arguments (in days) are given in Table 13.4. It happen that in order to determine the initial phase Π_0 (the epoch 2000.0 JD) and amplitude S, we have a system of eight equations with two unknowns, which should take the following initial values for the well known arguments of the theory of orbital motion of the Moon $F_0 = 93^0 27$ and $\Omega_0 = 125^0 045$ (Simon et al. 2004, Kudryavtsev 2007). These equations are derived from a comparison of some trigonometric perturbations of our theory with empirical formulas for libration of the Moon. Table 13.5 shows the values Π_0 (an epoch 2000,0 JD) and amplitude S (or the third and fourth columns) at each pair.

It is clearly seen that the initial phase of the libration Π_0 is well determined from laser ranging observations. The amplitude of the free libration is determined less precisely, but with a certain spread around the values $S = 0''02 \div 0''05$. Thus, all eight equations discussed above in Table 13.5 are approximately satisfied for the values of the amplitude and phase of the free core nutation of the Moon $S = 0''02 \div 0''05$ and $\Pi_0 = -128^0 \div -139^0$. Optimum values of these parameters are determined by the least squares as $S = 0''0395$ and $\Pi_0 = -134^0$. The numerical results are applicable to the theory of orbital motion of the Moon DE421, but the real value of flattening, the amplitude,

TABLE 13.5

Determination of the Initial Phases and Amplitudes of the Fourth Free Librations of the Moon, due to the Influence of the Ellipsoidal Liquid Core

1	$S \cos (\Omega_0+\Pi_0) = -0''052$	$S \sin (\Omega_0+\Pi_0) = 0''013$	$\Pi_0 = -139^0 1$	$S = 0''0536$
2	$S \cos (\Omega_0+\Pi_0) = -0''045$	$S \sin (\Omega_0+\Pi_0) = -0''002$	$\Pi_0 = -127^0 6$	$S = 0''047$
3	$S \cos (F_0+\Omega_0+\Pi_0) = 0''000$	$S \sin (F_0+\Omega_0+\Pi_0) = 0''016$	$\Pi_0 = -128^0 3$	$S = 0''016$
4	$S \cos (F_0+\Omega_0+\Pi_0) = -0''001$	$S \sin (F_0+\Omega_0+\Pi_0) = -0''016$	$\Pi_0 = -124.7^0$	$S = 0''016$

and period T_s of free libration of the Moon are very uncertain due to liquid core (Rambaux and Williams 2011).

In addition, we made one of the first attempts to determine the amplitude, phase and period of the fourth mode of libration of the Moon by a comparison of our analytic theory and the empirical model (laser) of the theory of physical libration of the Moon. For a more complete study of the problem of lunar rotation, new experimental data of the laser observations and telescope mounted on the surface of the Moon are required. Both these trends are developing intensively at present. Significant progress in the study of the Moon's rotation and its internal structure, selenodesy, and internal dynamics is expected by the Japanese project SELENA-2 (ILOM) with the telescope installed on the lunar surface (Hanada et al. 2004).

Periods of free libration of the Moon with liquid ellipsoidal core of 75133.87 d or 205.70 years correspond to the sum of dynamic compressions of the core such as $\varepsilon_D + \mu_D = 7.244 \times 10^{-4}$. Assuming a similarity of the ellipsoidal core and the whole Moon, the estimations of oblatenesses of the liquid core will be $\varepsilon_D = 4.419 \times 10^{-4}$, $\mu_D = 2.825 \times 10^{-4}$. Four librations of the Moon in the variables ρ and $I\sigma$ have been obtained with their mechanical explanation and interpretation. Also we have predicted one new libration in longitude (in variable τ) and estimated its period of 7449.89 d and a small amplitude about 0"001. In future, further studies of dynamic effects in the rotation of the Moon due to its liquid and solid cores, including studies relative to the dynamics of shells of celestial bodies (Barkin 2002; Barkin and Vilke 2004), are required. Much work is also needed addressed to the analyses of laser ranging and seismometer observations of the Moon as well as the new telescope on the lunar surface.

Acknowledgments

JMF work has been partially supported by grants AYA2010-22039-C02-01 from the Spanish Ministerio de Economia y Competitividad (MINECO) and ACOMP/2014/141 from Generalitat Valenciana.

References

Araki et al. (2009) Lunar global shape and polar topography derived from Kaguya-LALT laser altimetry. *Science* 13 February 2009: Vol. 323 no. 5916 pp. 897–900.
Arkhangelskii Yu.A. (1977) *Analytical Rigid Body Dynamics*. Nauka, Moscow 1977. 328s. In Russian.

Barkin, Yu. V. (2000) Perturbated rotational motion of weakly deformable celestial bodies, *Astron. Astrophys. Trans.* 19 (1), 19–65.

Barkin Yu.V. (1987) An analytical theory of the lunar rotational motion. In: *Figure and Dynamics of the Earth, Moon and Planets*, Ed. P. Holota. Proceedings of the Int. Symp. (Prague, Czechoslovakia, September 15–20, 1986). Monograph Series of UGTK, Prague, 1987, 657–677.

Barkin Yu.V. (1989) Dynamics of non-spherical celestial bodies and the theory of the Moon's rotation. Thesis for Professorship dissertation. Moscow, SAI MSU. 412c. In Russian.

Barkin Yu.V. (2002) Explanation of endogenous activity of planets and satellites and its cyclicity. *Proceedings of the Section of Earth Sciences of the Russian Academy of Natural Sciences*. M., VINITI. 2002. Issue. 9, C. 45–97. In Russian.

Barkin Yu.V. and Vilke V.G. (2004) Celestial mechanics of the planet shells. *Astronom. Astrophys. Trans.* 23, 533–554.

Barkin Y., and Ferrándiz J.M. (2004) New approach to development of moon rotation theory. 35th Lunar and Planetary Science Conference, March 15–19, League City, Texas, abstract no.1294.

Barkin Yu.V., Kudryavtsev S.M., Barkin M.Yu. (2009) Perturbations of the first order of the Moon rotation. Proceedings of the International Conference "Astronomy and World Heritage: Across Time and Continents" (Kazan, 19–24 August). KSU. pp. 161–164.

Barkin Y., Hanada H., Matsumoto K., Noda H., Petrova N., Sasaki S. (2012) The effects of the physical librations of the Moon, caused by a liquid core and their possible detection from the long-term laser observations and in the Japanese lunar project ILOM. The third Moscow Solar System Symposium (SM-S3), October 8–12, 2012, Moscow, Russia. Book of abstracts. Abstract 3MS3-MN-10. 18–19.

Barkin M. (2012) Report about visiting research period in Mizusawa astronomical observatory. Mizusawa, October – November 2012, Japan. (Available from the author)

Beletsky V.V. (1975) The motion of the satellite relative to the center of mass in a gravitational field. Moscow: Moscow State University, 1975. -308p. In Russian.

Calame O. (1976) Free librations of the Moon determined by an analysis of laser range measurements. *The Moon* 15, N3–4. p. 343–352.

Ferrándiz, J.M. and Barkin, Yu.V. (2000) Model of the Earth with the eccentric and the moveable liquid core. In: *Motion of Celestial Bodies, Astrometry and Astronomical Reference Frames*, Eds. M. Soffel and N. Capitaine, Journées 1999 & IX. Lohrmann-Kolloquium Proceedings (Dresden, September 13–15, 1999) , Dresden, Germany. 2000. P. 192.

Ferrándiz, J.M. and Barkin, Yu.V. (2001) On integrable cases of the Poincaré problem, *Astron. Astrophys. Transac.* 19, 769–780.

Getino, J. and Ferrándiz, J.M. (1991) A Hamiltonian theory for an elastic earth: Secular rotational acceleration. *Celest. Mech.* 52, 381–396.

Hanada, H., Heki, K., Araki, H., et al. (18 coauthors) (2004) Application of a PZT telescope to in situ lunar orentation measurements (ILOM), in *Proceedings of 25th General Assembly of the International Union of Geodesy and Geophysics*. Springer. pp. 163–168.

Jin, S.G., Arivazhagan, S., Araki, H. 2013 New results and questions of lunar exploration from SELENE, Chang'E-1, Chandrayaan-1 and LRO/LCROSS. *Adv. Space Res.*, 52(2), 285–305.

Kinoshita, H. (1977), Theory of the rotation of the rigid Earth, *Celest. Mech. Dyn. Astron.* 15, 277–326.

Kudryavtsev S.M. (2007) Long-term harmonic development of lunar ephemeris. *Astron. Astrophys.* 471, 1069–1075.

Lamb, G. (1947) *Hydrodynamics*. Gostekhizdat. 928p. In Russian. 6th edition in English by Cambridge Univ. Press (1932).

Matsumoto K., Goossens S., Ishihara Y., Liu Q., Kikuchi K., Iwata T., Namiki N., Noda H., Hanada H., Kawano N., Lemoine F.G. and Rowlands D.D. (2010) An improved lunar gravity field model from SELENE and historical tracking data: Revealing the far side gravity features, *J. Geophys. Res.*, Vol. 115, E06007.

Poincaré, H. (1910) Sur la précesion des corps déformables, *Bull. Astronom.* 27, 321–356.

Rambaux N., Williams J.G. (2011) The Moon's physical librations and determination of their free modes. *Celest. Mechan. Dyn. Astron.* 109(1), 85–10.

Sagitov M.U. (1979) *Lunar Gravimetry*. Moscow: Nauka Publishing House. 432 p. In Russian.

Sasao T., Okubo S. and Saito M. (1980) A simple theory on the dynamical effects of a stratified fluid core upon nutational motion of the Earth. In Fedorov, E. P., Smith, M. L., Bender, P. L. (eds) *Nutation and the Earth's Rotation*, Proc. IAU Symp. 78, pp 165–183.

Sevilla M. and Romero P. (1987) Polar motion for an elastic earth model with a homogeneous liquid core using a canonical theory. *Bull. Geod.* 61, 1–20.

Simon J.L, Bretagnon P, Chapront J, Chapront-Touzé M, Francou G, Laskar J (1994) Numerical expressions for precession formulae and mean elements for the Moon and planets. *Astron. Astrophys* 282, 663–683

Weber R.C., Lin Pei-Ying, Garnero E.J., Williams Q., Lognonné P. (2011) Seismic detection of the lunar core. *Science*, 331, 309.

Wei, E., Yan, W., Jin, S.G., Liu, J., Cai J. (2013), Improvement of Earth orientation parameters estimate with Chang'E-1 ΔVLBI observations. *J. Geodyn* 72, 46–52.

Williams J. G., Boggs, D. H., Yoder, C . F., Ratcliff, J. T., and Dickey, J. O. (2001) Lunar rotational dissipation in solid body and molten core, *J. Geophys. Res. Planets* 106, 27933–27968.

Williams J.G., Dale H. Boggs and Slava G. Turyshev (2010) LLR analysis – JPL model and data analysis. LLR Workshop, Harvard, Boston, MA December 9–10, 2010.

Williams J.G., Boggs D.H., Ratcliff J.T. (2011) Lunar moment of inertia and Love number. 42nd Lunar and Planetary Science Conference.

Williams J.G., Boggs D.H., Ratcliff J.T. (2012) Lunar moment of inertia, Love number and core. 43rd Lunar and Planetary Science Conference.

13.A1 Tables Forced Variations of the Projections of the Angular Velocity of the Moon at Its Principal Central Axis of Inertia

For variations of the projections of the angular velocity $z = (p, q, r)$ and the corresponding amplitudes of these variations $Z_v = (p_v, q_v, r_v)$ in Tables 13.A1–13.A3, the following notations are used:

Z_v (P, l) - the amplitudes of the libration of the Moon with a liquid core and the solid mantle (a method based on the use of the Poincaré variables);

Z_v (P, r) - the amplitudes of the libration of the Moon without a liquid core and with the solid mantle (a method based on the use of the Poincaré's variables);

Z_v (B, r) - the amplitude of the libration of the rigid Moon without a liquid core (a method based on the use of variables Andoyer);

Z_v (P, l) - Z_v (P, r) - the difference between the amplitudes of the perturbation models of the Moon rotation with a liquid core and a nonliquid core (a method based on the use of the Poincaré variables). Introduced characteristics are given in the three tables in this application.

In the tables for the variations of the components of angular velocity (columns 7–9):

$$\tilde{p}^{(1)} = \sum_{|v|>0} p_v^{(1)}(\rho)\sin\theta_v, \ \tilde{q}^{(1)} = \sum_{|v|>0} q_v^{(1)}(\rho)\cos\theta_v, \ \tilde{r}^{(1)} = \sum_{|v|>0} r_v^{(1)}(\rho)\cos\theta_v$$

The mentioned amplitudes and differences are given in arc seconds. Periods T_v are given in days (column 6). The columns 2–5 are the index values—factors in the arguments for the corresponding variables of the orbital motion of the Moon $\theta_v = v_1 l_M + v_2 l_s + v_3 F + v_4 D$. The tables are all showing the detected variations with amplitudes >0 "00049.

The results in Table 13.A1 with 76 variations show that the liquid core provides the main contribution to the variation of the equatorial components of the angular velocity of the Moon p with an amplitude of 0 "007 (corresponding period of 27.2122 days).

p: The main conclusion. Due to the influence of the ellipsoidal liquid core of the Moon projection p of the angular velocity of rotation of the Moon on its axis inertia directed toward Earth, are

$$\Delta p_v = 0.0074"\sin F - 0.0012"\sin(l - F).$$

The period of variation of the first term is 27.2122 days, and the second 2190.38 days or 5.9968 years. From the results obtained here (Table 13.4 with 94 variations of amplitude >0 "00049) it follows that the liquid core dominates

TABLE 13.A1

Amplitudes p_v of the Forced Variations of Component of the Angular Velocity p for a Model of the Moon with a Liquid Core and for Model without Core and the Difference between Their Values Δp_v in Arc Seconds

No.	v_1	v_2	v_3	v_4	T_v	$P_v\ (P,l)$	$P_v\ (P,r)$	$P_v\ (P,l)-P_v$ (P,r)
1	0	0	0	1	29.5306	0.0008	0.0008	0.00000
2	0	0	1	−4	−10.1312	−0.0010	−0.0010	0.00000
3	0	0	1	−2	−32.2807	0.2136	0.2136	0.00000
4	0	0	1	−1	346.6216	0.1129	0.1129	0.00000
5	0	0	1	0	27.2122	−0.1096	−0.1170	0.00744
6	0	0	1	1	14.1620	−0.0044	−0.0044	0.00000
7	0	0	1	2	9.5717	0.0846	0.0846	0.00000
8	0	0	1	4	5.8072	0.0021	0.0021	0.00000
9	0	0	2	−1	25.2314	0.0007	0.0007	0.00000
10	0	0	3	−2	23.5194	−0.0018	−0.0018	0.00000
11	0	0	3	0	9.0707	−0.0167	−0.0167	0.00000
12	0	1	−1	−2	−9.8293	−0.0058	−0.0058	0.00000
13	0	1	−1	0	−29.4028	−0.0487	−0.0487	0.00000
14	0	1	−1	1	−6793.3970	0.2241	0.2241	−0.00001
15	0	1	−1	2	29.6595	−0.0054	−0.0054	0.00000
16	0	1	1	−2	−35.4102	0.0148	0.0148	0.00000
17	0	1	1	0	25.3254	−0.0320	−0.0320	0.00000
18	0	1	1	1	13.6334	0.0010	0.0010	0.00000
19	0	1	1	2	9.3273	−0.0018	−0.0018	0.00000
20	0	2	−2	1	−29.2760	−0.0006	−0.0006	0.00000
21	0	2	1	−2	−39.2116	0.0008	0.0008	0.00000
22	1	−2	1	−2	−6168.4120	−0.0358	−0.0358	0.00000
23	1	−1	−1	−2	−14.1003	0.0022	0.0022	0.00000
24	1	−1	−1	0	−313.0548	0.1211	0.1211	0.00000
25	1	−1	1	−2	388.2514	0.0272	0.0272	0.00000
26	1	−1	1	0	14.2243	0.0083	0.0083	0.00000
27	1	−1	1	2	7.2449	0.0015	0.0015	0.00000
28	1	0	−3	−2	−7.0582	0.0005	0.0005	0.00000
29	1	0	−3	0	−13.5221	0.0023	0.0023	0.00000
30	1	0	−3	2	−160.6033	−0.0070	−0.0070	0.00000

(Continued)

TABLE 13.A1 (*Continued*)

Amplitudes p_v of the Forced Variations of Component of the Angular Velocity p for a Model of the Moon with a Liquid Core and for Model without Core and the Difference between Their Values Δp_v in Arc Seconds

No.	v_1	v_2	v_3	v_4	T_v	P_v (P,l)	P_v (P,r)	P_v (P,l)–P_v (P,r)
31	1	0	−1	−4	−7.3578	−0.0039	−0.0039	0.00000
32	1	0	−1	−2	−14.6664	−0.1569	−0.1569	0.00000
33	1	0	−1	0	−2190.3840	123.3921	123.3933	−0.00120
34	1	0	−1	1	29.9342	0.0012	0.0012	0.00000
35	1	0	−1	2	14.8655	0.0026	0.0026	0.00000
36	1	0	0	−1	411.7854	−0.0030	−0.0030	0.00000
37	1	0	1	−2	188.2021	−2.4988	−2.4988	0.00000
38	1	0	1	−1	25.5254	0.0023	0.0023	0.00000
39	1	0	1	0	13.6912	0.7752	0.7752	0.00000
40	1	0	1	1	9.3543	−0.0010	−0.0010	0.00000
41	1	0	1	2	7.1040	0.0193	0.0193	0.00000
42	1	0	3	−4	90.2249	0.0013	0.0013	0.00000
43	1	0	3	−2	12.6888	−0.0006	−0.0006	0.00000
44	1	0	3	0	6.8243	−0.0024	−0.0024	0.00000
45	1	1	−1	−2	−15.2800	−0.0077	−0.0077	0.00000
46	1	1	−1	0	438.3568	0.1053	0.1053	0.00000
47	1	1	−1	2	14.2842	−0.0006	−0.0006	0.00000
48	1	1	1	−2	124.2047	−0.0737	−0.0737	0.00000
49	1	1	1	0	13.1965	−0.0074	−0.0074	0.00000
50	1	2	1	−2	92.6871	−0.0010	−0.0010	0.00000
51	2	−4	2	−4	−3084.2060	0.0010	0.0010	0.00000
52	2	−1	−1	0	28.2139	−0.0015	−0.0015	0.00000
53	2	−1	1	0	9.3814	0.0014	0.0014	0.00000
54	2	0	−3	0	−26.5525	−0.0009	−0.0009	0.00000
55	2	0	−1	−4	−10.0384	−0.0035	−0.0035	0.00000
56	2	0	−1	−2	−31.3565	0.0211	0.0211	0.00000
57	2	0	−1	−1	507.1231	−0.0017	−0.0017	0.00000
58	2	0	−1	0	27.9056	−0.0997	−0.0997	0.00000
59	2	0	−1	2	9.6561	0.0009	0.0009	0.00000
60	2	0	1	−4	−38.2829	0.0060	0.0060	0.00000

(Continued)

TABLE 13.A1 (*Continued*)

Amplitudes p_v of the Forced Variations of Component of the Angular Velocity p for a Model of the Moon with a Liquid Core and for Model without Core and the Difference between Their Values Δp_v in Arc Seconds

No.	v_1	v_2	v_3	v_4	T_v	$P_v\ (P,l)$	$P_v\ (P,r)$	$P_v\ (P,l)-P_v$ (P,r)
61	2	0	1	−3	129.1675	0.0006	0.0006	0.00000
62	2	0	1	−2	24.0355	−0.0476	−0.0476	0.00000
63	2	0	1	0	9.1465	0.0883	0.0883	0.00000
64	2	0	1	2	5.6479	0.0028	0.0028	0.00000
65	2	1	−1	−2	−34.3012	0.0007	0.0007	0.00000
66	2	1	−1	0	25.9249	0.0011	0.0011	0.00000
67	2	1	1	−4	−42.7652	0.0006	0.0006	0.00000
68	2	1	1	−2	22.5515	−0.0018	−0.0018	0.00000
69	2	1	1	0	8.9231	−0.0012	−0.0012	0.00000
70	3	0	−3	0	−728.1281	−0.0009	−0.0009	0.00000
71	3	0	−1	−2	227.2543	−0.0039	−0.0039	0.00000
72	3	0	−1	0	13.8645	−0.0032	−0.0032	0.00000
73	3	0	1	−4	98.3252	−0.0008	−0.0008	0.00000
74	3	0	1	−2	12.8375	−0.0040	−0.0040	0.00000
75	3	0	1	0	6.8670	0.0082	0.0082	0.00000
76	4	0	1	0	5.4971	0.0007	0.0007	0.00000

the variation of equatorial angular velocity component q of the Moon with amplitude $0''012$ (corresponding to the period of 27.2122 days) and with an amplitude of $0''003$ (with a period of 2190.38 days).

q: The main conclusion. Due to the influence of the ellipsoidal liquid core of the Moon, projection q of the angular velocity of lunar rotation on its axis inertia directed to the east, are

$$\Delta q_v = 0.0116''\cos F + 0.00256''\cos(l-F).$$

The period of variation of the first term is 27.2122 days, and the second term is 2190.38 days or 5.9968 years. Thus, the liquid core leads to an additional wobble of the axis of rotation of the Moon (in its principal axes of inertia) with a period of 27.2122 days. The trajectory of the pole of the angular velocity of the Moon is an ellipse with an eccentricity $e_{l,c} = \sqrt{1-a^2/b^2} = 0.7701$. The ellipse is again stretched along an axis orthogonal to the radius vector of the Moon. The ratio of semiaxes of this ellipse is 1.568. Determination of this libration,

TABLE 13.A2

Amplitudes q_v of the Forced Variations of Component of the Angular Velocity q for a Model of the Moon with a Liquid Core and for Model without Core and the Difference between Their Values Δq_v in Arc Seconds

No.	v_1	v_2	v_3	v_4	T_v	q_v (P,l)	q_v (P,r)	q_v (P,l)−q_v (P,r)
1	0	0	0	1	29.5306	0.0012	0.0012	0.00000
2	0	0	1	−4	−10.1312	−0.0063	−0.0063	0.00000
3	0	0	1	−3	−15.4222	0.0007	0.0007	0.00000
4	0	0	1	−2	−32.2807	−0.7809	−0.7809	0.00000
5	0	0	1	−1	346.6216	0.0867	0.0867	0.00000
6	0	0	1	0	27.2122	−44.4282	−44.4398	0.01164
7	0	0	1	1	14.1620	0.0167	0.0167	0.00000
8	0	0	1	2	9.5717	−0.3434	−0.3434	0.00000
9	0	0	1	4	5.8072	−0.0038	−0.0038	0.00000
10	0	0	2	−1	25.2314	0.0010	0.0010	0.00000
11	0	0	3	−2	23.5194	0.0125	0.0125	0.00000
12	0	0	3	0	9.0707	−0.0239	−0.0239	0.00000
13	0	1	−3	0	−9.3017	−0.0007	−0.0007	0.00000
14	0	1	−1	−4	−5.9010	−0.0005	−0.0005	0.00000
15	0	1	−1	−2	−9.8293	−0.0233	−0.0233	0.00000
16	0	1	−1	0	−29.4028	−0.0185	−0.0185	0.00000
17	0	1	−1	1	−6793.3970	−0.3107	−0.3107	0.00001
18	0	1	−1	2	29.6595	0.0082	0.0082	0.00000
19	0	1	1	−4	−10.4203	−0.0007	−0.0007	0.00000
20	0	1	1	−2	−35.4102	−0.0272	−0.0272	0.00000
21	0	1	1	0	25.3254	0.0516	0.0516	0.00000
22	0	1	1	1	13.6334	−0.0019	−0.0019	0.00000
23	0	1	1	2	9.3273	0.0033	0.0033	0.00000
24	0	1	3	−2	22.0965	0.0005	0.0005	0.00000
25	0	2	−2	1	−29.2760	0.0009	0.0009	0.00000
26	0	2	−1	−2	−10.1011	−0.0005	−0.0005	0.00000
27	0	2	0	1	25.4202	−0.0008	−0.0008	0.00000
28	0	2	1	−2	−39.2116	−0.0006	−0.0006	0.00000
29	0	2	1	0	23.6834	0.0005	0.0005	0.00000
30	1	−2	1	−2	−6168.4120	−0.0476	−0.0476	0.00000
31	1	−1	−1	−2	−14.1003	0.0039	0.0039	0.00000

(Continued)

TABLE 13.A2 (*Continued*)

Amplitudes q_v of the Forced Variations of Component of the Angular Velocity q for a Model of the Moon with a Liquid Core and for Model without Core and the Difference between Their Values Δq_v in Arc Seconds

No.	v_1	v_2	v_3	v_4	T_v	q_v (P,l)	q_v (P,r)	q_v (P,l)−q_v (P,r)
32	1	−1	−1	0	−313.0548	−0.1221	−0.1221	0.00000
33	1	−1	−1	2	15.4962	−0.0018	−0.0018	0.00000
34	1	−1	1	−2	388.2514	0.0402	0.0402	0.00000
35	1	−1	1	0	14.2243	−0.0237	−0.0237	0.00000
36	1	−1	1	2	7.2449	−0.0042	−0.0042	0.00000
37	1	0	−3	−2	−7.0582	−0.0005	−0.0005	0.00000
38	1	0	−3	0	−13.5221	0.0125	0.0125	0.00000
39	1	0	−3	2	−160.6033	0.0035	0.0035	0.00000
40	1	0	−1	−4	−7.3578	−0.0110	−0.0110	0.00000
41	1	0	−1	−3	−9.7995	0.0007	0.0007	0.00000
42	1	0	−1	−2	−14.6664	−0.6216	−0.6216	0.00000
43	1	0	−1	0	−2190.3840	−92.5060	−92.5086	0.00255
44	1	0	−1	1	29.9342	−0.0008	−0.0008	0.00000
45	1	0	−1	2	14.8655	−0.0473	−0.0473	0.00000
46	1	0	−1	4	7.4076	−0.0010	−0.0010	0.00000
47	1	0	0	−1	411.7854	−0.0047	−0.0047	0.00000
48	1	0	1	−4	−16.0223	−0.0091	−0.0091	0.00000
49	1	0	1	−2	188.2021	−1.3023	−1.3023	0.00000
50	1	0	1	−1	25.5254	−0.0013	−0.0013	0.00000
51	1	0	1	0	13.6912	−3.0485	−3.0485	0.00000
52	1	0	1	1	9.3543	0.0026	0.0026	0.00000
53	1	0	1	2	7.1040	−0.0550	−0.0550	0.00000
54	1	0	1	4	4.7963	−0.0008	−0.0008	0.00000
55	1	0	3	−4	90.2249	0.0013	0.0013	0.00000
56	1	0	3	−2	12.6888	0.0023	0.0023	0.00000
57	1	0	3	0	6.8243	−0.0035	−0.0035	0.00000
58	1	1	−1	−4	−7.5091	−0.0013	−0.0013	0.00000
59	1	1	−1	−2	−15.2800	−0.0286	−0.0286	0.00000
60	1	1	−1	0	438.3568	−0.1589	−0.1589	0.00000
61	1	1	−1	2	14.2842	0.0010	0.0010	0.00000
62	1	1	1	−4	−16.7574	−0.0007	−0.0007	0.00000

(*Continued*)

TABLE 13.A2 (*Continued*)

Amplitudes q_v of the Forced Variations of Component of the Angular Velocity q for a Model of the Moon with a Liquid Core and for Model without Core and the Difference between Their Values Δq_v in Arc Seconds

No.	v_1	v_2	v_3	v_4	T_v	q_v (P,l)	q_v (P,r)	q_v (P,l)−q_v (P,r)
63	1	1	1	−2	124.2047	−0.0371	−0.0371	0.00000
64	1	1	1	0	13.1965	0.0190	0.0190	0.00000
65	1	1	1	2	6.9684	0.0008	0.0008	0.00000
66	1	2	−1	0	199.2423	−0.0007	−0.0007	0.00000
67	1	2	1	−2	92.6871	0.0008	0.0008	0.00000
68	2	−4	2	−4	−3084.2060	−0.0007	−0.0007	0.00000
69	2	−1	−1	0	28.2139	0.0019	0.0019	0.00000
70	2	−1	1	0	9.3814	−0.0034	−0.0034	0.00000
71	2	−1	1	2	5.7366	−0.0006	−0.0006	0.00000
72	2	0	−1	−4	−10.0384	−0.0097	−0.0097	0.00000
73	2	0	−1	−2	−31.3565	−0.0045	−0.0045	0.00000
74	2	0	−1	−1	507.1231	0.0027	0.0027	0.00000
75	2	0	−1	0	27.9056	0.1086	0.1086	0.00000
76	2	0	−1	2	9.6561	−0.0038	−0.0038	0.00000
77	2	0	1	−4	−38.2829	0.0074	0.0074	0.00000
78	2	0	1	−2	24.0355	0.0364	0.0364	0.00000
79	2	0	1	0	9.1465	−0.2386	−0.2386	0.00000
80	2	0	1	2	5.6479	−0.0068	−0.0068	0.00000
81	2	1	−1	−4	−10.3221	−0.0009	−0.0009	0.00000
82	2	1	−1	0	25.9249	−0.0013	−0.0013	0.00000
83	2	1	1	−4	−42.7652	0.0008	0.0008	0.00000
84	2	1	1	−2	22.5515	0.0015	0.0015	0.00000
85	2	1	1	0	8.9231	0.0027	0.0027	0.00000
86	3	0	−3	0	−728.1281	0.0014	0.0014	0.00000
87	3	0	−1	−2	227.2543	−0.0036	−0.0036	0.00000
88	3	0	−1	0	13.8645	0.0032	0.0032	0.00000
89	3	0	1	−4	98.3252	−0.0018	−0.0018	0.00000
90	3	0	1	−2	12.8375	0.0058	0.0058	0.00000
91	3	0	1	0	6.8670	−0.0189	−0.0189	0.00000
92	3	0	1	2	4.6871	−0.0007	−0.0007	0.00000
93	4	0	1	−2	8.7575	0.0006	0.0006	0.00000
94	4	0	1	0	5.4971	−0.0015	−0.0015	0.00000

TABLE 13.A3

Forced Variations in the Amplitude of the Duration of the Day of the Moon for the Models with the Liquid Core and without the Liquid Core LOD_v (P, l) and LOD_v (P, r) and Their Difference ΔLOD_v (in Seconds of Time)

No.	v_1	v_2	v_3	v_4	T_v	$LOD_v (P,l)$	$LOD_v (P,r)$	ΔLOD_v
1	0	0	0	1	29.5306	1.0209	1.0204	0.00044
2	0	0	0	2	14.7653	−12.5222	−12.5169	−0.00536
3	0	0	0	3	9.8435	0.0102	0.0102	0.00000
4	0	0	0	4	7.3826	−0.1572	−0.1571	−0.00007
5	0	0	2	−2	173.3108	2.9392	2.9379	0.00126
6	0	0	2	−1	25.2314	−0.0084	−0.0084	0.00000
7	0	0	2	0	13.6061	2.7495	2.7483	0.00118
8	0	0	2	2	7.0810	0.0203	0.0203	0.00001
9	0	1	−2	0	−14.1326	−0.0061	−0.0061	0.00000
10	0	1	−2	2	−329.7944	−0.1229	−0.1228	−0.00005
11	0	1	0	−4	−7.5349	−0.0211	−0.0211	−0.00001
12	0	1	0	−2	−15.3873	−0.8897	−0.8894	−0.00038
13	0	1	0	−1	−32.1281	0.0079	0.0079	0.00000
14	0	1	0	0	365.2596	67.4929	67.4639	0.02895
15	0	1	0	1	27.3217	−0.1384	−0.1383	−0.00006
16	0	1	0	2	14.1916	0.1706	0.1705	0.00007
17	0	1	2	−2	117.5396	0.0855	0.0855	0.00004
18	0	2	0	−2	−16.0640	−0.0450	−0.0450	−0.00002
19	0	2	0	0	182.6298	0.3713	0.3711	0.00016
20	1	−3	2	−3	−67048.7100	0.9813	0.9809	0.00042
21	1	−2	0	−2	−27.0927	0.0193	0.0193	0.00001
22	1	−2	0	0	32.4506	−0.0310	−0.0310	−0.00001
23	1	−2	0	2	10.1479	−0.0084	−0.0083	0.00000
24	1	−1	0	−4	−9.8136	0.0081	0.0081	0.00000
25	1	−1	0	−2	−29.2633	0.3119	0.3118	0.00013
26	1	−1	0	−1	−3232.6970	−0.9259	−0.9255	−0.00040
27	1	−1	0	0	29.8028	−1.6172	−1.6165	−0.00069
28	1	−1	0	2	9.8736	−0.1564	−0.1564	−0.00007
29	1	0	−2	−2	−9.5301	0.0380	0.0379	0.00002
30	1	0	−2	0	−26.8783	1.0815	1.0810	0.00046
31	1	0	−2	1	−299.2639	−0.0065	−0.0065	0.00000

(Continued)

TABLE 13.A3 (*Continued*)

Forced Variations in the Amplitude of the Duration of the Day of the Moon for the Models with the Liquid Core and without the Liquid Core LOD_v (P, l) and LOD_v (P, r) and Their Difference ΔLOD_v (in Seconds of Time)

No.	v_1	v_2	v_3	v_4	T_v	$LOD_v (P,l)$	$LOD_v (P,r)$	ΔLOD_v
32	1	0	−2	2	32.7636	0.0433	0.0433	0.00002
33	1	0	0	−6	−5.9921	−0.0067	−0.0067	0.00000
34	1	0	0	−4	−10.0846	−0.4072	−0.4071	−0.00017
35	1	0	0	−3	−15.3144	0.0310	0.0310	0.00001
36	1	0	0	−2	−31.8119	−39.7207	−39.7037	−0.01703
37	1	0	0	−1	411.7854	−2.1979	−2.1970	−0.00094
38	1	0	0	0	27.5546	−172.8478	−172.7739	−0.07398
39	1	0	0	1	14.2542	0.1100	0.1099	0.00005
40	1	0	0	2	9.6137	−2.0063	−2.0054	−0.00086
41	1	0	0	4	5.8226	−0.0332	−0.0332	−0.00001
42	1	0	2	−2	23.7746	0.1949	0.1949	0.00008
43	1	0	2	0	9.1085	0.1533	0.1533	0.00007
44	1	1	0	−4	−10.3709	0.0474	−0.0474	−0.00002
45	1	1	0	−2	−34.8469	−2.0211	−2.0202	−0.00087
46	1	1	0	0	25.6217	1.2033	1.2028	0.00052
47	1	1	0	1	13.7188	−0.0167	−0.0167	−0.00001
48	1	1	0	2	9.3672	0.0372	0.0372	0.00002
49	1	1	2	−2	22.3217	0.0072	0.0072	0.00000
50	1	2	0	−2	−38.5220	−0.0816	−0.0816	−0.00004
51	1	2	0	0	23.9422	0.0098	0.0098	0.00000
52	2	−4	2	−4	−3084.2060	−0.0180	−0.0180	−0.00001
53	2	−2	0	−2	−1616.4040	0.1038	0.1038	0.00004
54	2	−1	0	−4	−15.2422	0.0063	0.0063	0.00000
55	2	−1	0	−2	471.8964	0.4777	0.4775	0.00020
56	2	−1	0	0	14.3173	−0.1600	−0.1599	−0.00007
57	2	−1	0	2	7.2689	−0.0209	−0.0209	−0.00001
58	2	0	−2	−2	−14.5689	0.0171	0.0171	0.00001
59	2	0	−2	0	−1095.1920	0.6306	0.6303	0.00027
60	2	0	0	−6	−7.6572	−0.0100	−0.0100	0.00000
61	2	0	0	−4	−15.9060	−0.4229	−0.4227	−0.00018
62	2	0	0	−3	−34.4753	0.0116	0.0116	0.00000

(*Continued*)

TABLE 13.A3 (*Continued*)

Forced Variations in the Amplitude of the Duration of the Day of the Moon for the Models with the Liquid Core and without the Liquid Core LOD_v (P, l) and LOD_v (P, r) and Their Difference ΔLOD_v (in Seconds of Time)

No.	v_1	v_2	v_3	v_4	T_v	LOD_v (P,l)	LOD_v (P,r)	ΔLOD_v
63	2	0	0	−2	205.8927	14.3133	14.3072	0.00613
64	2	0	0	−1	25.8264	−0.0326	−0.0326	−0.00001
65	2	0	0	0	13.7773	−9.9586	−9.9543	−0.00427
66	2	0	0	1	9.3944	0.0121	0.0121	0.00001
67	2	0	0	2	7.1271	−0.2473	−0.2472	−0.00011
68	2	0	2	−2	12.7627	0.0147	0.0147	0.00001
69	2	0	2	0	6.8456	0.0097	0.0097	0.00000
70	2	1	−2	0	548.0334	0.0063	0.0063	0.00000
71	2	1	0	−4	−16.6302	−0.0396	−0.0396	−0.00002
72	2	1	0	−2	131.6710	0.3627	0.3625	0.00016
73	2	1	0	0	13.2765	0.1297	0.1296	0.00006
74	2	1	0	2	6.9907	0.0058	0.0058	0.00000
75	2	2	0	−2	96.7824	0.0093	0.0093	0.00000
76	3	−4	1	−4	−1280.7810	−0.0112	−0.0112	0.00000
77	3	−4	2	−5	475.2364	0.0083	0.0083	0.00000
78	3	−1	0	−2	26.0344	0.0067	0.0067	0.00000
79	3	−1	0	0	9.4218	−0.0164	−0.0164	−0.00001
80	3	0	0	−6	−10.6040	−0.0063	−0.0063	0.00000
81	3	0	0	−4	−37.6253	0.0109	0.0109	0.00000
82	3	0	0	−2	24.3022	0.4214	0.4213	0.00018
83	3	0	0	0	9.1848	−0.7462	−0.7459	−0.00032
84	3	0	0	2	5.6625	−0.0269	−0.0268	−0.00001
85	3	1	0	−2	22.7861	0.0137	0.0137	0.00001
86	3	1	0	0	8.9596	0.0136	0.0136	0.00001
87	4	0	0	−4	102.9464	−0.0120	−0.0120	−0.00001
88	4	0	0	−2	12.9132	0.0349	0.0349	0.00001
89	4	0	0	0	6.8886	−0.0574	−0.0573	−0.00002
90	5	2	−6	1	3247.7810	0.1897	0.1896	0.00008

$$\Delta p_v = 0.0074" \sin F - 0.0012" \sin(l - F),$$

$$\Delta q_v = 0.0116" \cos F + 0.00256" \cos(l - F)$$

and eccentricity of the pointed ellipse is an actual problem of lunar astrometry.

Table 13.5 shows the variations of the duration of the day of the Moon. It shows the rotation characteristics of the Moon. LOD_v (P, l) are variations in the amplitude of the length of the day of the Moon (the model with a liquid core and the solid mantle) and LOD_v (P, r) are amplitude variations of the length of the day of the Moon (model of the solid the Moon without a liquid core),

$\Delta LOD_v = LOD_v$ $(P, l) - LOD_v$ (P, r) are the difference between these amplitudes (the effects of variations in the length of the lunar day, due to the effect of the liquid core). There are a total of 90 variations and 14 were found in the duration of the lunar day with amplitudes ≥ 0.001 s due to the influence of the liquid core. The most significant contributions of the liquid core are characterized by amplitudes: 0.074 s with a period of 27.56 days, with 0.029 s (corresponding to the period 365.26 days), with -0.017 s (period of days -31.81).

In the first column of Table 13.5 are given numbers of corresponding variations. The values of the periods (the sixth column) are given in days. The values of the amplitudes are given in seconds of time. The columns 2–5 are the index values—factors in the arguments for the corresponding variables of the orbital motion of the Moon: $\theta_v = v_1 l_M + v_2 l_s + v_3 F + v_4 D$. The table shows all the detected variations LOD with amplitudes $> 0"00049$ s.

If you highlight the main variations in the amplitude of >0.00044 sec, then the variation of length of the day of the Moon are determined by the formula (where the amplitude is given in milliseconds, ms).

$$\Delta(LOD_v) = -73.98 \cos l + 28.95 \cos l' - 17.03 \cos(l - 2D) + 6.13 \cos 2(l - D)$$

$$- 5.36 \cos 2D - 4.27 \cos 2l + 1.26 \cos 2(F - D) + 1.18 \cos 2F$$

$$- 0.94 \cos(l - D) - 0.86 \cos(l + 2D) - 0.87 \cos(l + l' - 2D)$$

$$- 0.69 \cos(l - l') + 0.52 \cos(l + l') + 0.46 \cos(l - 2F) + 0.44 \cos D$$

We emphasize that the variation determines the effect of the liquid core to the axial rotation of the Moon. The effects are quite significant and probably will be obtained from future observations, that is, Japanese mission SELENA-2.

Index

A

Alpha particle x-ray spectrometer (APXS), 178
Andoyer–Poincaré variables, 342–345, 350–352
Andoyer variables, 320–321, 353–355
Apollo 15, 16, and 17 laser altimeters, 57–58
Apollo 15 landing area
    Chang'E-1 and LRO DTMs, 103–106
    SELENE and LRO DTMs, 106–108
APXS, *see* Alpha particle x-ray spectrometer

B

Back-projection, CE-1 and CE-2 images, 82–83
Barycentric celestial reference system (BCRS), 23
BCRS, *see* Barycentric celestial reference system
BELA, *see* BepiColombo laser altimeter
BepiColombo laser altimeter (BELA), 68
Brightness temperature (TB), 9
    hour angle correction, 211–213
    and instrument, 210–211
    map of different hour angle, 217
    methodology and model, 211–216
    microwave radiative transfer simulation, 214–216
    moon by SVD model, 218–220
    permanently shadowed regions, 216–217
    results and analysis, 216–220
    simulation of detecting depth, 217–218
    singular value decomposition (SVD) model, 216
    solar solar altitude angle, 213
    TB data sets, 210–211
    terrain screen angle, 213–214

C

Cassini's laws
    determination of, 339–341
    with liquid core, 338–339
    of moon rotation, 341–342
    periodic intermediate solution, 339
CE-1 and CE-2, *see* Chang'E-1 and Chang'E-2
Chandrayaan-1, 60
Chang'E-1, 59 60
Chang'E-1 (CE-1) and Chang'E-2 (CE-2)
    back-projection, 82–83
    experimental results, 87–94
    exterior orientation, 82–83
    interior orientation, 79–82
    intratrack adjustment of, 87–94
    sensor model refinements, 83–87
    space intersection, 82–83
Chang'E-2, 61–62
ChemCam, *see* Chemistry and camera complex
CheMin, *see* Chemistry and mineralogy
Chemistry and camera complex (ChemCam), 179–180
Chemistry and mineralogy (CheMin), 180
China National Space Administration (CNSA), 77
Chinese Lunar Exploration Program (CLEP), 77
Clementine LIDAR model, 58, 59
Clementine ultraviolet-visible (UV-VIS), 158
CLEP, *see* Chinese Lunar Exploration Program
CMB, *see* Core–mantle boundary
CNSA, *see* China National Space Administration
Command and service module (CSM), 58
Compact reconnaissance imaging spectrometer for Mars (CRISM), 178

Context camera (CTX), 178
Coregistration of multisource lunar
   DTMs
   in Apollo 15 landing area, 103–108
   multifeature-based surface matching
      method, 103
   in Sinus Iridum Area, 108–113
Core–mantle boundary (CMB), 223
Crater size frequency distributions
      (CSFD), 11
CRISM, *see* Compact reconnaissance
      imaging spectrometer for Mars
Cross-calibration, of IIM NIR bands,
      130–133
CSFD, *see* Crater size frequency
      distributions
CSM, *see* Command and service
      module
CTX, *see* Context camera

D

DAN, *see* Dynamic albedo of neutrons
Deep space network (DSN), 285
DEM, *see* Digital elevation model
Differential VLBI
   adjustment model, 24
   CE-1 transfer orbit, 22–23
   entire transfer orbit, 30–32
   observations, 24–25
   orbit determination strategy, 25
   results, 26–32
   strategies and methods, 25–26
   time delay observations of CE-1, 27–30
Digital elevation model (DEM), 78
Digital ortho map (DOM), 78
Digital terrain models (DTMs), 98
Diviner lunar radiometer experiment
      (DLRE), 9
DLRE, *see* Diviner lunar radiometer
      experiment
DOM, *see* Digital ortho map
DRT, *see* Dust removal tool
DSN, *see* Deep space network
DTMs, *see* Digital terrain models
Dust removal tool (DRT), 180
ΔVLBI, principle of, 2–3
Dynamic albedo of neutrons (DAN), 180

E

Elemental abundances, 141, 142
Ellipsoidal liquid core, 336–338
Euler angles, 320–321
Exterior orientation, CE-1 CCD, 82–83

F

FeO
   inversion of, 133–137
   remote measurements of, 121
FeO and TiO_2
   inversion of, 133–141
   prediction of, 133–141
Field of view (FOV), 128
Force function
   in Andoyer variables, 331–332
   expansions of spherical functions,
      329–331
   Poincaré's variables, 332–336
FOV, *see* Field of view
Free oscillations, pole of moon, 352–359
Full width at half maximum (FWHM),
      9, 126
FWHM, *see* Full width at half maximum

G

Gamma ray spectrometer, 178, 181
Gamma ray spectroscopy (GRS), 120
Geoid-to-topography ratios (GTRs), 295
Global positioning system (GPS), 285
GPS, *see* Global positioning system
GRACE mission, *see* Gravity recovery
      and climate experiment
      mission
GRAIL mission, *see* Gravity recovery
      and interior laboratory mission
Gravity recovery and climate experiment
      (GRACE) mission, 5
Gravity recovery and interior laboratory
      (GRAIL) mission, 5
γ-ray (GRS), 9
Ground Segment for Data, Science and
      Application (GSDSA), 125
GRS, *see* Gamma ray spectrometer;
      Gamma ray spectroscopy; γ-ray

GSDSA, *see* Ground Segment for Data, Science and Application
GTRs, *see* Geoid-to-topography ratios

H

Hamiltonian function, 323
Hamiltonian, noninertial part of, 328–329
Hapke's model, 168
Hazard avoidance cameras (hazcams), 180
Hazcams, *see* Hazard avoidance cameras
HEND, *see* Highenergy neutron detector
High-energy neutron detector (HEND), 181
High-resolution imaging science experiment camera (HiRISE), 178
High-resolution stereo camera (HRSC), 178
HiRISE, *see* High-resolution imaging science experiment camera
HRSC, *see* High-resolution stereo camera
HST, *see* Hubble Space Telescope
Hubble Space Telescope (HST), 176

I

ICRS, *see* International Celestial Reference System
IERS, *see* International Earth Rotation and Reference Systems Service
IIM, *see* Interference Imaging Spectrometer
IIM data
    calibration of, 124–126
    data selection, 127–128
    in-flight calibration of, 126–133
    inhomogeneity of spatial response, 128–129
    laboratory calibration of, 126
    NIR bands, 130–133
    performance characteristics of, 124
    photometric model, 126–128
    systematic artifacts, 129–130
Interference Imaging Spectrometer (IIM), 9
Interior orientation, CE-1 CCD, 79–82

International Celestial Reference System (ICRS), 23
International Earth Rotation and Reference Systems Service (IERS), 25

K

Kaguya/SELENE spacecraft, 58–59
Kinetic energy of planet, 321

L

LALT, *see* Laser ALTimeter
LAM, *see* Laser altimeter
LAMP, *see* Lyman Alpha Mapping Project
Laser altimeter, 78, 98
Laser ALTimeter (LALT), 98
Laser altimetry (LAM)
    BepiColombo mission, 68–69
    Ganymede laser altimeter (GALA), 69–71
    Jupiter Icy Moons Explorer (JUICE), 69–71
    mars, 62–63
    measurement principle and scientific objectives, 52–57
    mercury, 63–65
    moon, 57–62
    near-earth asteroids, 65–67
Laser ranging instrument (LIDAR), 58
LCROSS, *see* Lunar crater observation sensing satellite
LECS, *see* Lunar equatorial coordinate system
LGRS, *see* Lunar gravity ranging system
LIDAR, *see* Laser ranging instrument; Light detection and ranging instrument
Light detection and ranging instrument (LIDAR), 66
Liquid core
    in Andoyer variables, 323–326
    dynamic compressions of, 318–319
    dynamic oblateness of, 316–317
    in Eulerian variables, 321–323
    of moon, 314
    polar and equatorial axes, 317–318

LLRI, *see* Lunar laser range instrument
LMO, *see* Lunar magma ocean
LO, *see* Lunar orbiter missions
LOLA, *see* Lunar orbiter laser altimeter
Long-lived volcanism, 10–11
Long-wave infrared camera (LWIR), 6
LP, *see* Lunar Prospector
LP-GRS, *see* Lunar Prospector gamma
    ray spectroscopy
LRO, *see* Lunar Reconnaissance Orbiter
LROC, *see* Lunar Reconnaissance
    Orbiter Camera
LRO/LCROSS, *see* Lunar
    Reconnaissance Orbiter/
    Lunar Crater Observation and
    Sensing Satellite
LRO spacecraft, *see* Lunar
    reconnaissance orbiter
    spacecraft
LSCC, *see* Lunar soil characterization
    consortium
Lunar core
    Axial rotation of, 349–350
    pole of, 349
Lunar crater observation sensing
    satellite (LCROSS), 9
Lunar day, duration of, 348–349
Lunar dynamics, 13
Lunar equatorial coordinate system
    (LECS), 212
Lunar geodesy and sensing
    lunar missions and explorations, 6–7
    selenodetic methods and techniques,
       2–6
    selenodetic questions and missions,
       12–14
    selenodetic results, 7–12
Lunar gravity field, 8–9
    CEGM-01, 272–277
    Chang'E-1, 277–284
    Chang'E-1 precision orbit
       determination, 271–272
    GRAIL mission, 284–287
    model determination, 265–270
    SELENE, 277–284
Lunar gravity ranging system (LGRS), 5
Lunar interior dynamics, 11–12
Lunar laser range instrument (LLRI), 60
Lunar magma ocean (LMO), 120

Lunar major elements
    of FeO, 121
    of nontransition elements, 122–123
    remote measurements of, 120–123
    TiO_2, 121–122
Lunar mineral inversion methods, 158
Lunar minerals, 9, 158
Lunar orbiter laser altimeter (LOLA),
    60–61, 98
Lunar orbiter (LO) missions, 6
Lunar Prospector (LP), 264
Lunar Prospector gamma ray
    spectroscopy (LP-GRS), 123
Lunar quakes, 13
Lunar Reconnaissance Orbiter (LRO), 98
Lunar Reconnaissance Orbiter Camera
    (LROC), 129
Lunar Reconnaissance Orbiter/
    Lunar Crater Observation
    and Sensing Satellite (LRO/
    LCROSS), 7
Lunar reconnaissance orbiter (LRO)
    spacecraft, 60
Lunar rotation, axis of, 345–347
Lunar sandmeier model
    angle i_s and emergence angle i_v,
       163–164
    control cast shadow, binary
       coefficient to, 167
    horizontal surface, direct component
       of irradiance, 165
    horizontal surface, total irradiance
       on, 165–166
    parameters of, 162–167
    slope e and aspect φ, 162–163
    terrain-view factor V_t, 164–165
Lunar soil characterization consortium
    (LSCC), 9, 158
Lunar system
    dynamic compressions of core, 318–319
    dynamic oblateness of, 316–317
    geodynamic parameters of, 312–314
    liquid core, on moon rotation, 316
    liquid core, polar and equatorial axes
       of, 317–318
    moon, liquid core of, 314
    moon, shell of, 314–316
Lunar topographic data sets
    from Chang'E-1, 99–103

LRO missions, 99–103
SELENE, 99–103
Lunar topography, 7–8, 158
LWIR, *see* Long-wave infrared camera
Lyman Alpha Mapping Project
    (LAMP), 9

M

MAG, *see* Magnetometer
Magnetometer (MAG), 176
MAHLI, *see* Mars hand lens imager
MARDI, *see* Mars descent imager
MaRS, *see* Mars Radio Science
    Experiment
Mars climate sounder (MCS), 178
Mars descent imager (MARDI), 179, 180
Mars Exploration Rover (MER), 178
Mars Express lander communications
    (MELACOM), 178
Mars Global Surveyor (MGS), 61, 294
Mars hand lens imager (MAHLI), 180
Mars orbiter camera (MOC), 176
Mars orbiter laser altimeter (MOLA),
    62–63, 176, 294
Mars Pathfinder (MPF), 178
Mars Radio Science Experiment
    (MaRS), 178
Mars Reconnaissance Orbiter (MRO),
    178, 293
Mars Science Laboratory (MSL), 179
Martian crust thickness and structure
    anomalies correction, 299–300
    crustal dichotomy, 303
    crustal thickness variations, 303–306
    gravity field, 296–297
    inversion method, 297–299
    inversion with constraint, 300
    localized admittance analysis,
        300–303
    localized analysis, 306
    topography, 295–296
Martian minerals and rock components
    clay mineral, 193–194
    early and telescopic observations,
        175–176
    geologic setting, 188–190
    kaolinite, 194–195
    methods of mineral analysis, 180–188

mineral classes distribution, 197–201
mineral information from, 185–188
MRO CRISM images, 188–201
oxide, 197
reflection spectroscopy remote
    sensing, 182–184
remote sensing methods, 185
smectite, 195
spacecraft missions, 176–180
sulfate and carbonate, 195–197
MCS, *see* Mars climate sounder
MECA, *see* Microscopy,
    electrochemistry, conductivity
    analyzer
MELACOM, *see* Mars Express lander
    communications
MER, *see* Mars Exploration Rover
Mercury laser altimeter (MLA), 63–65
Mercury's external magnetic field,
    231–235
Mercury's internal dynamo, numerical
    models of
    alternatives, 255–257
    dynamo theory, 241–246
    inhomogeneous boundary
        conditions, 252–255
    stably stratified outer layer, dynamos
        with, 248–252
    standard earth-like dynamo models,
        246–248
Mercury's internal magnetic field,
    235–241
Mercury's internal structure, 225–230
MESSENGER mission, 224
MESSENGER offset dipole model
    (MODM), 224, 225
MGS, *see* Mars Global Surveyor
MI, *see* Microscopic imager
Microscopic imager (MI), 178
Microscopy, electrochemistry,
    conductivity analyzer
    (MECA), 179
Mineral classes distribution, Martian
    mineral
    hydrated minerals, 200–201
    mafic minerals, 197–199
    sulfate and carbonate, 201
MLA, *see* Mercury laser altimeter
MOC, *see* Mars orbiter camera

MODM, *see* MESSENGER offset dipole
model
MOLA, *see* Mars orbiter laser altimeter
Moon
fourth free librations of, 359
global elemental abundances,
141–150
global Mg#, 150–152
liquid core of, 314
masses and moments of, 314–316
physical libration of, 319–329
rotation, liquid core on, 316
Moon's rotation theory, 355–356
MPF, *see* Mars Pathfinder
MRO, *see* Mars Reconnaissance Orbiter
MSL, *see* Mars Science Laboratory
Multifeature-based surface matching
method, 103
Multisource lunar topographic data sets
synergistic use of, 113–115

N

Navcams, *see* Navigation cameras
Navigation cameras (navcams), 180
Near-earth asteroids (NEAs), 67
Near infrared (NIR) absorbance, 10
NEAR-Shoemaker Laser Rangefinder
(NLR), 65
NEAs, *see* Near-earth asteroids
Neutron spectrometer (NS), 181
NIPALS, *see* Nonlinear iterative partial
least squares
NIR absorbance, *see* Near infrared
absorbance
NLR, *see* NEAR-Shoemaker Laser
Rangefinder
Nonlinear iterative partial least squares
(NIPALS), 139
Nontransition elements
inversion of, 139–141
remote measurements of, 122–123
NS, *see* Neutron spectrometer

O

ODY, *see* Orbit–Mars Odyssey
Orbit–Mars Odyssey (ODY), 293
OSIRIS-REx, 71

P

Partial least squares regression (PLSR)
method, 139
PCA, *see* Principal component analysis
Physical libration of moon
in Andoyer variables, 323–327
coordinate system and variables,
319–320
Euler angles and Andoyer's
variables, 320–321
Hamiltonian, noninertial part of,
328–329
kinetic energy and canonical
equations, 321–323
Poincaré's variables, 327–328
PLSR method, *see* Partial least squares
regression method
Poincaré's variables
Hamiltonian of, 332–336
physical libration of, 327–328
Principal component analysis (PCA), 158

R

RAC, *see* Robotic arm camera
RAD, *see* Radiation assessment detector
RADF, *see* Radiance factor
Radiance factor (RADF), 126
Radiation assessment detector (RAD), 180
RAT, *see* Rock abrasion tool
Refinement
on exterior orientation, 83–84
on IO, 85
Reflectance experiment laboratory
(RELAB) spectra, 9
RELAB spectra, *see* Reflectance
experiment laboratory spectra
REMS, *see* Rover environmental
monitoring station
RMSE, *see* Root mean square error
RMSEP, *see* Root-mean-square error of
prediction
Robotic arm camera (RAC), 179
Rock abrasion tool (RAT), 178
Root mean square error (RMSE), 133
Root-mean-square error of prediction
(RMSEP), 140
Rover environmental monitoring station
(REMS), 180

S

SAM, *see* Sample analysis at Mars;
    Spectral angle mapper
Sample analysis at Mars (SAM), 180
Sandmeier model, 159–162
SAR, *see* Synthetic aperture radar
SCS, *see* Sun-canopy-sensor
SELENE, *see* SELenological and
    ENgineering Explorer
Selenodetic methods and techniques
    lunar laser altimetry, 5
    lunar laser ranging (LLR), 3–5
    lunar remote sensing, 6
    satellite gravimetry, 5–6
    very long baseline interferometry
    (VLBI), 2–3
SELenological and ENgineering
    Explorer (SELENE), 158
Shallow subsurface radar
    (SHARAD), 178
SHARAD, *see* Shallow subsurface radar
Signal-to-noise ratios (SNR), 64
Sinus Iridum area
    Chang'E-1 and SELENE DTMs,
    108–112
    SELENE and LRO DTMs, 112–113
SMA, *see* Spectral mixing analysis
SNR, *see* Signal-to-noise ratios
Spacecraft missions, 176–180
Space intersection, CE-1 and CE-2
    images, 82–83
Space VLBI
    adjustment model, 37–39
    for CE-1 transfer orbit, 35–37
    experiment, 39–46
    optimal conditions for observation,
    46–47
    unknown parameters, estimation
    of, 39
Spectral angle mapper (SAM), 193
Spectral mixing analysis (SMA), 158,
    168–169
SSI, *see* Surface stereo imager
STS, *see* Sun-terrain-sensor
Sun-canopy-sensor (SCS), 159
Sun–Moon distance, 126
Sun-terrain-sensor (STS), 159
Surface components mapping, 9–10

Surface stereo imager (SSI), 179
Synthetic aperture radar (SAR), 10

T

TB, *see* Brightness temperature
TC, *see* Terrain camera
TDI, *see* Time delay integration
TECP, Thermal and electrical
    conductivity probe
Tectonics of moon, 12
TEGA, *see* Thermal and evolved gas
    analyzer
Terrain camera (TC), 11
TES, *see* Thermal emission
    spectrometer; Thermal
    spectrometer
THEMIS, *see* Thermal emission imaging
    system
Thermal and electrical conductivity
    probe (TECP), 179
Thermal and evolved gas analyzer
    (TEGA), 179
Thermal emission imaging system
    (THEMIS), 178
Thermal emission spectrometer (TES), 176
Thermal spectrometer (TES), 183
Time delay integration (TDI), 78
TiO_2
    inversion of, 138–139
    remote measurements of, 121–122
Topographic correction, 167–168
Topographic correction models, 159

U

Unified S-band (USB)
    monitoring system, 2
    system, 270
USB, *see* Unified S-band
UV-VIS, *see* Clementine
    ultraviolet-visible

V

Very long baseline interferometry
    (VLBI), 270, 271
Visible and near infrared wavelength
    (VNIR), 120

VLBI, *see* Very long baseline
    interferometry
VNIR, *see* Visible and near infrared
    wavelength

W

WAC, *see* Wide Angle Camera
WCL, *see* Wet chemistry lab

Wet chemistry lab (WCL), 179
Wide Angle Camera (WAC), 129

X

X-ray spectrometer (XRS), 9
XRS, *see* X-ray spectrometer